Ecological Morphology

Ecological Morphology

*Integrative
Organismal
Biology*

Edited by
Peter C. Wainwright

and
Stephen M. Reilly

The University of Chicago Press

Chicago and London

PETER C. WAINWRIGHT is assistant professor of biology at Florida State University;
STEPHEN M. REILLY is assistant professor of biology at Ohio University

The University of Chicago Press, Chicago 60637
The University of Chicago Press, Ltd., London
© 1994 by The University of Chicago
All rights reserved. Published 1994
Printed in the United States of America
03 02 01 00 99 98 97 96 95 94   1 2 3 4 5
ISBN: 0-226-86994-6 (cloth)
        0-226-86995-4 (paper

Library of Congress Cataloging-in-Publication Data

Ecological morphology : integrative organismal biology / edited by Peter C. Wainwright and
   Stephen M. Reilly.
        p.   cm.
     Includes index.
     1. Morphology.   2. Ecology.   3. Adaptation (Biology).   I. Wainwright, Peter
   Cam, 1957–   .   II. Reilly, Stephen M.
   QH351.E37   1994
   591.5—dc20   93-40993
                                                                         CIP

Contents

Preface

The discipline of morphology has a rich history of interest in evolutionary questions in recent years increasingly focused on patterns and processes occurring in natural populations. By its nature, this movement has drawn researchers from a range of disciplines, including morphology, physiology, ethology, ecology, population biology, and systematics. While this breadth has added tremendously to our understanding of organismal design, it has also made it difficult for a single researcher to keep up with all relevant developments. Thus, we decided in 1990 to organize a multiauthor book on the integration of functional morphology and ecology. We hope that this book can introduce beginners to the major concepts and approaches in ecological morphology and that even experienced practitioners will find new ideas and perspectives in the contributions of a diverse authorship.

Most of the chapters included here were presented at a symposium held at the annual meetings of the American Society of Zoologists in San Antonio, Texas, in December 1990. The symposium gave authors an opportunity to present their ideas and to take advantage of subsequent discussions before preparing written manuscripts. Most of the speakers who presented oral contributions at the symposium have written chapters for the book, and one chapter (chap. 4, Losos and Miles) was added following the meetings.

Many people contributed to the development of this book, and we have been fortunate in enjoying the best that the academic community has to offer. In the planning stages of this project we benefited from the thoughtful advice of several people, including Stevan Arnold, Willi Bemis, Albert Bennett, Zoe Eppley, Raymond Hudey, Bruce Jayne, George Lauder, and Stephen Wainwright. Rhonda Reilly provided many hours of expert editorial assistance. Each chapter of the book was reviewed by at least two experts in the field, and we are deeply grateful to those readers for their considerable time and energy spent in helping to improve the final product. The efforts of two readers who reviewed the entire book manuscript were invaluable in helping to integrate the parts of the whole and in improving various sections. Funds in support of the symposium were very generously supplied by the Cocos Foundation. We thank James Hanken and Burt Bogitsh and the local committee of the American Society of Zoologists for their assistance in organizing the symposium. Susan Abrams assisted with every phase of the book; we owe her a great debt of gratitude for her help and constant encouragement. Travel funds and additional support necessary to complete the book were provided by the Department of Biological Science at Florida State University and the Department of Biological Sciences at Ohio University.

1

Introduction

Peter C. Wainwright and Stephen M. Reilly

Understanding the basic structure and function of organisms is one of the oldest and still most active areas of biology. The continued success of morphological and physiological research is due both to our natural fascination with how organisms work, and to the emergence of new perspectives and conceptual frameworks within which biologists couch questions about the nature of organismal design. Two of the major themes that have emerged from the past two decades of research in organismal functional morphology, physiology, and biomechanics are (1) the need for making observations and interpretations on how organisms function within an environmentally and historically relevant context, and (2) the need for understanding the ecological and evolutionary consequences of organismal form. These themes are central to this volume. Our overall goal is to present a variety of modern viewpoints on these topics in hopes of stimulating thought and provoking new avenues of integrative research on the nature of organismal design.

No feature so uniquely characterizes life as the process of evolution. This mechanism of change in organismal form sets apart the functional analysis of biological systems from attempts to understand inanimate or man-made structures. Thus, a complete understanding of organismal design by nature should involve a functional analysis, a historical analysis, and an ecological analysis. How have organisms come to be designed the way they are, and what role does their current design play in shaping their modern ecology? Such a holistic approach to studying organismal form and function clearly requires the integration of information from what are normally considered to be separate disciplines of biology. Phylogenetic systematics, quantitative genetics, behavioral biology, and ecology are broad areas that can directly contribute to studies of morphology and show great potential for enriching our understanding of the nature of organismal design. Not surprisingly, one of the major messages emerging from this volume is the need for collaborative efforts between specialists from two or more disciplines to address general issues in ecological morphology.

We have purposely sought diverse opinions on our central topic. Among the disciplines represented are community ecology, population biology, functional morphology, comparative physiology, biomechanics, systematics, and paleontology. Although the emphasis of the book is on vertebrate animals, several chapters include invertebrate or plant examples, and two (chapter 8, by Denny; chapter 12, by Bradley) focus on invertebrate taxa. Most of the discussions in the book should generalize readily to studies on any life form.

A Brief History of Ecological Morphology

We visualize ecological morphology as a field with no current conceptual distinctions from physiological ecology (Feder et al., 1987), ecomechanics (Denny, 1988; Denny et al., 1985; S. Wainwright et al., 1976), or any discipline that seeks to link some form of the analysis of how organisms work with questions about the ecological context of the function. Nevertheless, ecological morphology has a history that is distinct from physiological ecology, and while the two disciplines may be largely indistinguishable today (at least to us), this has not always been the case. In both fields, periodic contributions of ecologists and evolutionary biologists have played a key role in shaping the modern discipline.

The general notion that the morphology of organisms is indicative of the way in which they make their living goes at least as far back as the recorded thoughts of natural historians such as Aristotle, but the formalization of the term "adaptation" by Darwin (1859) provided the central paradigm of ecological morphology that dominates the field to this day. Numerous European and American anatomists of the early part of this century attempted to relate detailed studies of animal morphology to the ways in which animals utilized the environment during their day-to-day life (e.g., Gregory, 1913; Howell, 1925; Jordan, 1923; Nobel, 1930). These workers were concerned mostly with understanding and describing the morphology of animals, and how those parts functioned, but often they attempted to place their observations into the context of the known natural history of species. Somewhat later Van der Klaauw (1948) coined the term "ecological morphology" (which is frequently shortened to "ecomorphology"). Van der Klaauw's (1948) definition of ecological morphology as the study of the relationship between the morphology of the organism and its environment is, in its essence, still the most common definition (Alexander, 1988; Barel et al., 1989; Bock, 1990; Losos, 1990; Motta and Kotrschal, 1992; Norton, 1991; P. Wainwright, 1991).

Beginning in the 1960s, several papers written by ecologists began to explore the relationship between the morphology of species in communities and the ecological space that they occupy (Gatz, 1979; Grant, 1968; Karr and James, 1975; Hespenheide, 1973; Werner, 1974). These studies pursued one of two general goals. In the first, the precise functional meaning of morphological variables was not emphasized so much as the patterns of dispersion of species' morphologies and how they were correlated with features of the environment (Gatz, 1979; Karr and James, 1975; Felley, 1984; Findley and Black, 1983; Miles and Ricklefs, 1984; Ricklefs and Travis, 1980; Winemiller, 1991). Most of the questions addressed in these studies are ecological in nature: When species are added to communities, is the average morphological distance between members increased, or is the total morphological space occupied by the community left unchanged? How even or clumped is the morphological and ecological spread of species in communities? Do similar morphologies associate with similar environmental features, irrespective of phylogenetic affinities?

A second theme that developed in ecomorphological studies has its roots in morphology and focuses on the causal consequences of anatomical differences between species (or individuals within species) for the ecology of the organism. A series of papers by Bock (Bock, 1977, 1980, 1988, 1990; Bock and von Wahlert, 1965) were pivotal in clarifying several important points and articulating a conceptual framework for interpreting the ecological and evolutionary implications of functional morphological observations. Bock defined the function of structures as their action, or how they work, and made the useful distinction between function and biological role—the latter referring to how the structure is used in the life of the organism (Bock and von Wahlert, 1965). A central theme of Bock's papers has been a call for active collaborations between functional morphologists and ecologists, natural historians, and ethologists (Bock, 1977, 1980, 1990).

During the late 1970s and early 1980s a crucial connection was made that linked population genetics and the mechanism of natural selection to the paradigm that had emerged in functional morphology and physiological ecology for studying adaptation (Arnold, 1983; Huey and Stevenson, 1979; Lande and Arnold, 1983). The role of organismal design in determining how a creature interacts with its environment comes about through the effect that design has on the ability of the organism to perform various tasks and behaviors. Arnold (1983) formalized this relationship by noting that morphological, performance, and fitness gradients can be constructed for the members of a population, aiding in both the visualization and the quantification of the connection between morphology and fitness. The clear role of disciplines such as functional morphology and physiology in this framework is in understanding exactly how or why specific morphological variation causes performance differences. One important consequence of this contribution has been the realization that "performance testing" is a key component of studies that seek to link morphology to ecology. Some of the earliest studies that related the functional design of animals to performance capacity and showed how variation in performance within and among species shape fitness distributions or community ecological phenomena include Werner's work with sunfishes (Werner, 1974, 1977) and studies by Grant and his colleagues on Galapagos finches (Grant, 1968; 1986).

The development of methodologies for studying ecomorphological relationships within populations from the standpoint of fitness was paralleled by the emergence of a modern comparative method (Brooks and McLennan, 1991; Harvey and Pagel, 1991) that focuses on phyletic diversity and offers a methodology for testing whether a particular feature of design may be termed an adaptation. Lauder (1982) distinguished the "extrinsic," or environmentally generated, selective determinants of design, from "intrinsic" factors. Intrinsic factors are features inherited from ancestors that may limit or determine future directions of evolution in a lineage. New species tend to look most like their direct ancestors, and some aspects of design may increase the probability of certain directions of future change. Phylogenetic hypotheses provide our best estimate of the evolutionary history of life and provide a rigorous basis for testing hypotheses about the adaptive nature of specific features. One of the chief lessons from this area has been that most traits of species are features inherited from an ancestor, and any hypothesis testing about how a particular condition came to be must focus on the specific phyletic transition where the trait evolved. Recently, aspects of the morphology/performance/fitness paradigm have been incorporated into the phylogenetic approach (Lauder and Liem, 1989; Schluter, 1989) and may be used to examine patterns of correlated evolution among different levels of analysis (Garland et al., 1992; Martins and Garland, 1991).

There is no doubt that ecological morphology has emerged as an active, contributing subdiscipline of morphology. Several review papers have been published on the topic in the past few years (Alexander, 1988; Barel et al., 1989; Bock, 1990; Motta and Kotrschal, 1992; Wainwright, 1991), and a new panel was created at the National Science Foundation in 1991 with the title "Physiological and Functional Ecology." Nevertheless, the excitement in this area does not reflect the emergence of an entirely new approach in organismal biology but, rather, the current conviction of those engaged in functional morphology that to understand the diversity of organismal design and its ecological and evolutionary consequences one must explicitly build field studies and historical analyses into the research program.

STRUCTURE AND OVERVIEW OF THIS VOLUME

Part 1 introduces the reader to many of the major concepts of ecological morphology and the approaches that have been used (and in some cases have yet to be used) to address central research questions. The chapters in this part present few data but instead emphasize discussions of conceptual issues that lie at the heart of ecological morphology. Contributors were urged to be creative and to present new or untested ideas. Contributors to part 2 summarize research on the ecomorphology of a particular system. Their contributions serve not only as a reference and review sources for the topics they address, but provide insights into some of the general conclusions that have emerged from ecomorphological research on well-studied systems and point in the direction of some of the major challenges that continue to face researchers.

Part 1, "Concepts, Issues, and Approaches," begins with an ecological perspective by Ricklefs and Miles that summarizes research on the relationship between the morphological and ecological space occupied by members of communities. A key contribution of their chapter is to formalize the linkage that is often assumed to occur between morphology and ecology and to question some of the underlying assumptions involved. Chapter 3, by Wainwright, discusses the potential role of experimental functional morphology in ecological research and presents a framework for implementing this research program. He emphasizes that this is a greatly neglected area of research in spite of the apparently predominant role that organismal design plays in shaping patterns of resource use. Chapter 4, by Losos and Miles, reviews current methodology for using phylogenetic hypotheses as a basis for inferring patterns of evolution in functional characters. This has been an active area in recent years and rigorous statistical techniques are being developed for examining the phylogenetic distribution of characters in connection with such hypotheses as adaptation and coevolution. Travis, in chapter 5, presents a discussion of the adaptive value of phenotypic plasticity. He distinguishes between developmental plasticity and genetically based variation that allows the individual to take advantage of diverse environmental settings. Chapter 6, by Emerson, Greene, and Charnov, formalizes some theoretical considerations on the role of functional morphology of the feeding mechanism in shaping predator diet. Central to their thesis is the role of changing body size that occurs during the life of the predator. In chapter 7, Van Valkenburgh presents a paleontologist's view of using morphological characters to infer ecological patterns in fossil organisms and communities. Her findings suggest that a key tool in understanding patterns in extinct lineages involves applying the results of experimental functional analyses that are performed with living creatures.

In part 2, "Model Systems," Denny (chap. 8) reviews research examining the role of organismal design in shaping the ecology of rocky intertidal animal and plant communities. Denny emphasizes the remarkable ability of organisms to effectively avoid the seemingly severe constraints placed on organismal form by the continued presence of strong wave action in this environment. Norberg, in chapter 9, examines the biomechanical basis of flight performance in bats and the influence of flight performance on ecological parameters such as diet. Bats exhibit a large range of flight capabilities that seem to very closely match the behavioral tendencies of species as they forage. Chapter 10, by Garland and Losos, summarizes research on the functional basis of locomotor performance in reptiles and the ecological consequences of variation in performance. A key revelation of their investigation is that in spite of a tremendous data base on the functional basis of locomotor performance, there exists a remarkable lack of knowledge about the ecological ramifications of differences among species or among individuals in locomotor capacities. Chapter 11 summarizes work done by Bradley and his colleagues on the functional basis of salt tolerance in larval mosquitoes. These data are interpreted in the light of a phylogeny for mosquitoes, and they show that saline tolerance has evolved several times and does not always involve the same suite of modifications. In chapter 12 Reilly offers a reinterpretation of the ecological circumstances surrounding heterochronic developmental changes. He uses salamanders as a model system and points out that, while many heterochronic changes become fixed during evolution, others are developmentally plastic and permit the existence of multiple life history strategies, even within a single population.

While our aim was to give broad coverage to the perspectives that dominate research in ecological morphology, we would not presume to have covered all important topics. For example, the use of optimal design in functional analysis has a rich history in research on the nature of organismal design. Mechanical and physiological optima can be identified and used as benchmarks against which to compare the actual design of biological systems (Alexander, 1982; Sibly and Calow, 1986; Taylor and Weibel, 1981). Although the utility of optimal criteria as a heuristic aid to interpreting form and function cannot be disputed, optimal design as an expectation for the end product of morphological evolution has been brought into question (Bennett, 1987; Dudley and Gans, 1991; Garland and Huey, 1987; Gould and Lewontin, 1979). This is an interesting and active area of debate, and one that promises to generate numerous insights into how complex functional systems evolve.

Another major contributor to ecological morphology has been the perspective of phylogenetic systematics. The modern comparative method (Brooks and McLennan, 1991; Harvey and Pagel, 1991) is a tool for identifying the pitfalls of assuming that variation among taxa is adaptive, particularly given that most characteristics of any species are features inherited from an ancestor. This lesson provided the impetus for new techniques that permit one to actively account for phylogeny while studying ecomorphological relationships (see summary in chapter 4). The reader may notice that the chapters of this book vary in the degree to which they address the issue of phylogeny directly in their comparative analyses. Two reasons may be given for this heterogeneity. First, it is unfortunately the case that for the vast majority of taxa no adequate phylogenetic hypothesis exists. This paucity of information greatly restricts the range of organisms for which we are able to address historical questions and is a major challenge for the study of diversity (Brooks and McLennan, 1991). The second point is that it is not always necessary to have a phylogeny in hand in order to draw significant ecomorphological conclusions. An example is provided by Norberg in chapter 9. She is able to offer a clear biomechanical interpretation of the aerodynamic consequences of wing morphology in bats. These interpretations and their implications for flight performance and possible habitat and food use can be expected to hold true regardless of the specific evolutionary history of wing design in bats. Physics offers an absolute scale with which to gauge the consequences of wing design.

This book is aimed especially at graduate students and others who are developing research programs in this area. However, we hope that by bringing together researchers with different backgrounds to present their perspectives and opinions on studying the evolution and ecology of organismal design, stimulating ideas will be exchanged between fields. Perhaps one of the greatest strengths of ecological morphology is that it is characterized by such a diversity of approaches. Integrated efforts hold much promise for answering new and exciting questions.

There are three principle goals for this volume. First, in part 1 we hope to present an overview of many of the major concepts in ecological morphology in a fashion that can reach a broad audience. Second, by presenting in part 2 summaries of some well-studied systems in ecological morphology we hope to illustrate the fruits that are available and to point to some of the apparent, and perhaps surprising, limitations and shortcomings of research to date. Third, we want to emphasize the value of an integrative approach and present both a conceptual framework and a practical guide so that individuals working on the mechanistic basis of organismal performance may link that performance to evolutionary and ecological processes. Above all, we hope this book stimulates thought and encourages researchers to venture into interesting and productive new areas of investigation.

REFERENCES

Alexander, R. McN. 1982. *Optima for Animals*. London: Edward Arnold Publishers.
Alexander, R. McN. 1988. The scope and aims of ecological morphology. *Neth. J. Zool.* 38:3–22.
Arnold, S. J. 1983. Morphology, performance and fitness. *Amer. Zool.* 23:347–361.
Barel, C. D. N., G. C. Anker, F. Witte, R. J. C. Hoogerhoud, and T. Goldschmidt. 1989. Constructional constraint and its ecomorphological implications. *Acta Morphol. Neerl.-Scand.* 27:83–109.

Bartholomew, G. A. 1986. The role of natural history in contemporary biology. *Bioscience* 36:324–329.

Bennett, A. F. 1987. The accomplishments of ecological physiology. In *New Directions in Ecological Physiology*, ed. M. E. Feder, A. F. Bennett, W. W. Burggren, R. B. Huey, 1–8. Cambridge: Cambridge University Press.

Bock, W. J. 1977. Toward an ecological morphology. *Vogelwarte* 29:127–135.

Bock, W. J. 1980. The definition and recognition of biological adaptation. *Amer. Zool.* 20:217–227.

Bock, W. J. 1988. The nature of explanations in morphology. *Amer. Zool.* 28:205–215.

Bock, W. J. 1990. From biologische anatomie to ecomorphology. *Neth. J. Zool.* 40:254–277.

Bock, W. J., and G. von Wahlert. 1965. Adaptation and the form-function complex. *Evolution* 19:269–299.

Brooks, D. R., and D. A. McLennan. 1991. *Phylogeny, Ecology, and Behavior*. Chicago: University of Chicago Press.

Darwin, C. D. 1859. *On the Origin of Species*. New York: Penguin Books.

Denny, M. W. 1988. *Biology and the Mechanics of the Wave-Swept Environment*. Princeton: Princeton University Press.

Denny, M. W., T. L. Daniel, and M. A. R. Koehl. 1985. Mechanical limits to size in wave swept organisms. *Ecol. Mon.* 55:69–102.

Dudley, R., and C. Gans. 1991. A critique of symmorphosis and optimality models in physiology. *Physiol. Zool.* 64:627–637.

Feder, M. E., A. F. Bennett, W. W. Burggren, and R. B. Huey, eds. 1987. *New Directions in Ecological Physiology*. Cambridge: Cambridge University Press.

Felley, J. D. 1984. Multivariate identification of morphological-environmental relationships within the Cyprinidae (Pisces). *Copeia* 1984:442–455.

Findley, J. S., and H. Black. 1983. Morphological and dietary structuring of a Zambian insectivorous bat community. *Ecology* 64:625–630.

Garland, T., Jr., and R. B. Huey. 1987. Testing symmorphosis: Does structure match functional requirements? *Evolution* 41:1404–1409.

Garland, T., Jr., P. H. Harvey, and A. R. Ives. 1992. Procedures for the analysis of comparative data using phylogenetically independent contrasts. *Syst. Biol.* 41:18–32.

Gatz, A. J., Jr. 1979. Community organization in fishes as indicated by morphological features. *Ecology* 60:711–718.

Gould, S. J., and R. C. Lewontin. 1979. The spandrels of San Marcos and the Panglossian paradigm. *Proc. R. Soc.* (London) 205B:581–598.

Grant, P. R. 1968. Bill size, body size and the ecological adaptations of bird species to competitive situations on islands. *Syst. Zool.* 17:319–333.

Grant, P. R. 1986. *Ecology and Evolution of Darwin's Finches*. Princeton: Princeton University Press.

Gregory, W. K. 1913. Locomotory adaptations in fishes illustrating "Habitus" and "Heritage." *Ann. N.Y. Acad. Sci.*, Feb. 10:266–268.

Harvey, P. H., and M. D. Pagel. 1981. *The Comparative Method in Evolutionary Biology*. New York: Oxford University Press.

Hespenheide, H. A. 1973. Ecological inferences from morphological data. *Ann. Rev. Ecol. Syst.* 4:213–229.

Howell, A. B. 1925. On the alimentary tracts of squirrels with diverse food habits. *J. Washington Acad. Sci.* 15:145–150.

Huey, R. B., and R. D. Stevenson. 1979. Integrating thermal physiology and ecology of ectotherms: A discussion of approaches. *Amer. Zool.* 19:357–366.

Jordan, D. S. 1923. A classification of fishes, including families and genera as far as known. *Stanford Univ. Publ., Univ. Ser. Biol. Sci.* 3:79–243.

Karr, J. R., and F. C. James. 1975. Eco-morphological configurations and convergent evolution of species and communities. In *Ecology and Evolution of Communities,* ed. M. L. Cody and J. M. Diamond, 258–291. Cambridge: Harvard University Press.

Lande, R., and S. J. Arnold. 1983. The measurement of selection on correlated characters. *Evolution* 37:1210–1226.

Lauder, G. V. 1982. Historical biology and the problem of design. *J. Theor. Biol.* 97:57–67.

Lauder, G. V., and K. F. Liem. 1989. The role of historical factors in the evolution of complex organismal functions. In *Complex Organismal Functions: Integration and Evolution in Vertebrates,* ed. D. Wake and G. Roth, 63–78. New York: John Wiley and Sons.

Losos, J. B. 1990. The evolution of form and function: Morphology and locomotor performance in West Indian *Anolis* lizards. *Evolution* 44:1189–1203.

Martins, E. P., and T. Garland. 1991. Phylogenetic analyses of the correlated evolution of continuous characters. *Evolution* 45:534–557.

Miles, D. B., and R. E. Ricklefs. 1984. The correlation between ecology and morphology in deciduous forest passerine birds. *Ecology* 65:1629–1640.

Motta, P. J., and K. M. Kotrschal. 1992. Correlative, experimental, and comparative experimental approaches in ecomorphology. *Neth. J. Zool.* 42:400–415.

Nobel, G. K. 1930. *The Biology of the Amphibia.* New York: McGraw-Hill.

Norton, S. F. 1991. Capture success and diet of cottid fishes: The role of predator morphology and attack kinematics. *Ecology* 72:1807–1819.

Ricklefs, R. E., and J. Travis. 1980. A morphological approach to the study of avian community organization. *Auk* 97:321–338.

Schluter, D. 1989. Bridging population and phylogenetic approaches to the evolution of complex traits. In *Complex Organismal Functions: Integration and Evolution in the Vertebrates,* ed. D. B. Wake and G. Roth, 79–95. New York: John Wiley and Sons.

Sibly, R. M., and P. Calow. 1986. *Physiological Ecology of Animals: An Evolutionary Approach.* Oxford: Blackwell Scientific.

Taylor, C. R., and E. R. Weibel. 1981. Design of the mammalian respiratory system. I. Problem and strategy. *Respir. Physiol.* 44:1–10.

Van der Klaauw, C. J. 1948. Ecological studies and reviews. IV. Ecological morphology. *Bibliotheca Biotheoretica* 4:27–111.

Wainwright, P. C. 1991. Ecological morphology: Experimental functional anatomy for ecological problems. *Amer. Zool.* 31:680–693.

Wainwright, S. A., W. D. Biggs, J. D. Curry, and J. M. Gosline. 1976. *Mechanical Design on Organisms.* London: Edward Arnold.

Werner, E. E. 1974. The fish size, prey size, handling time relation and some implications. *J. Fish. Res. Board Can.* 31:1531–1536.

Werner, E. E. 1977. Species packing and niche complementarity in three sunfishes. *Amer. Nat.* 111:553–578.

Winemiller, K. O. 1991. Ecomorphological diversification in lowland freshwater fish assemblages from five biotic regions. *Ecol. Mon.* 61:343–365.

I. Concepts, Issues, and Approaches

2

Ecological and Evolutionary Inferences from Morphology: An Ecological Perspective

Robert E. Ricklefs and Donald B. Miles

INTRODUCTION

From the perspective of the ecologist, ecomorphological analyses have three distinct goals: (1) estimation of ecological relationships among species from their positions in morphological space, that is, to make ecological inferences from morphological pattern; (2) measurement of the ecology-morphology correlation as a means of validating the first goal; and (3) elucidation of the functional relationship between morphology and ecology as it is mediated by the behavior and performance (ethotype) of the organism. The latter involves the combination of ecological studies with functional morphology.

Within this general perspective, ecologists are interested in a number of issues for which morphological studies have played, and will continue to play, an important role. Among these are the association between values and character states of phenotypic traits and those of environmental variables, the distribution and abundance of species, the coexistence of species within communities, and the role of historical and regional processes in the development of local ecological communities. With respect to structure-function correlations, studies of morphology may reveal selective factors in the environment and constraints on the response of the phenotype to these factors (Arnold, 1983; Lande and Arnold, 1983; Lauder, 1981, 1982). To the degree that morphology determines the ecological range of the phenotype, it may limit the ecological (and geographical) distribution of a population and the coexistence of populations in local communities (Ricklefs, 1987, 1989; Schluter, 1988a). The influence of history is revealed by phylogenetically conservative morphology (Emerson, 1984, 1988; Lauder and Liem, 1989; Livezey, 1989; Paton and Collins, 1989; Schluter, 1989).

In each case, morphology represents aspects of the relationship between the organism and its environment. Indeed, the concept of ecomorphology is based on the premise that the phenotype provides useful information about this relationship; that is, ecology and morphology provide alternative, but mutually consistent expressions of the outcome of ecological and evolutionary adjustments between phenotype and environment (Bock, 1977; Findley, 1976; Hespenheide, 1973; Karr and James, 1975; Losos, 1990a,b; Losos and Miles, chap. 4; Ricklefs and Travis, 1980; Schluter, 1988a; Smith, 1990; Williams, 1972, 1983). The environment itself is complex and impinges upon morphology through many processes related to physical factors (such as climate), habitat structure (which relates, for example, to locomotory structures), diet breadth, and the partitioning of resources among species (which are expressed as variation in trophic characteristics). Ecologists have employed a variety of data and analytical techniques in studying these relationships. Among those relevant to the present discussion are characterization of niche and habitat relationships by measurement of resource and substrate use (MacArthur, 1958; Pianka, 1973), functional analysis of morphology to deduce engineering constraints on the phenotype (Alexander, 1968, 1985; Emerson, 1978; Gans, 1974; Vogel, 1988), statistical correlation of phenotype and environment in comparative studies (Cody and Mooney, 1978; Karr and James, 1975; Miles and Ricklefs, 1984), prediction of the outcome of adaptive radiation by quantifying the morphology-resource dependence (e.g., Schluter and Grant, 1984a), and assessing the effects of adaptive plasticity on ecomorphological relationships (e.g., Sinervo and Huey, 1990; James and Nesmith, 1991; Travis, chap. 5). Each of these approaches may be complemented by analyses based upon the ecomorphological premise that the association between morphology and ecology represents the expression of the phenotype-environment interaction.

Our goal in this paper is to assess, and advocate, the application of morphological analysis to the description of community structure. We validate this approach by demonstrating strong, functionally based correlations between the distribution of species in morphological space and their distributions in ecological space. We discuss morphological convergence in the adaptations of species and the structure of communities. Finally, we summarize the results of ecomorphological analyses which have established some generalized principles of community structure.

ECOMORPHOLOGICAL ENDEAVORS

The application of ecomorphological techniques in the ecological literature of the past thirty years may be illustrated by an example. Ecomorphological analysis has been employed widely to assess certain hypotheses concerning the regulation of community organization and levels of interspecific competition between coexisting species. The work of Hutchinson (1959), MacArthur (1965), MacArthur and Levins (1967), May (1975), Schoener (1974), Vandermeer (1972), and others suggested that competition between species is a major organizing, principle in community ecology. Accordingly, many ecologists expected competitive interactions to result in resource partitioning, by which the impact of interspecific competition was reduced and coexistence made more likely. In particular, one expected to find more-or-less regular distribution of species over resource space (the community niche). In a complementary fashion, independently derived communities (for example, on different continents) encountering similar resources and other ecological conditions should exhibit similar patterns of organization, including niche partitioning. This is the principle of convergence.

The first indications that niche relationships among competing species could be embodied by morphology came from Lack (1947) and Hutchinson (1959), who suggested that the sizes of consumers (or of their trophic appendages) reflect the size array of consumed resources. Hutchinson developed the generalization that closely related species (usually congeners) exhibited a regular size hierarchy that progressed from one species to the next by a factor of 2 in body mass or about 1.3 in linear dimensions. Hutchinsonian ratios became an obsession of community ecology for several decades (e.g., Schoener, 1965; Hespenheide, 1971; Brown, 1975; among others) until Strong et al. (1979) and Simberloff and Boecklen (1981) challenged ecologists to demonstrate the regularity of size ratios statistically; the general failure of these tests threw community ecology into a state of disarray from which it has not completely recovered (Gilpin and Diamond, 1984; Wiens, 1984; Schoener, 1986).

Regardless of the controversy over the nonrandomness of size distributions, the diversification of species with respect to body size and shape (Grant, 1986; Schluter and Grant, 1984b) are well established. Convergence is also a conspicuous phenomenon (e.g., Cody and Mooney, 1978; Fuentes, 1976; Mares, 1980; Schluter, 1986; but see Blondel et al., 1984). In ecologically similar but geographically distant communities, similar morphology associated with shared locomotory behavior, mediation of the physical and structural environment, and handling of food resources suggests a rough correspondence in the ecomorphological relationships in each place.

Because competition occurs over a broad range of habitats and localities that are only partially shared by each pair of interacting species, one should not expect to find regular morphological spacing of size or shape within a local community; instead, competitive interactions are accommodated by morphological adjustment of each species to interactions summed over the whole of its range. In addition, differences between communities in the phylogenetic relationships of their respective species pools, which may reflect the geological history of the environment, may affect patterns of morphological similarity independently of interspecific relationships. The commonly perceived failure of morphological and ecological data to confirm a misconceived ecological hypothesis has unfortunately and mistakenly tarnished the broader application of ecomorphological analysis. We believe in the general utility of this approach for reasons that we outline in the following section.

THE VIRTUES AND LIMITATIONS OF ECOMORPHOLOGICAL ANALYSIS

Variables of the phenotype and of the environment constitute the axes of complementary multivariate spaces. The two are linked by the behavior and performance of the organism at an ecologically relevant task (e.g., Arnold, 1983), or the ethotype described by Ricklefs (1991), which maps each space onto the other. The general correspondence between ecology and morphology may be modified by variation in behavior or performance. We suggest that particular morphologies may accentuate or limit flexibility in the ethotype. As described by Lauder (1982), the structural complexity of morphological design may determine the breadth of performance by the individual, in the sense that a structurally simple feeding system might restrict the range of prey chosen by an individual. Consequently, the mapping of the phenotype on the ecological space is constrained by morphology through an intervening variable, be it performance or the ethotype (e.g., see Pounds, 1988; Losos, 1990b). The range of covariation between morphology and behavior indelibly affects the relationship between morphology and ecology and produces predictably correlated patterns in the two.

This relationship between morphology and ecology is portrayed in a single dimension in figure 2.1. Each species may be represented by a distribution, perhaps a bell-shaped curve in both ecological (niche) space and morphological space. The arrows represent an ecologically relevant performance variable, or set of behaviors (the ethotype) that link morphology to ecology. Each aspect of this relationship can be described by a set of movements and behavioral decision rules, whose outcome (speed, endurance, rate of prey capture, predator escape) constitutes the performance of the individual along each arrow.

Each point in morphological and ecological space can be linked by a specific behavior path, although many of these are precluded by morphological constraints (mice can't fly) or by considerations of efficiencies and energetics; that is, behavior is contingent on performance. Lizard morphological characteristics, such as width of the pelvis, length of the femur, and insertion of the caudifemoralis muscle, may directly affect such performance variables as sprint speed in a particular environment. Variation in sprint speed, in turn, may limit the range of habitats exploited by a species, just as variation in jumping or climbing performance may affect substrate and habitat choice. Thus, we might expect partitioning of the substrate or habitat niche space to correlate with those mor-

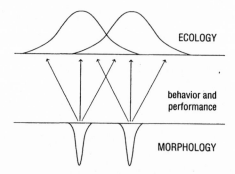

FIGURE 2.1 The relationship between morphology and ecology. Two axes are portrayed, one which reflects variation in an arbitrary ecological trait and a second which pertains to variation in a morphological trait. We assume the distribution of the phenotype may be adequately represented by a bell-shaped curve (although the precise form of the distribution is not critical). The connection between morphology and ecology (arrows) is accomplished through an intervening variable, namely behavior and performance.

phological traits associated with the relevant performance variable. We suppose that a particular morphology maps onto a specific point in ecological space with greatest utilization efficiency (perhaps measured in terms of energetic return to the organism), and that efficiency decreases with distance from that point; that is, only certain links between morphology and ecology are realized for a given set of energetically favorable conditions. We also suppose that for a set of quantitatively distributed traits (size or shape), points of greatest utilization are arranged in the ecological space in approximately the same order as in morphological space. Thus, although the utilization curves of different phenotypes may overlap broadly, each is unique and is constrained by morphology.

If one accepts the correspondence between ecology and morphology, then the advantages of ecomorphological analysis become readily apparent, as do its weaknesses. First, morphological characters are unambiguously defined as measurements of recognizable landmarks or homologous structures; they are organism-centered, and their choice by ecologists presumably reflects biomechanical principles. Morphological characters are organized into functional complexes that pertain to ecologically identifiable and relevant tasks. Second, morphological characters have high repeatability of measurement, they are measured independently of the environmental background, and they facilitate broad comparison. The last two features are especially important because measurements of behavior, diet, substrate use, or habitat selection are highly specific to particular environmental structures and cannot be compared readily between different habitats.

The limitations of ecomorphological analysis are equally apparent. First, morphology provides little information about breadth of niche utilization by a single individual or even a population, or about overlap of niche utilization between individuals or populations. In general, morphology indicates the general dispersion of phenotypes over niche space, but provides no information about within-phenotype distribution of niche occupancy. Second, morphology is less responsive than physiology and behavior to short-term temporal change in the environment during the life span of a species. Morphology integrates ecological relationships over time and space. Thus, ecomorphology presumes that such generalized patterns of correspondence between the phenotype and the environment have evolutionary significance. In many cases, this limitation will restrict the application of ecomorphological studies. Third, while morphology is measured independently of the habitat background, it is highly taxon-dependent. One cannot identically measure birds, rodents, and ants because of their different morphologies, even though they may share some resources; that is, ecomorphology depends on measurement of homologous structures. Moreover, when analogous measurements may be made, different behavioral transformations relating phenotype to the environment may preclude direct comparison.

Inferences concerning ecological relationships based on ecomorphological analyses presume a direct, consistent relationship between ecology and morphology. Validation of this relationship has taken two forms: correlation of ecological and morphological measurements (Cody and Mooney, 1978; Emerson, 1985; Karr and James, 1975; Losos, 1990b; Miles and Ricklefs, 1984) and concordance of ecomorphological correlations between different species assemblages (Miles et al., 1987). The statistical methods employed have varied tremendously among authors, but generally involve univariate or multivariate positioning of species in morphological and ecological niche spaces and calculating a correlation between these positions, often using canonical correlation techniques. Other statistical procedures may be used to assess concordance in the patterns of ecomorphological relationships; for example, Miles et al. (1987) applied analysis of covariance techniques to assess the concordance of ecomorphological relationships between avian assemblages from different regions. Functional morphological analyses and univariate analyses may serve to refine the ecological interpretation of the species positions in morphological space. In general, these analyses have supported the basic premise of ecomorphological inference. We should point out, however, that historically such analyses have not taken into account the phylogenetic structure of the data (Felsenstein, 1985).

The hierarchical nature of species relationships present two problems. First, species may show similarity in ecomorphological pattern as a consequence of phylogenetic legacy. Traditionally, species located at adjacent positions in the ecomorph space were interpreted to be examples of a common relationship between morphology and ecology. However, this may also be evidence of shared ancestry. Second, because of the varying levels of relatedness, species should not be used as independent data points in statistical analyses. Considerable analytical advances have been made in accounting for the influence of phylogeny in inter-specific comparisons (Brooks and McLennan, 1991; Harvey and Pagel, 1991; Losos and Miles, chap. 4; Miles and Dunham, 1993). Application of these new methods for removing the influence of phylogeny in ecomorphological analyses has generally validated the derived association. Therefore, we advocate that phylogenetic considerations be addressed in all future work (Garland et al., 1992; see Losos and Miles, chap. 4).

Why have ecologists generally shied away from using morphology, particularly in multivariate applications, to assess such ecological attributes as niche separation and community organization? Ecologists may trust direct measures more than indirect ones; they may doubt that morphology provides sufficient information to fully describe the subtleties of ecological relations; they may be suspicious of multivariate descriptions that combine variables into derived axes of variation whose biological meanings are inferred from examination of the position of species along those axes and the direction and magnitude of "loadings." But each of these perceived difficulties also underscores strengths of the approach. Ecologically based measurements vary so much over time of day and season as activity patterns, habitat structure, and diet change that morphology may provide a better-generalized description of ecological position. Where morphological and ecological measurements have been compared, morphology generally predicts a greater proportion of the variation in ecology than ecology does of morphology. In a study that employed canonical correlation analysis to measure the correspondence between the morphology and foraging ecology of eastern deciduous forest passerine birds, Miles and Ricklefs (1984) found a strong correlation ($R^2 = .94$). However, morphology explained a greater proportion of variation in ecology (55%) than ecology did for morphology (36%). This result suggests that morphological variables, as they are typically measured in ecological studies, contain more information pertaining to ecology—for example, habitat use, diet breadth, and foraging behavior—than the corresponding ecological measurements contain regarding morphology. Alternatively, because many possible morphological pathways may connect to a given ecological feature (Bock's multiple pathway model), such a result is not unexpected. Finally, fear of multivariate analysis generally reflects misunderstanding. Morphological spaces may be structured in such a way that factor rotations and other transformations do not distort the positions of species with respect to each other. When log-transformed variables are used and principal components analyses are based on covariance matrices, not only are the original positions of species preserved but the dimensionality and internal spacing of different samples of species are directly comparable.

Still, many ecologists fail to appreciate the ability of morphological variation to resolve subtle differences in ecology, which are believed to arise from behavioral differences between species independently of morphology. For example, MacArthur (1958) demonstrated that closely related warblers of the genus *Dendroica* (Parulinae) partitioned habitat by the different manners in which they exploited the substrate provided by spruce trees in boreal forest. MacArthur emphasized the similar morphologies of the species and represented the warblers as an example of behavioral niche partitioning. However, Leisler et al. (1989) demonstrated a close relationship between detailed measurements of suites of morphological characters from three functional complexes (comprising forty-eight skeletal and external variables) and twelve breeding habitat variables that captured subtle aspects of behavior and habitat selection in six species of European warblers in the genus *Acrocephalus* (Sylviidae). Traditionally, ecologists have used simple measures of morphology, such as beak dimensions and lengths of appendages. The success of ecomorphological analysis will depend to a large degree on selection of morphological characters based upon an understanding of their consequences for the functioning of the organism and their relationship to other characters of the phenotype.

GENERALIZATIONS FROM ECOMORPHOLOGICAL ANALYSES

Adaptation and Convergence

Much has been written concerning the general relationship of morphology to the environment. Correlations of structures with ecology and behavior emphasize the intimate link between phenotype and environment. In some cases, such correlations have been elevated to the status of ecomorphological "rules" (e.g., Bergmann's, Allen's, and Gloger's rules) although the functional basis of these patterns has been the subject of considerable controversy (Stevenson, 1986). One aspect of ecomorphological analysis—elucidation of the adaptive nature of the phenotype, in the sense that particular phenotypes maximize fitness in particular environments—has been demonstrated by studies of survival and reproductive success of alternative phenotypes (Endler, 1986), and by engineering analyses of the function of certain structures (Lauder, 1989, 1990) and the extrapolation of considerations of mechanical and physiological efficiencies to natural settings (e.g., Jayne, 1988; Jayne and Bennett, 1989, 1990; Wainwright, 1987, 1988).

Interpretation of the phenotype-environment correlation as having resulted from evolutionary adaptation of the phenotype to the particular environment—as opposed to selection of environments by phenotypes that have arisen by nonadaptive mechanisms (e.g., see Lauder, 1981)—has been based in part on the phenomenon of convergence. Convergence of form and function in similar environments from dissimilar ancestry implies selection by environmental factors, although habitat choice cannot be excluded. Convergence may be inferred from the similarity of unrelated organisms in the same place, such as the common foliage structure of many species from different families in desert and Mediterranean-climate environments (Cody and Mooney, 1978; Mooney and Dunn, 1970; Mooney, 1977; Orians and Solbrig, 1977). More frequently, convergence is inferred from similarity in appearance in organisms living in geographically isolated places under similar climates, such as the intercontinental comparisons of vertebrates in rainforest (Keast, 1972; Bourlière, 1973) and desert (Pianka, 1971; Mares, 1976).

Often convergence of function is not particularly well matched by convergence in form (Collins and Paton, 1989; Paton and Collins, 1989). For example, Lack (1947) contrasted the adaptations of various bark-feeding birds in places from which woodpeckers are absent. Woodpeckers trench and chip bark with their beaks and probe into holes and crevices with their long tongues. The Hawaiian honeycreeper *Heterorhynchus* taps with its short lower mandible and probes with its long upper mandible; the Galapagos woodpecker finch trenches with its short beak and probes with a cactus spine held in its beak; in the extinct New Zealand huia, males excavated with their short beaks and females probed with their long decurved beaks (the two usually fed in pairs). Thus, behavior appears to have allowed divergent morphologies to converge on rather more similar ecological positions. We presume that the woodpecker morphology utilizes this particular region of the ecological space most efficiently, and alternatives arise and persist only in the absence of this widespread group from isolated islands.

Electric fish from the families Mormyridae and Gymnotidae provide an example of convergence of form and function. Species in the former family occur in Africa, and the latter in South America. Both groups inhabit river systems that contain high levels of particulate matter, obscuring vision. The two families have converged in morphological form, perhaps reflecting constraints in the production of electric fields from muscle-derived tissues. In addition, their feeding and social behavior are also convergent (Lannoo, pers. comm.; Hopkins, 1974).

Other studies of convergence which have applied ecomorphological data have shown that similarity in form may reflect a common phyletic heritage (e.g., Blondel et al., 1984; Mares, 1976; Niemi, 1985). These studies attempted to ascertain the degree of ecomorphological similarity among species within a particular habitat type, say desert, relative to their closest congeners in habitats of different structure and generally have revealed a strong phylogenetic component to morphology independent of habitat (Blondel et al., 1984; Niemi, 1985).

Schluter (1986) made the useful distinction between similarity of appearance and convergence from dissimilar ancestors. Comparisons among different communities may reveal a greater degree of similarity in some characteristic, say mean body size, than expected under a null model of random variation. Such evidence of similarity does not support the hypothesis of convergence. Schluter (1986) suggested that unequivocal documentation of convergence must demonstrate that species from different communities in similar habitats show greater similarity than their ancestors. He also discussed analysis of variance techniques that allows one to quantify an index of convergence. When one characterizes an array of species utilizing different habitats or niche space within habitats (ecology) in two or more regions having taxonomically distinct biotas (geography, history), one may relate variation in morphology to variation in ecology and region. The ecology effect represents the "convergence" fraction of variation, that is, that part of the variation which is correlated with the environment. The regional effect represents the imprint of history and the failure of adaptation to fully mask ancestry. The roles of natural selection by the environment and habitat choice by the phenotype cannot be separated by this analysis, but this is immaterial to the purposes of ecomorphology.

In applying the ANOVA model for assessing convergence, Schluter (1986) compiled finch species lists from twenty-four localities representing nine habitat types and five biogeographic regions. Patterns of covariation in five external morphological traits were summarized using principal components analysis, and the species scores along the first two axes (representing "size" and "shape") were used in the model. Significant convergence (about 60% of the total body size variation) was demonstrated for size and shape in the finch communities. Historical causes, that is, a significant effect of geographic region, were evident only in shape variation (40%).

Community Convergence

Tests of convergence have also been applied at the community level to examine whether the niche space in different localities with similar environments is occupied to the same extent and whether similar positions are filled within the niche space as a whole. Several studies have applied morphological data to address these questions. Keast (1972) pointed out that the South American avian family Tyrannidae is distributed over a morphological space occupied by a diverse array of families in Africa that fill most of the same ecological niches. In a community-wide analysis, Karr and James (1975) suggested that morphological distributions of passerine birds in similar habitats in Panama and Liberia occupied similar space, although they did not provide a statistical evaluation of similarity. In contrast to the general finding of morphological comparability, strikingly different morphological spaces are occupied by lizards in desert habitats in Australia, Africa, and North America (Ricklefs et al., 1981) and by passerine birds in temperate forests in Australia and North America (Ricklefs, Ford, and Miles, unpubl.).

Several authors have suggested a one-to-one matching of species between different regions, indicating the presence of discrete niches that are filled by particular phenotypes in each region, or of similar extrinsic selective pressures that yield a common morphology. For example, Cody (1975) pointed out ecological and morphological counterparts between unrelated sets of birds in Mediterranean habitats in Chile and southern California, as did Fuentes (1976) and Pianka (1986) for assemblages of lizards. However, in none of these cases were statistical criteria for species-matching employed. Ricklefs and Travis (1980) suggested that when species-for-species convergence has occurred, the nearest neighbor of a species in morphological space will be its counterpart in a convergent fauna rather than another species in the same fauna. In an analysis of Cody's (1974) communities, with species distributed in an eight-dimensional morphological space based on external measurements, Ricklefs and Travis (1980) determined that the faunas as a whole occupied statistically indistinguishable morphological spaces but that nearest neighbors occurred as often in the same fauna as in the corresponding fauna in the other region. Thus, they rejected the notion of species-for-species matching, as did Wiens (1990), using similar analytical techniques, for birds of shrub-steppe habitats in western North America and Australia. Schluter (1990) has recently applied a general statistical technique for assessing the degree to which species exhibit species-for-species matching.

While studies of convergence, both ecological and morphological, date back to the beginnings of ecological and evolutionary study, they have been mostly anecdotal and unstatistical. Recent quantitative analyses reveal both convergent and historical (ancestry) effects on morphology, at the levels of both the individual organism and the biotal assemblage (Schluter and Ricklefs, 1993), but fail to support the notion of species-for-species matching and the implication of discrete niches. These results will depend, of course, on the particular communities and traits examined, as some dimensions of ecological space are organized as discrete, noncontinuous variables, at least within a locality (e.g., hosts for parasites and herbivores, resting backgrounds and other hiding or nesting places, predators). We feel that ecologists have hardly begun to explore the phenomenon of convergence and that ecomorphological approaches should play an important role in the development of this area.

Community Organization

Patterns of morphological variation and similarity have been used extensively to study community organization since Hutchinson's (1959) remarks on the constancy of size ratios among congeners within assemblages of species. The principal concern of these studies has been the regularity of spacing between species along morphological dimensions as a measure of the evolutionary effects of competitive interactions within communities. Although, as we have pointed out above, the prediction of regular spacing does not necessarily follow from local competition, studies of spacing have emphasized the utility of morphological analyses in community structure and have led to the development of statistically valid approaches to the problem (Ricklefs and Travis, 1980; Simberloff and Boecklen, 1981).

The Hutchinsonian tradition of ecomorphology has restricted analyses primarily to comparisons within groups of closely related species, as these are more likely to compete with each other, and their similarity of structure lends some degree of confidence to ecological interpretations of size ratios. Two other directions of inquiry have substantially expanded the scope of community analysis. The first of these centers about the concept of the guild (Root, 1967; Simberloff and Dayan, 1991), namely, that assemblages of species are subdivided into subsets that may be distinguished by particular suites of ecological characteristics, regardless of taxonomic identity. Thus within an avian community, one may distinguish a foliage-gleaning guild, bark-gleaning guild, ground-foraging guild, and so on. Among other groups, guilds have been distinguished by prey type as well as microhabitat selection, among other factors. It is reasonable to ask whether guilds represent artificial boundaries placed within a continuous distribution of species in ecological or morphological space, or discrete clusters of species separated ecologically perhaps by discontinuities in the resource space, with corresponding clustering in morphological space.

The usual techniques for analyzing guild structure are various ordination analyses used in conjunction with a clustering algorithm (e.g., Holmes et al., 1979; Poysa, 1983), which groups species according to their similarities, but does not address the discreteness of the groups in ecological or morphological space. In general, the problem of clustering must be explored by various resampling procedures—for example, Monte Carlo approaches, bootstrapping or jackknifing—which have yet to be applied to community analyses (e.g., MacNally and Doolan, 1986). While species sometimes appear to occupy discrete clusters in morphological space, particularly in small assemblages where some species have distinct morphological roles (e.g., stream-fish clades [Douglas, 1987]; coniferous-forest passerines [Norberg, 1979, 1981]; insectivorous bats [McKenzie and Rolfe, 1986]; bone-crushing carnivores [Van Valkenburgh, 1988]), more general appraisals remain to be done.

The second direction is an appraisal of community-wide occupancy of morphological and ecological space combined with estimates of species packing within the community volume. This approach arose out of considerations of the regulation of species diversity but was developed more in the tradition of descriptive (albeit quantitative, multivariate) natural history, rather than the testing of particular hypotheses about community structure. Before describing the results of such studies and generalizations that have emerged, we must discuss various methods of portraying the structure of the morphological space itself.

MULTIVARIATE DESCRIPTION OF MORPHOLOGICAL SPACE

The portrayal of species in morphological space depends upon the axes used. In general, the primary data are various linear measurements, although body mass occasionally is included in the data set. In this discussion we shall ignore discrete variables, such as number of teeth or digits, and binary variables, such as the presence or absence of a particular trait; these traits may be included in alternative analyses. Rather, we shall concentrate solely upon continuously distributed variables, although this requires the setting aside of potentially important information. By restricting comparisons to relatively small taxonomic groups, which generally are uniform with respect to discrete and binary variables, one sidesteps this problem while accepting the limitation of taxonomic scope. As in any descriptive analysis, the portrayal of species in morphospace entails the resolution of certain compromises.

Morphological spaces have been established by using raw measurements as axes (Leisler, 1980; Douglas, 1987; Ricklefs and Travis, 1980), combinations of measurements and ratios of measurements to depict shape (Karr and James, 1975; James and Boecklen, 1984), a single measurement to depict size and the ratios of other measurements to the size variable to depict shape (Mosimann and James, 1979; Leisler, 1980; Niemi, 1985), linear combinations of the original variables (especially principal components) to reduce redundancy and the dimensionality of the space (Ricklefs and Travis, 1980), and various transformations of the original variables, particularly the logarithmic transformation. Many of these techniques are reviewed by James and McCulloch (1990) and Rohlf (1990, and references therein).

We favor the approach used by Ricklefs and Travis (1980), in which axes are the logarithmically transformed original variables. The log transformation tends to produce multivariate normal distributions of species in morphospace. Furthermore, distances correspond to factorial differences between points rather than additive differences, which we feel is biologically realistic and conforms to the general independence of factorial differences between species and the average value of their measurements. Vectors in the morphological space correspond to products and ratios of original measurements and may be interpreted roughly as size and shape.

Regardless of the species or measurements used, samples of species can be compared directly with regard to the dimensionality and volume of morphologi-

TABLE 2.1. Dispersion of species on the first three principal component axes of morphological spaces

Axis	Lizards[a]	Birds[b]
I	0.577[c]	0.380
II	0.175	0.115
III	0.104	0.102

[a] Pooled sample of eighty-three species from deserts of Australia, North America and Africa (Ricklefs et al., 1981). Nine external measurements were taken: PC Axis I characterized a composite measure of body size; PC Axis II contrasted head and limb measurements; PC Axis III contrasted forelimb and hindlimb measurements.
[b] Pooled sample of seventy-six species of passerine birds from scrub communities in temperate North and South America (Ricklefs and Travis, 1980). Eight external measurements were taken: PC Axis I may be interpreted as a measure of overall size; PC Axis II represented a contrast between tarsus and culmen length versus culmen depth and width; PC Axis III depicted a contrast between bill depth versus wing length.
[c] Square root of the eigenvalue, equal to the standard deviation of the orthogonal projections of species onto each of the axes, a measure of the size of each dimension.

cal space occupied, provided the same number of measurements are used to construct the space. Thus, although the morphological spaces occupied by lizards and birds do not overlap, one may nonetheless directly compare the dimensions of the spaces (table 2.1). In this example, the morphological volume occupied by lizards exceeds that of birds, suggesting greater variety of size and body shape associated with the greater variety of ecological relationships exhibited by lizards. Finally, we advocate the use of principal components analysis (or any eigenanalysis) based on the covariance matrix to explore the pattern of the morphological space. This technique preserves the original positions of species in morphological space without distortion and is amenable to direct statistical comparison of the variance-covariance matrices. Standardization of the variables (as in calculating a correlation matrix) distorts the position of species in the morphological space, thereby preventing comparison among studies. A common practice in most ecomorphological analyses is to determine community structure of the morphological space by interpreting the loadings of the original variables on the principal components; often, however, only the first two or three PC axes are retained (see Gibson et al., 1984). Because Ricklefs and Travis (1980) and Travis and Ricklefs (1983) found that many of the smaller principal components contained ecologically relevant information, we recommend examination and retention of all the principal components for subsequent analysis.

SPECIES PACKING AND SPECIES DIVERSITY

One of the important ways in which ecologists have utilized morphological space is to explore the pattern of spacing between species and the relationship of niche packing to community diversity (Fenton, 1972; Findley, 1976; Gatz, 1979). If species evolve to reduce the deleterious population consequences of interspecific competition within local assemblages, one would expect to find significantly nonrandom (hyperdispersed) distribution of species in ecological (hence morphological) space. If similarity between species is constrained by competitive interactions (MacArthur and Levins, 1967; May, 1975) and communities are generally saturated with species, then the addition of species to a community can be accomplished by increasing the total volume of niche (= morphological) space occupied while maintaining as a constant the average distance between species. Alternatively, diverse communities may be built up by inserting species within the established ecological and morphological space, reducing the average distance between species centroids either by reducing average niche size or increasing niche overlap (MacArthur, 1972). Niche volume has been calculated in a variety of ways; species packing is often equated with the average distance between each species and its nearest neighbor or as the average segment length of a Prim network (minimum spanning tree) connecting all species.

A variety of studies have addressed the relationship between community volume and average nearest-neighbor distance (NND, the density of species packing) in morphological space (table 2.2). In all cases, community volume either increases with diversity or is not significantly correlated with diversity. In the case of Ricklefs and Travis's (1980) study of Mediterranean bird communities, most of the increase in community volume occurred along the fifth and seventh of eight principal components of the log-transformed morphological space. The survey of studies in table 2.2 also clearly demonstrates that NND does not vary with species diversity in comparisons of similar communities. Thus, species do not appear to be added at random to the community. Rather, increased diversity is accommodated at the periphery of existing morphological space, and internal spacing appears to be more or less conserved. A few exceptions to this general pattern indicate that it is not an artifact of some mathematical manipulation. This pattern can also be checked by randomization tests in which communities of different sizes are assembled at random from pools of potential species or synthesized from the morphological space occupied by the entire pool of species (Ricklefs and Travis, 1980). These randomly assembled communities consistently differ from the real communities on which they are based by showing no relationship between community morphological volume and diversity and a negative relationship between NND and diversity. These comparisons strengthen the notion that communities have an inherent structure that is determined at least in part by interactions between their members.

Randomized communities, in which species are assigned to communities without regard to other species present (hence they are noninteractive commu-

TABLE 2.2. Ecomorphological assessments of the relationship between niche packing and species diversity

Study	Taxon	N	Vol	NND	Null model
			Relation to diversity		
Random assemblage	—	—	NS	neg	—
Karr and James (1975)	birds	3	—	—	—
appendage length	—	—	pos	pos	—
bill dimension	—	—	pos	NS	—
Bierregaard (1978)	raptorial birds	35	pos	NS	yes
Ricklefs and Travis (1980)	passerine birds	11	pos	NS	yes
Travis and Ricklefs (1983)	island birds	5	pos	neg	—
Blondel et al. (1984)	bird guilds	6	pos	—	—
Miles (1985)	passerine birds	5	pos	NS	yes
Fenton (1973)	bats	3	(pos)	—	—
among continent	—	3	NS	NS	—
among habitat	—	32	NS	neg	yes
Findley (1973)	bats	6	pos	NS	—
Shum (1984)	bats	—	pos	NS	—
Schiebe (1987)	lizards	20	pos	NS	yes
Strauss (1987)	fish	7	pos	NS	yes
Gatz (1979)	fish	3	pos	NS	—
Winemiller (1991)	fish	5	(pos)	NS	—

Note: NND = nearest-neighbor distance; vol = morphological volume; pos = positive relationship; (pos) = marginally significant positive relationship ($0.05 < p < 0.10$); neg = negative relationship; NS = nonsignificant.

nities), also allow one to assess the regularity of species packing within the morphological space. In real communities average distance between nearest neighbors is generally conserved, suggesting a limiting similarity established by interactions between species. Even so, the variation in NNDs (low standard deviations representing regular spacing, i.e., hyperdispersion) within real communities and randomly constructed ones is similar. This does not strike us as inconsistent because, as we mentioned earlier, interactions that select upon morphology occur throughout the range and habitat distribution of each species, which are rarely congruent for all species within a local community. Indeed, evidence supporting nonrandom spacing has come primarily from small, depauperate islands where most species are broadly distributed across habitats (MacArthur et al., 1972; Cox and Ricklefs, 1977) and congruence of populations is high (Travis and Ricklefs, 1983).

FUTURE ECOLOGICAL APPLICATIONS OF MORPHOLOGICAL DATA

We emphasize two important areas of endeavor. The first is validation of the ecomorphological relationship and the seeking of a greater understanding of how behavior maps the phenotype onto ecological space. The second is the continued application of the ecomorphological paradigm to explore basic questions of ecology and evolution, particularly those related to adaptation and community structure. With respect to the first goal, we advocate exploration of the links between morphology and such ecological characteristics as habitat choice, substrate use, occupancy of biophysical space, diet breadth, and so on. While comparative studies may be used to establish general patterns, intraspecific studies of ecomorphological correlation (e.g., Arnold and Bennett, 1988; Herrera, 1978) should allow investigators to refine these relationships. These detailed ecomorphological analyses may provide extensive descriptions of the behavioral filter between ecology and morphology. For example, Miles (in preparation) estimated the relationship between components of locomotory performance—that is, maximum velocity, burst speed, mean velocity, stride frequency, and stride length—and eight external morphological traits in two populations of *Urosaurus ornatus*. Sprint speed was highly correlated with morphology (fig. 2.2). Maximum velocity was significantly related to hind-leg (thigh) and crus (shank) length.

The ecological relations of a particular phenotype at different seasons (e.g., Carrascal et al., 1990) and in different habitats should provide clues to the flexibility of the ecomorphological relationship as it is determined by behavior, including physiological response and habitat selection. Geographical comparisons

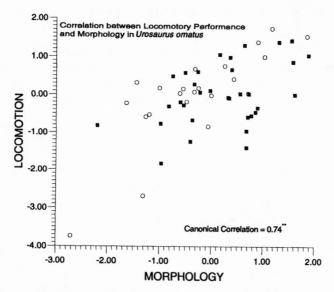

FIGURE 2.2 Summary of a canonical correlation analysis testing the association between locomotory performance and morphology in *Urosaurus ornatus*. *Circles* represent the position of males in the ecomorphological space and *squares* represent females. Axes are linear combinations of the original variables and are centered on zero. ** $P < .01$.

of ecomorphological relationships should demonstrate habitat-related correlations between morphology and ecology.

Measures of survival and reproductive success of individuals in different parts of the morphological space (Arnold, 1983), combined with ecomorphological correlations and functional analyses, will enable us to gain a better understanding of the fitness consequences of different morphologies and their evolutionary implications. So far, few analyses have adequately demonstrated the operation of selection on morphology by means of the performance and fitness consequences of morphological variation (Benkman and Lindholm, 1991; Boag and Grant, 1981; Grant, 1986; Grant and Grant, 1989; Jayne and Bennett, 1990; Miles, in preparation).

Ecomorphologists must also confront the issues of how many and which variables to measure. Generally, more is better, but too many independent variables reduce error degrees of freedom in statistical analyses, forcing one to use principal components or other measures to reduce the intercorrelations among the variables and dimensionality of the data set. These issues will probably be best resolved by some combination of functional analysis of traits (e.g., Bock, 1977; Norberg, 1979) and empirical work with large data sets. We wish to stress that meaningful interpretations of the morphological space requires a judicious selection of morphological variables reflecting a priori biomechanical function. Thus, we reiterate Bock's (1977) plea for integration between ecologists and functional morphologists in the design and analysis of ecomorphological studies.

Another problem is the confounding of phylogenetic history and adaptation to local environmental conditions (e.g., Cadle and Greene, 1993; Gould and Vrba, 1982; Lauder, 1981, 1982; Losos and Miles, chap. 4). Each species carries with it vestiges of its history of adaptation. For example, the dispersion of avian species in the morphological space characterized by Ricklefs and Travis (1980) was affected, in part, by phylogenetic affinity. This is evident in phylogenetically informative traits, but may also be true of characters that directly influence various aspects of ecological relationships. Because morphology, behavior, and ecology are directly linked, historical and contemporary aspects of this complex are inseparable. Indeed, the distinction between phylogenetic constraint and local adaptation may be a nonissue, in part, in the context of ecomorphological analysis, because the phenotype of a species limits the variety of environments in which it can function competitively (i.e., in the face of varying suites of predators, competitors, and range of resource availability), regardless of its evolutionary conservatism. However, we must be cautious in our interpretation of ecomorphological pattern, for the structure elucidated may more accurately reflect the contribution of a subset of related taxa than that of the community as a whole. For example, examination of the species included in Felley's (1984) analysis of community organization in fish reveals that out of forty-three species, twenty-six belonged to the genus *Notropis*. Thus, the analysis may suggest more about phenotype-environment relationships within *Notropis* than the Cyprinidae as a whole. The phylogenetic constraint versus adaptation argument may pertain largely to the evolution of phenotype-environment correlations and the evolutionary flexibility of certain characters, not the basic adaptive premise of the ecomorphological relationship.

At the risk of belaboring obvious points, we emphasize once more that (1) ecologists must pay attention to the functional implications of phenotypic characters and morphologists should consider the ecological context of characters through behavior in a natural setting; (2) morphology, physiology, behavior, and ecology constitute a single suite of intimately interrelated and basically interpredictable characters that define the organism in a particular environmental setting; and (3) morphological and ecological data are multivariate and must be dealt with as such, implying that ecomorphologists must use multivariate techniques to embrace the complexity of their systems rather than simplify them by reducing dimensions.

UNREALIZED POTENTIAL OF ECOMORPHOLOGICAL ANALYSIS

Finally, we have thought about a set of questions that reflect our own interests in the phenotype-environment relationship and indicate some of the potential of ecomorphological analysis. This list is by no means complete, but we hope it will be suggestive. The discipline of ecomorphology attempts to integrate morphological traits with aspects of performance in an ecological context. Yet, we are largely ignorant of how to characterize the fit between morphology and ecology and how to determine the goodness of fit. Many studies have found strong correlations between the morphological characteristics of an organism and aspects of its ecology, such as diet, foraging behavior, foraging height, and locomotory performance (table 2.3). However, in many instances the nature of the behavioral transformations (ethotype) that relate them is relatively unexplored (e.g., see Losos, 1990b). Indeed, may the ethotype itself be considered as a set of characters having its own multidimensional space sandwiched between morphology and ecology? Borrowing from Arnold (1983), we may ask how strong is the path from morphology to behavior/performance, and from performance to ecology. Interspecific comparisons that employ canonical correlation analyses to match variation in morphology with ecology, as referred to in this paper (e.g., Losos, 1990b; Miles and Ricklefs, 1984; Miles et al., 1987), provide a tentative beginning to answer these questions, but our understanding here is presently quite primitive.

Most ecomorphological analyses compare the variation in morphology

TABLE 2.3. Summary of multivariate analyses demonstrating phenotype-environment correlations among vertebrates

Study	Number of species	Number of sites	Number of morph. variables	Ecological variables	Correlation
			FISH		
Felley (1984)	138	43	17	Diet	—[a]
Wikramanayake (1990)	12	1	10	Diet	0.61
			LIZARDS		
Losos (1990b)	15	2	6	Substrate	0.80
				Performance	0.98
Scheibe (1987)	29	20	13	Habitat	0.80
			BIRDS		
Karr and James (1975)	136	4	17	Foraging behavior	0.81
Leisler and Winkler (1985)	17	1	6	Foraging behavior	0.91
Cody and Mooney (1978)	97	3	7	Foraging height	0.75
				Foraging behavior	0.81
				Ecology	0.75
Miles and Ricklefs (1984)	19	1	8	Foraging behavior	0.97
Miles et al. (1987)	19	2	8	Foraging behavior	0.95
	38	3	8	Foraging behavior	0.93

Note:
[a] The correlation between morphology and ecology was found to be significant, but the correlation coefficient was not given.

among a set of species with variation in ecology. The morphological variables are assumed to have some functional role in the environmental context (e.g., Moermond, 1979). However, estimates of the form-function complex are mainly derived from intraspecific functional analyses and rarely interspecific comparison. Thus, can we predict that the relationship between intraspecific and interspecific ecomorphological relationships is uniform? Do both sets occupy similar canonical correlations, with the intraspecific sets of data having smaller variance? Or do the two types of data have fundamentally different organization? If they do differ, how is this organization changed during the course of the evolutionary diversification of species? An initial analysis comparing intraspecific and interspecific morphology-performance relationships among an assemblage of rhacophorid frogs suggests that ecomorphological associations are not uniform (Emerson, pers. comm.). The statistical relationships between performance and morphology, based upon biomechanical predictions, were evaluated for each species. Comparison of the correlations generated within species with those generated among species revealed inconsistencies in the ecomorphological associations. This pattern is all the more intriguing for the predicted correlations, as derived from biomechanical principles, should have been equivalent.

What is the role of history in generating morphological diversity (Lauder and Liem, 1989)? Does the contemporary morphological space depend upon phylogenetic history, that is, the particular taxa represented in the sample of species? Analyses performed by Strauss (1987) were designed to evaluate the influence of phylogenetic constraint on the morphological structure of assemblages. In a comparison of stream fish assemblages between North America and South America, Strauss (1987) found significant clustering in the morphological space. He attributed this clustering to phylogenetic relatedness of the taxa included in the study. Future studies should link comparative and phylogenetic approaches to the analysis of ecomorphological relationships (Schluter, 1989). The analysis of ecological differences among Old World leaf warblers (genus *Phylloscopus*) by Richman and Price (1992) exemplifies such an approach. In this analysis, they applied Felsenstein's (1985) method of independent contrasts to determine the evolution of the association between morphology and ecology among eight species of *Phylloscopus* warblers. Strong correlations between morphology and feeding ecology and habitat selection were detected after removing the influence of phylogeny. Their analysis also revealed that the foraging behaviors and feeding sites characterizing the warblers probably evolved quite rapidly and early during the diversification of the clade. Do ecomorphological relationships differ between conservative and evolutionarily more flexible characters?

What is the dimensionality of morphological space and does this correspond to the dimensionality of the environment? How does dimensionality increase with the number of characters included in the analysis? To what degree is the phenotype organized into suites of characters from ecomorphological, functional, and genetic viewpoints? How can one assess the dimensionality of the environment independently of canonical correlations with morphological traits? That is, will the apparent dimensionality of the environment increase with the number of morphological traits distinguished, or can one identify its intrinsic structure? Does the environment limit the volume and dimensionality of the morphological space occupied by a community? Are there gaps in morphological space, representing possible phenotypes that are not realized? What produces such gaps and how do these map onto ecological space? Conversely, are there clusters of species in morphological space, and do these represent ecological guilds or does behavior transform morphological clusters into an ecological continuum (e.g., Norberg, 1986).

Another area of interest to both ecologists and morphologists concerns the evolution of specialization. Can one identify a habitat or resource specialist from its position in morphological space? Conversely, does specialization in the ecological space necessarily allow predictions of morphological specialization? Are such traits as specialization organized along certain dimensions of the morphological space or are they distinguished by central versus peripheral position? Preliminary analyses by Miles (1985) demonstrated a significant difference between habitat generalists and habitat specialists in an assemblage of Sonoran Desert passerine birds. Generalists tended to exhibit larger body sizes relative to specialists as evidenced by the position of these species in a morphological space defined by the first two axes of a principal component space (fig. 2.3). In addition, the ecomorphological relationships differed between specialists and generalists. The volume occupied by generalists and specialists also seems to differ. Leisler (1980) showed greater amounts of morphological differentiation related to habitat preferences among genera of sylviid warblers with broad habitat preferences. The morphological differentiation was also described to affect coexistence mechanisms among the species' morphological space. Of related interest, what is the ecological "cost" of morphological specialization? For example, Andrews et al. (1987) found that modifications of the head in a fossorial species of skink (*Chalcides ocellatus*) reduced its feeding performance (increase in handling time and limits to the size of prey) compared to a less modified, nonfossorial skink, *Eumeces inexpectatus*.

Does distance along each dimension in morphological space define equivalent ecological separation? Total morphological space occupied by a community is a multivariate ellipse, but is the same true of the ecological space of the community or of the individual species that constitute the community?

How similar are the morphological spaces, and ecomorphological relationships, of different groups of species? Are levels of species packing conserved among disparate groups, or does this bear a consistent relationship to ecology?

How do the morphological space and ecomorphological relationship compare at different taxonomic levels, such as species within families versus family means? Can one infer anything about evolutionary diversification from these patterns (e.g., Schluter, 1984)? Can we infer levels of morphological differentiation associated with speciation? Morphological data have been used to distinguish between modes of evolution—for example, rectangular versus gradual (e.g.,

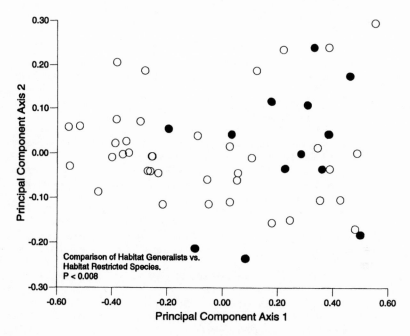

FIGURE 2.3 Plot of the positions of fifty-one passerine bird species from five habitat types in the Sonoran Desert along the first two principal component axes based on eight external characters. *Open circles:* habitat specialists species, i.e., species that are restricted to a single habitat type; *solid circles:* habitat generalists, i.e., species that are found in two or more habitats. Axes are linear combinations of the original variables and are centered on zero.

Douglas and Avise, 1982; Lemen and Freeman, 1989)—but the absence of an appropriate null model has hindered the general applicability of such analyses (e.g., Lynch, 1990).

CONCLUSION

Ecomorphology has been an integral tool of ecologists and evolutionary biologists in elucidating pattern and interpreting processes. The next decade should witness expanded studies of the mapping of morphology onto ecology and of the underlying studies of functional morphology necessary to understanding this close coupling. Growing utilization of multivariate methods will more closely match the complexity of analyses to the complexity of natural systems. By combining experimental, modeling, and observational studies on restricted taxa with more broadly comparative analyses, the ecomorphological approach should clarify fundamental issues in evolutionary adaptation and the development of biological communities.

REFERENCES

Alexander, R. McN. 1968. *Animal Mechanics*. London: Sidgwick and Jackson.

Alexander, R. McN. 1985. Body support, scaling and allometry. In *Functional Vertebrate Anatomy,* ed. M. D. Hildebrand, B. Bramble, K. F. Liem, and D. B. Wake, 26–37. Cambridge: Belknap Press.

Andrews, R. M., F. H. Pough, A. Collazo, and A. de Queiroz. 1987. The ecological cost of morphological specialization: Feeding by a fossorial lizard. *Oecologia* 73:139–145.

Arnold, S. J. 1983. Morphology, performance and fitness. *Amer. Zool.* 23:347–361.

Arnold, S. J., and A. F. Bennett. 1988. Behavioural variation in natural populations. V. Morphological correlates of locomotion in the garter snake (*Thamnophis radix*). *Biol. J. Linnean Soc.* 34:175–190.

Benkman, C. W., and A. K. Linholm. 1991. The advantages and evolution of a morphological novelty. *Nature* 349:519–520.

Bierregaard, R. O. 1978. Morphological analyses of community structure in birds of prey. Unpublished Ph.D. dissertation. Philadelphia, Pennsylvania, University of Pennsylvania.

Blondel, J., F. Vuilleumier, L. E. Marcus, and E. Terouanne. 1984. Is there ecomorphological convergence among mediterranean bird communities of Chile, California and France? *Evol. Biol.* 18:141–213.

Boag, P. T., and P. R. Grant. 1981. Intense natural selection in a population of Darwin's Finches (*Geospizinae*) in the Galapagos. *Science* 214:82–85.

Bock, W. J. 1977. Toward an ecological morphology. *Die Vogelwarte* 29:127–135.

Bourlière, F. 1973. The comparative ecology of rainforest mammals in Africa and tropical America: Some introductory remarks. In *Tropical Forest Ecosystems in Africa and South America: A Comparative Review,* ed. B. J. Meggars, E. S. Ayensu, and W. D. Duckworth, 279–292. Washington, D.C.: Smithsonian Institution Press.

Brooks, D. R., and D. A. McLennan. 1991. *Phylogeny, Ecology, and Behavior: A Research Program in Comparative Biology.* Chicago: University of Chicago Press.

Brown, J. H. 1975. Geographical ecology of desert rodents. In *Ecology and Evolution of Communities,* ed. M. L. Cody and J. M. Diamond, 315–341. Cambridge: Belknap Press.

Cadle, J. E., and H. W. Greene. 1993. Phylogenetic patterns, biogeography, and the ecological structure of neotropical snake assemblages. In *Species Diversity in Ecological Communities: Historical and Geographical Perspectives,* ed. R. E. Ricklefs and D. Schluter. Chicago: University of Chicago Press.

Carrascal, L. M., E. Moreno, and J. L. Tellaria. 1990. Ecomorphological relationships in a group of insectivorous birds of temperate forests in winter. *Hol. Ecol.* 13:105–111.

Cody, M. L. 1974. *Competition and the Structure of Bird Communities*. Princeton: Princeton University Press.

Cody, M. L. 1975. Towards a theory of continental species diversities: Bird distributions over Mediterranean habitat gradients. In *Ecology and Evolution of Communities,* ed. M. L. Cody and J. M. Diamond, 214–257. Cambridge: Belknap Press.

Cody, M. L., and H. A. Mooney. 1978. Convergence versus nonconvergence in mediterranean-climate ecosystems. *Ann. Rev. Ecol. Syst.* 9:265–321.

Collins, B. G., and D. C. Paton. 1989. Consequences of differences in body mass, wing length and leg morphology for nectar-feeding birds. *Aust. J. Ecol.* 14:269–289.

Cox, G. W., and R. E. Ricklefs. 1977. Species diversity, ecological release, and community structuring in Caribbean land bird faunas. *Oikos* 29:60–66.

Douglas, M. E. 1987. An ecomorphological analysis of niche packing and niche dispersion in stream fish clades. In *Community and Evolutionary Ecology of North American Stream Fishes,* ed. W. J. Matthews and D. C. Heins, 144–149. Norman: University of Oklahoma Press.

Douglas, M. E., and J. Avise. 1982. Speciation rates and morphological divergence in fishes: Tests of gradual versus rectangular modes of evolutionary change. *Evolution* 36:224–232.

Emerson, S. B. 1978. Allometry and jumping in frogs: Helping the twain to meet. *Evolution* 32:551–564.

Emerson, S. B. 1984. Morphological variation in frog pectoral girdles: Testing alternative to a traditional adaptive explanation. *Evolution* 38:376–388.

Emerson, S. B. 1985. Skull shape in frogs: Correlations with diet. *Herpetologica* 41:177–188.

Emerson, S. B. 1988. Testing for historical patterns of change: A case study with frog pectoral girdles. *Paleobiology* 14:174–186.

Endler, J. 1986. *Natural Selection in the Wild.* Princeton: Princeton University Press.

Felley, J. D. 1984. Multivariate identification of morphological-environmental relationships within the Cyprinidae (Pisces). *Copeia* 1984:442–455.

Felsenstein, J. 1985. Phylogenies and the comparative method. *Amer. Nat.* 125:1–15.

Fenton, M. B. 1972. The structure of aerial-feeding bat faunas as indicated by ears and wing elements. *Can. J. Zool.* 50:287–296.

Findley, J. S. 1976. The structure of bat communities. *Amer. Nat.* 110:129–139.

Fuentes, E. R. 1976. Ecological convergence of lizard communities in Chile and California. *Ecology* 57:3–17.

Gans, C. 1974. *Biomechanics.* Philadelphia: Lippincott.

Garland, T., Jr., P. H. Harvey, and A. R. Ives. 1992. Procedures for the analysis of comparative data using phylogenetically independent contrasts. *Syst. Biol.* 41:18–32.

Gatz, A. J., Jr. 1979. Community organization in fishes as indicated by morphological features. *Ecology* 60:711–718.

Gibson, A. R., A. J. Baker, and A. Moeed. 1984. Morphometric variation in introduced populations of the common myna (*Acridotheres tristis*): An application of the jackknife to principal component analysis. *Syst. Zool.* 33:408–421.

Gilpin, M. E., and J. M. Diamond. 1984. Are species co-occurences on islands non-random, and are null hypotheses useful in community ecology? In *Ecological Communities: Conceptual Issues and the Evidence,* ed. D. R. Strong, Jr., D. Simberloff, L. G. Abele, and A. B. Thistle, 297–315. Princeton: Princeton University Press.

Gould, S. J., and E. S. Vrba. 1982. Exaptation: A missing term in the science of form. *Paleobiology* 8:4–15.

Grant, P. R. 1986. *Ecology and Evolution of Darwin's Finches.* Princeton: Princeton University Press.

Grant, B. R. and P. R. Grant. 1989. *Evolutionary Dynamics of a Natural Population: The Large Cactus Finch of the Galápagos.* Chicago, IL: University of Chicago Press.

Harvey, P. H., and M. Pagel. 1991. *The Comparative Method in Evolutionary Biology.* Oxford: University of Oxford Press.

Herrera, C. M. 1978. Ecological correlates of residence and non-residence in a mediterranean passerine bird community. *J. Anim. Ecol.* 47:871–890.

Hespenheide, H. A. 1971. Food preference and the extent of overlap in some insectivorous birds, with special reference to the Tyrannidae. *Ibis* 113:59–72.

Hespenheide, H. A. 1973. Ecological inferences from morphological data. *Ann. Rev. Ecol. Syst.* 4:213–229.

Holmes, R. T., R. E. Bonney, and S. W. Pacala. 1979. Guild structure of the Hubbard Brook bird community: A multivariate approach. *Ecology* 60:512–520.

Hopkins, C. D. 1974. Electrical communication in fish. *Amer. Sci.* 62:426–437.

Hutchinson, G. E. 1959. Homage to Santa Rosalia, or Why are there so many kinds of animals? *Amer. Nat.* 93:145–159.

James, F. C., and W. J. Boecklen. 1984. Interspecific morphological relationships and the densities of birds. In *Ecological Communities: Conceptual Issues and the Evidence,* ed. D. R. Strong, Jr., D. Simberloff, L. G. Abele, and A. B. Thistle, 458–477. Princeton: Princeton University Press.

James, F. C., and C. E. McCulloch. 1990. Multivariate analysis in ecology and systematics: Panacea or Pandora's box? *Ann. Rev. Ecol. Syst.* 21:129–166.

James, F. C., and C. Nesmith. 1991. Nongenetic effects in geographic differences among nestling populations of redwinged blackbirds. *Acta XIX Congr. Intern. Ornithol.* 2:1424–1433.

Jayne, B. C. 1988. Muscular mechanisms of snake locomotion: An electromyographic study of the sidewinding and concertina modes of *Crotalus cerastes, Nerodia fasciata* and *Elaphe obsoleta. J. Exp. Biol.* 140:1–33.

Jayne, B. C., and A. F. Bennett. 1989. The effect of tail morphology on locomotor performance of snakes: A comparison of experimental and correlative methods. *J. Exp. Zool.* 252:126–133.

Jayne, B. C., and A. F. Bennett. 1990. Selection on locomotor performance capacity in a natural population of garter snakes. *Evolution* 44:1204–1229.

Karr, J. R., and F. C. James. 1975. Ecomorphological configurations and convergent evolution in species and communities. In *Ecology and Evolution of Communities,* ed. M. L. Cody and J. M. Diamond, 258–291. Cambridge: Belknap Press.

Keast, A. 1972. Ecological opportunities and dominant families, as illustrated by the Neotropical Tyrannidae (Aves). *Evol. Biol.* 5:229–277.

Lack, D. 1947. *Darwin's Finches.* Cambridge: Cambridge University Press.

Lande, R., and S. J. Arnold. 1983. Measuring selection on correlated characters. *Evolution* 37:1210–1226.

Lauder, G. V. 1981. Form and function: Structural analysis in evolutionary morphology. *Paleobiology* 7:430–442.

Lauder, G. V. 1982. Historical biology and the problem of design. *J. Theor. Biol.* 97:57–67.

Lauder, G. V. 1989. Caudal fin locomotion in ray-finned fishes: Historical and functional analyses. *Amer. Zool.* 29:85–102.

Lauder, G. V. 1990. Function, morphology and systematics: Studying functional patterns in an historical context. *Ann. Rev. Ecol. Syst.* 21:317–340.

Lauder, G. V., and K. F. Liem. 1989. The role of historical factors in the evolution of complex organismal functions. In *Complex Organismal Functions: Integration and Evolution in Vertebrates,* ed. D. B. Wake and G. Roth, 63–78. Chichester: John Wiley and Sons.

Leisler, B. 1980. Morphological aspects of ecological specialization in bird genera. *Okologische Vogel* 2:199–220.

Leisler, B. and H. Winkler. 1985. Ecomorphology. Curr. Ornithol. 2:155–186.

Leisler, B., H.-W. Ley, and H. Winkler. 1989. Habitat, behavior and morphology of *Acrocephalus* warblers: An integrated analysis. *Ornis Scan.* 20:181–186.

Lemen, C. A., and P. W. Freeman. 1989. Testing macroevolutionary hypotheses with cladistic analysis: Evidence against rectangular evolution. *Evolution* 43:1538–1554.

Livezey, B. C. 1989. Morphometric patterns in recent and fossil penguins (Aves, Sphenisciformes) *J. Zool.* (London) 219:269–307.

Losos, J. B. 1990a. The evolution of form and function: Morphology and locomotor performance in West Indian *Anolis* lizards. *Evolution* 44:1189–1203.

Losos, J. B. 1990b. Ecomorphology, performance capability, and scaling of West Indian *Anolis* lizards: An evolutionary analysis. *Ecol. Mon.* 60:369–388.

Lynch, M. L. 1990. The rate of morphological evolution in mammals from the standpoint of the neutral expectation. *Amer. Nat.* 136:727–741.

MacArthur, R. H. 1958. Population ecology of some warblers in northeastern coniferous forests. *Ecology* 39:599–619.

MacArthur, R. H. 1965. Patterns of species diversity. *Biol. Rev.* 40:510–533.

MacArthur, R. H. 1972. *Geographical Ecology.* New York: Harper & Row.

MacArthur, R. H., J. M. Diamond, and J. R. Karr. 1972. Density compensation in island faunas. *Ecology* 53:330–342.

MacArthur, R. H., and R. Levins. 1967. The limiting similarity, convergence, and divergence of coexisting species. *Amer. Nat.* 101:377–385.

MacNally, R. C., and J. M. Doolan. 1986. Patterns of morphology and behaviour in a cicada guild: A neutral model analysis. *Aust. J. Ecol.* 11:279–294.

Mares, M. A. 1976. Convergent evolution of desert rodents: Multivariate analysis and zoo-geographic implications. *Paleobiology* 2:39–63.

Mares, M. A. 1980. Convergent evolution among desert rodents a global perspective. *Bull. Carnegie Mus. Nat. Hist.* 16:1–51.

May, R. M. 1975. *Stability and Complexity in Model Ecosystems,* 2d ed. Princeton: Princeton University Press.

McKenzie, N. L., and J. K. Rolfe. 1986. Structure of bat guilds in the Kimberly mangroves, Australia. *J. Anim. Ecol.* 55:401–420.

Miles, D. B. 1985. An eco-morphological comparison of avian communities in contrasting habitats. Ph.D. diss., University of Pennsylvania, Philadelphia.

Miles, D. B., and A. E. Dunham. 1993. Historical perspectives in ecology and evolutionary biology: The use of phylogenetic comparative analyses. *Ann. Rev. Ecol. Syst.* In press.

Miles, D. B., and R. E. Ricklefs. 1984. The correlation between ecology and morphology in deciduous forest passerine birds. *Ecology* 65:1629–1640.

Miles, D. B., R. E. Ricklefs, and J. Travis. 1987. Concordance of eco-morphological relationships in three assemblages of passerine birds. *Amer. Nat.* 129:347–364.

Moermond, T. C. 1979. Habitat constraints on the behavior, morphology, and community structure of *Anolis* lizards. *Ecology* 60:152–164.

Mooney, H. A., ed. 1977. *Convergent Evolution in Chile and California.* Stroudsburg, PA: Dowden, Hutchinson & Ross.

Mooney, H. A., and E. L. Dunn. 1970. Convergent evolution of Mediterranean-climate evergreen sclerophyll shrubs. *Evolution* 24:292–303.

Mosimann, J. E., and F. C. James. 1979. New statistical methods for allometry with application to Florida red-winged blackbirds. *Evolution* 33:444–459.

Niemi, G. J. 1985. Patterns of morphological evolution in bird genera of New World and Old World peatlands. *Ecology* 66:1215–1228.

Norberg, U. M. 1979. Morphology of the wings, legs, and tail of three coniferous forest tits, the goldcrest, and the treecreeper in relation to locomotor pattern and feeding station selection. *Phil. Trans. R. Soc.* (London) B 287:131–165.

Norberg, U. M. 1981. Flight, morphology and the ecological niche in some birds and bats. *Symp. Zool. Soc. Lond.* 48:173–197.

Norberg, U. M. 1986. Evolutionary convergence in foraging niche and flight morphology in insectivorous aerial-hawking birds and bats. *Ornis Scand.* 17:253–260.

Orians, G. H., and O. T. Solbrig, eds. 1977. *Convergent Evolution in Warm Deserts.* Stroudsburg, PA: Dowden, Hutchinson & Ross.

Paton, D. C., and B. G. Collins. 1989. Bills and tongues of nectar-feeding birds: A review of morphology, function and performance, with intercontinental comparisons. *Aust. J. Ecol.* 14:473–506.

Pianka, E. R. 1971. Comparative ecology of two lizards. *Copeia* 1971:129–138.

Pianka, E. R. 1973. The structure of lizard communities. *Ann. Rev. Evol. Syst.* 4:53–74.

Pianka, E. R. 1986. *Ecology and Natural History of Desert Lizards.* Princeton: Princeton University Press.

Pounds, J. A. 1988. Ecomorphology, locomotion, and microhabitat structure: Patterns in a tropical mainland *Anolis* community. *Ecol. Mon.* 58:299–320.

Poysa, H. 1983. Morphology-mediated niche organization in a guild of dabbling ducks. *Ornis Scand.* 14:317–326.

Richman, A. D. and T. Price. 1992. Evolution of ecological differences in the Old World leaf warblers. *Nature* 355:817–821.

Ricklefs, R. E. 1987. Community diversity: Relative roles of local and regional processes. *Science* 235:167–171.

Ricklefs, R. E. 1989. Speciation and diversity: Integration of local and regional processes. In *Speciation and Its Consequences,* ed. D. Otte and J. Endler, 599–622. Sunderland, MA: Sinauer Associates.

Ricklefs, R. E. 1991. Structures and transformations of life histories. *Func. Ecol.* 5:174–183.

Ricklefs, R. E., D. Cochran, and E. R. Pianka. 1981. A morphological analysis of the structure of communities of lizards in desert habitats. *Ecology* 62:1447–1483.

Ricklefs, R. E., and J. Travis. 1980. A morphological approach to the study of avian community organization. *Auk* 97:321–338.

Rohlf, F. J. 1990. Morphometrics. *Ann. Rev. Ecol. Syst.* 21:299–316.

Root, R. B. 1967. The niche exploitation pattern of the blue-gray gnatcatcher. *Ecol. Mon.* 37:317–350.

Scheibe, J. S. 1987. Climate, competition, and the structure of temperate zone lizard communities. Ecology 68:1424–1436.

Schluter, D. 1984. Morphological and phylogenetic relations among the Darwin's finches. *Evolution* 38:921–930.

Schluter, D. 1986. Tests for similarity and convergence of finch communities. *Ecology* 67:1073–1085.

Schluter, D. 1988. Character displacement and the adaptive divergence of finches on islands and continents. *Amer. Nat.* 131:799–824.

Schluter, D. 1989. Bridging population and phylogenetic approaches to the evolution of complex traits. In *Complex Organismal Functions: Integration and Evolution in Vertebrates,* ed. D. B. Wake and G. Roth, 79–95. Chichester: John Wiley and Sons.

Schluter, D. 1990. Species-for-species matching. *Amer. Nat.* 136:560–568.

Schluter, D., and P. R. Grant. 1984a. Determinants of morphological patterns in communities of Darwin's finches. *Amer. Nat.* 123:175–196.

Schluter, D., and P. R. Grant. 1984b. Ecological correlates of morphological evolution in a Darwin's finch, *Geospiza difficilis. Evolution* 38:856–869.

Schluter, D., and R. E. Ricklefs. 1993. Convergence and the regional component of species diversity. In *Species Diversity in Ecological Communities: Historical and Geographical Perspectives,* ed. R. E. Ricklefs and D. Schluter. Chicago: University of Chicago Press.

Schoener, T. W. 1965. The evolution of bill size differences among sympatric congeneric species of birds. *Evolution* 19:189–213.

Schoener, T. W. 1974. Resource partitioning in ecological communities. *Science* 185:27–39.

Schoener, T. W. 1986. Patterns in terrestrial vertebrate versus arthropod communities: Do systematic differences in regularity exist? In *Community Ecology,* ed. J. M. Diamond and T. J. Case, 556–586. New York: Harper & Row.

Schum, M. 1984. Phenetic structure and species richness in North and Central American bat faunas. Ecology 65:1315–1324.

Simberloff, D. S., and W. Boecklen. 1981. Santa Rosalia reconsidered: Size ratios and competition. *Evolution* 35:1206–1228.

Simberloff, D., and T. Dayan. 1991. The guild concept and the structure of ecological communities. *Ann. Rev. Ecol. Syst.* 22:115–143.

Sinervo, B., and R. B. Huey. 1990. Allometric engineering: An experimental test of the causes of interpopulational differences in performance. *Science* 248:1106–1109.

Smith, T. B. 1990. Resource use by bill morphs of an African finch: Evidence for intraspecific competition. *Ecology* 71:1246–1257.

Stevenson, R. D. 1986. Allen's rule in North American rabbits (*Sylvilagus*) and hares (*Lepus*) is an exception, not a rule. *J. Mammal.* 67:312–316.

Strauss, R. E. 1987. The importance of phylogenetic constraints in comparisons of morphological structure among fish assemblages. In *Community and Evolutionary Ecology of North American Stream Fishes,* ed. W. J. Matthews and D. C. Heins, 136–143. Norman: University of Oklahoma Press.

Strong, D. R., L. A. Szyska, and D. S. Simberloff. 1979. Tests of community-wide character displacement against null hypotheses. *Evolution* 33:897–913.

Travis, J., and R. E. Ricklefs. 1983. A morphological comparison of island and mainland assemblages of neotropical birds. *Oikos* 41:434–441.

Vandermeer, J. H. 1972. Niche theory. *Ann. Rev. Ecol. Syst.* 3:107–132.

Van Valkenburgh, B. 1988. Trophic diversity in past and present guilds of large predatory mammals. *Paleobiology* 14:155–173.

Vogel, S. 1988. *Life's Devices*. Princeton: Princeton University Press.

Wainwright, P. 1987. Biomechanical limits to ecological performance: Mollusc-crushing by the Caribbean hogfish, *Lachnolaimus maximus* (Labridae). *J. Zool.* (London) 213:283–297.

Wainwright, P. 1988. Morphology and ecology: Functional basis of feeding constraints in Caribbean labrid fishes. *Ecology* 69:635–645.

Wiens, J. A. 1984. On understanding a non-equilibrium world: Myth and reality in community patterns and processes. In *Ecological Communities: Conceptual Issues and the Evidence,* ed. D. R. Strong, Jr., D. Simberloff, L. G. Abele, and A. B. Thistle, 439–457. Princeton: Princeton University Press.

Wiens, J. A. 1990. *The Ecology of Bird Communities,* vols. 1 and 2. Cambridge: Cambridge University Press.

Wikramanayake, E. D. 1990. Ecomorphology and biogeography of a tropical stream fish assemblage: evolution of assemblage structure. Ecology 71:1756–1764.

Winemiller, K. O. 1991. Ecomorphological diversification in lowland freshwater fish assemblages from five biotic regions. Ecol. Monogr. 61:343–365.

Williams, E. E. 1972. The origin of faunas. Evolution of lizard congeners in a complex island fauna: A trial analysis. *Evol. Biol.* 6:47–89.

Williams, E. E. 1983. Ecomorphs, faunas, island size, and diverse end points in island radiations of *Anolis*. In *Lizard Ecology: Studies of a Model Organism,* ed. R. B. Huey, E. R. Pianka, and T. W. Schoener, 326–370. Cambridge: Harvard University Press.

3

Functional Morphology as a Tool in Ecological Research

Peter C. Wainwright

INTRODUCTION

It is axiomatic in biology that phenotypic differences among individuals and species are related to differences in their ecology. Indeed, this notion is a cornerstone of our understanding of the nature of organismal diversity. In general biology texts it is common to see pictures of birds with bills of different sizes and shapes and to read that this variation is related to differences among species in feeding behavior and diet. Thus, the inference is that the morphological differences determine ecological attributes. To the extent that morphology does, in fact, determine ecological patterns, understanding the mechanisms of this relationship can be a powerful explanatory tool in ecological research. But, what exactly is the role of functional morphology in shaping ecological patterns? What major questions in ecology can be addressed through research in functional morphology? And how can the impact of morphology on ecology be assessed rigorously? This chapter will address these questions in an evaluation of the utility of functional morphological approaches in ecology.

Performance: The Link Between Morphology and Ecology

One of the central paradigms in ecomorphology focuses on the role of organismal performance as a crucial link between the organism's phenotype and its ecology (fig. 3.1; Arnold, 1983; Emerson and Arnold, 1989; Huey and Stevenson, 1979; Wainwright, 1991). Here, and throughout this chapter, I use "performance" to refer to an organism's ability to carry out specific behaviors and tasks (e.g., capture prey, escape predation, obtain mates). An individual's phenotype (the way it is constructed) will determine the limits of its performance, because the ability to perform many behaviors is rooted in the design of underlying functional systems. For example, many aspects of flight performance in bats are determined by the aerodynamic consequences of wing shape (Norberg and Rayner, 1987; Norberg, chap. 9, this volume). Thus, design of the locomotor system

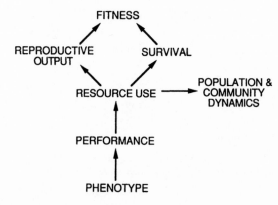

FIGURE 3.1 Flow diagram showing the paths through which phenotypic variation influences individual fitness and population and community ecology. Organismal design (the phenotype) affects ecology because it determines the limits of an individual's ability to perform day-to-day behaviors. Performance capacity interacts with the environment to constrain and shape patterns of resource use. Resource use is the key internal factor determining the two components of fitness, reproductive output and survival. In addition, patterns of resource use play a central role in determining patterns of ecology at the individual level. Individual ecology is compounded to produce population dynamics and community structure. Functional morphology (and other functional sciences) provide an understanding of the causal relationship between phenotype and performance and thus, can play an important role in providing mechanistic explanations for classical ecological questions. Note that the levels of this hierarchy can also exert forces in a downward direction and that direct connections between most levels are possible. For example, locomotor performance can directly influence survival by determining escape-success in predator encounters.

places limits on sprint speed, design of the visual system places limits on visual acuity, and design of the jaw places limits on biting force. Understanding the relationship between the phenotype and organismal performance is the realm of functional morphology and other fields concerned with how organisms function (note that physiology, biomechanics, biochemistry, and molecular biology fit into the scheme discussed in this chapter in much the same way that functional morphology does). The crucial role played by the functional sciences in eco-morphological research is that of elucidating the *causal* relationships between organismal design and behavioral performance.

Once the relationship between morphology and performance has been determined the latter may be related to aspects of ecology. Performance capacity affects ecology in two major ways. First, limitations on performance will constrain the range of environmental resources that individuals can exploit (fig. 3.1). For example, bats with relatively short wings and high wing loading exhibit relatively poor hovering performance and are therefore ineffective hovering nectarivores (Norberg and Rayner, 1987). Morphology shapes ecological patterns by determining the behavioral capacity to exploit resources. Patterns of resource use by individuals combine to produce patterns at the levels of population dynamics and community structure.

A second impact of performance on ecology is its influence on the individual's fitness (fig. 3.1; Arnold, 1983; Emerson and Arnold, 1989; Lande, 1979). Organismal performance affects fitness with varying degrees of directness. Of the two major components of fitness, reproductive output and survival, the latter may be affected directly by predator-escape performance, but, the major path by which performance capacity influences fitness is through its role in shaping patterns of resource use by individuals. Reproductive output is determined by a complex assortment of factors such as the energy available to devote to gamete production and the ability to obtain mates. Performance capacity may play a major role in determining energy intake or success in sexual competition for mates and provides a link between the phenotype and fitness through such channels.

Performance testing is a crucial step toward understanding the ecological and fitness consequences of morphological variation (fig. 3.1) and is a central component of research in ecological morphology. The importance of the functional basis of organismal performance has long been appreciated in comparative physiology (Bartholomew, 1987), but the recognition of its role as a link to ecological and evolutionary issues is more recent (Arnold, 1983; Bock, 1981; Huey and Stevenson, 1979; Lande, 1979). In the following sections I examine the link between morphology and ecology by means of behavioral performance testing and emphasize the power of this approach in resolving ecological questions.

The Role of Functional Morphology

The importance of functional morphology (or any of the functional sciences) in the scheme of linking the phenotype to ecology lies in its use in establishing a causal connection between morphology and performance (Lauder, 1990; Wainwright, 1991; Zweers, 1979). A common approach is to measure the correlation between morphology and performance. Univariate and multivariate correlative analyses are used to explore the relationship between morphology and performance, among species or among individuals within species (Arnold and Bennett, 1988; Ehlinger and Wilson, 1988; Losos, 1990a, b), but a difficulty arises in the interpretation of observed correlations between morphology and performance. Significant correlations do not demonstrate causation, and because so many morphological features tend to covary, both ontogenetically and phylogenetically, the danger of finding spurious correlations is usually high (Lauder, 1990; Wainwright, 1987). For example, larger mammals run faster than smaller mammals

(Garland, 1983), but which of the many functional features that change with body size is the basis of this phenomenon?

To determine the effect of morphology on performance and to separate causal features from spurious correlation it is necessary to understand how a particular functional system operates. Through a functional/biomechanical or physiological analysis of the relevant system, one can generate specific predictions about how variation in particular morphological features will influence performance capacity. When such predictions are supported by the results of laboratory or field performance experiments, the interpretation of the causal nature of the interaction between morphology and performance is greatly enhanced by the biomechanical or physiological principles that underlie the functional analysis. Thus, our belief that the long, slender wings of some bat species cause their relatively poor hovering performance is rooted in our confidence in the aerodynamic theory that underlies the interpretation of wing design.

From Phenotype to Performance

The functional basis of behavioral performance is a poorly investigated area of ecomorphology, yet it is an area crucial to developing a more general understanding of the role of morphological variation in shaping ecological patterns. Identifying the performance consequences of morphological variation involves two major steps: (1) a functional analysis, which is used to predict the consequences of variation in morphology for performance, and (2) tests of these predictions with performance experiments, contrived laboratory or field situations usually designed to test the limits of an individual's ability to perform a specific act.

In successful analyses, quantitative predictions of performance can be made on the basis of functional analyses, but only if the functional basis of performance in the behavior under study is clearly understood. However, instances where precise predictions of behavioral performance have been made from only a few morphological variables are rare, usually only occurring in biomechanically simple systems (Kiltie, 1982; Lawrence, 1957; Wainwright, 1987, 1988; Werner, 1974). An example is the pharyngeal jaw apparatus of labrid fishes (Wainwright, 1987, 1988), which is used to crack mollusc shells and other hard prey. The pharyngeal jaw apparatus consists primarily of a single pair of muscles originating from either side of the skull and inserting on either side of a robust lower jaw bone that is suspended below the skull. Mollusc prey are crushed between the pharyngeal jaws as the muscles contract and lift the lower jaw bone up against stationary upper jaw bones. This simple prey-crushing system involves no lever arms, and mollusc-crushing performance is a direct function of the force-producing capacity of the paired crushing muscles (Wainwright, 1987), a feature that can be independently measured and has a clear morphological basis in the size and the architecture of the muscles. In this case, variation among species and among size classes within species in the morphology of the crushing muscle accurately predicted differences in mollusc crushing performance (Wainwright, 1987, 1988).

Performance studies can be used to dissect systems with complex functional bases. In most cases the exact contribution of any one morphological feature to overall performance in some behavior is not clearly known. Further, performance at most behaviors (e.g., sprint speed, locomotor endurance, auditory acuity) has a complex functional basis, and performance may depend not on a single variable, such as leg length or locomotor muscle mass, but rather some combination of features. In these situations it is possible to address the relative importance of several variables to performance through multivariate analyses such as multiple regression and path analysis. In multiple regression, several independent variables (e.g., body length, hind limb length) can be measured and simultaneously regressed on the performance measure (Garland, 1984; Lauder et al., 1986; Losos, 1990a). To the extent that the independent variables under consideration do not entirely covary, multiple regression can be used to identify those variables most tightly correlated with performance. In cases where suites of performance and morphological variables have been measured, canonical correlation analysis can be used to examine more complex relationships among multiple independent and dependent variables. However, because multiple regression and canonical correlation analyses have no mechanism for distinguishing direct from indirect effects, spurious correlations may plague them just as in univariate cases. The strength of causal interpretations in multiple regression or any statistical analysis depends on a sound understanding of the biomechanical or physiological mechanisms that underlie the inferred causal relationship between phenotype and performance.

An example of such an analysis comes from work on locomotor endurance in the lizard *Ctenosaurus similus* (Garland, 1984). In an ontogenetic series of seventeen lizards, 90% of the variation in endurance capacity (body size–corrected variation) was explained by the combined effects of two morphological variables (thigh and heart mass), citrate synthase activity, and maximum oxygen consumption. This analysis simultaneously ruled out the predictive power of several additional variables, and the results are consistent with the prevalent view that endurance capacity is limited primarily by the maximal capacity of organisms for aerobic metabolism (Bennett, 1984, 1989; Davies et al., 1982; Garland, 1984). Because the capacity for aerobic metabolism cannot be assessed by measurement of a single morphological or physiological variable, but rather is a complex attribute with many components, it was logical to use an approach where the combined effects of several potentially important factors were examined simultaneously.

An interesting twist on the multivariate regression approach has recently been illustrated with work by Emerson and her coworkers on the functional basis of "flying" in frogs (Emerson and Koehl, 1990; Emerson et al., 1990). "Flying" frogs have specialized morphology, enlarged webbed feet and body fringes, that have been thought to be the key components of "flying" ability in these animals, but the consequences of this morphology for performance depend on the postural orientation of the limbs such that some orientations actually decrease performance relative to the unmodified morphology. Further, the suite of morphological modifications do not affect "flying" performance in a simple additive manner; rather, specific combinations produce positively nonadditive effects on performance (Emerson et al., 1990). This case study illustrates the crucial role of behavior in mediating morphological determinants of performance and the potentially complex relationship between morphology and performance.

Performance studies can also be used to test competing functional hypotheses where the functional basis of performance is poorly understood (see also Emerson et al., chap. 6, this volume). An example is provided by Emerson and Diehl (1980), who examined the basis of sticking ability in tree frogs. Several mechanisms of sticking by tree frogs had previously been proposed, including suction, capillarity, and adhesion. These mechanisms suggest very different outcomes of sticking performance of individuals under various conditions. No change in sticking performance was observed when animals were placed in pressure-controlled chambers and exposed to reduced air pressure, thus suggesting that suction is not a major mechanism of sticking. Animals immersed in water immediately slipped from their perch, ruling out capillarity as a sticking mechanism. In these cases the performance tests were used to provide critical tests for hypothesized sticking mechanisms and permitted the researchers to rule out several mechanisms of sticking by tree frogs.

Measuring Performance

The choice of laboratory performance measures is a crucial step in ecomorphological analyses and depends on the goal of the study. In general, when ecological consequences of morphological variation are being studied, performance tests should be designed to mimic the ecologically relevant behavior. However, realistic behaviors are usually complex and may not isolate the function of a single system. Consider the functional basis of sprint speed and its role in determining the ability of individuals to escape predation. Sprint speed can be measured on a race track, but a more ecologically relevant performance measure might entail a staged encounter between predator and prey. Escaping predation involves more than just sprint speed, however. Prey can choose when to begin their escape movement, and they may differ in their ability to detect and respond to an oncoming predator. In some systems prey rely on a lack of motion to escape detection (e.g., McPeek, 1990). The functional basis of sprint speed may be easier to determine than the functional basis of escape performance, but if staged encounters show that rapid locomotion plays a significant role in avoiding capture, it may be possible to integrate the two tests. A more complex approach, where more than one performance measure is used and in which the performance measures form a natural hierarchy, could'help forge a stronger link between morphology and ecology. For example, by measuring both sprint speed and performance in the predator encounter, one can look for patterns of correlation between these two measures that will strengthen the ultimate argument for a causal link between phenotype and ecological patterns.

Performance tests may emphasize an individual's maximal capabilities within a single tightly controlled set of conditions, or the aim may be to determine the effect of some experimentally controlled variable. For example, one of the most heavily studied environmental variables is temperature. Body temperature has pervasive effects on biological rate processes, including behavioral performance (Bennett, 1980, 1990; Prosser, 1973; Wardle, 1975) and their functional basis (Bennett, 1984; Jayne et al., 1990; Josephson, 1981). Similarly, the effects of prey size or type on a predator's capture success and handling time are frequently explored themes in feeding studies (Osenberg and Mittelbach, 1989; Reilly and Lauder, 1988; Stephens and Krebs, 1986, and references therein; Wankowski, 1979; Werner, 1974, 1977).

In most performance tests, experimental individuals are assumed to make a maximal effort, and this assumption is critical in efforts to relate performance to functional design, but it is unlikely that individuals are maximally, or even similarly, motivated to perform during tests. Variation among individuals or among species in motivation level may obscure the relationship between functional determinants and performance. Standardized research protocols may reduce variation in motivation, but problems of motivation are likely to continue to be a key concern in performance studies.

Variation is Fundamental

A key feature of the ecomorphological paradigm illustrated in figure 3.1 is that morphological variation among individuals or among species can be causally linked to variation in performance and ultimately to variation in resource use and fitness. This observation is especially important for two reasons. First, variation is the attribute that gives the paradigm great potential as an explanatory tool in ecology. We seek the consequences of individual variation in morphology for variation in ecology. Variation in performance that has its roots in morphological differences may account for a tremendous portion of the variation in ecological patterns. Thus, the interaction between the individual's functional capacity and its environment can play a prominent role in shaping ecological patterns at the level of the individual (Schoener, 1986).

Second, naturally occurring morphological variation is convenient for experimental purposes. Several sources of natural variation can be exploited: variation among species, intraspecific body size–related variation, and variation among individuals that is independent of body size. The choice of the source of morphological variation is usually determined by the ecological questions that are of interest. In species comparisons and comparisons among size classes of a single species, body size becomes a complicating factor because nearly all anatomical features covary strongly with body size. Removing the confounding effects of body size by calculating residuals from regressions of the morphological variable on some proxy for body size (usually body length or body mass) has become routine and can be used in both inter- and intraspecific studies (e.g., Garland, 1984; Jayne and Bennett, 1990a, b). In comparisons among size classes of a single species in which the effect of body size is not removed, it may be necessary to rely on the quantitatively different predictions of the relationship between body size and performance produced by the biomechanical models of different components of the functional system (e.g., Wainwright, 1987).

In some situations it may not be possible to isolate completely the effects of a single morphological factor, even by removing the effects of body size, because of the feature's tight correlation with other anatomical structures. In these instances experimental manipulations of morphology can provide tests of the causal connection between morphology and performance (Emerson and Koehl, 1990; Lauder and Reilly, 1988; Nishikawa and Roth, 1991; Reilly and Lauder, 1991; Webb, 1977). If one or a few aspects of morphology are isolated and manipulated, hypotheses about the functional significance of that particular feature can be tested directly. One ingenious approach has been applied by Sinervo and Huey, who manipulated egg yolk of snake embryos to isolate the effects of body size on locomotor performance of newborns (Sinervo, 1990; Sinervo and Huey, 1990). Other manipulative approaches have provided unexpected results and emphasize the need for critical tests of conventional wisdom. For example, Jayne and Bennett (1989) explored the effect of tail length on locomotor performance in the garter snake *Thamnophis sirtalis fitchi*. They found that naturally occurring intermediate tail length was significantly correlated with burst speed in a wild population of snakes. They then tested the implied influence of tail length on burst speed by artificially removing one third of the tail in sixteen snakes and two thirds of the tail in sixteen additional snakes. Loss of one third of the tail had no influence on burst speed, whereas loss of two thirds of the tail decreased performance by only 5%. These results show that the original correlation between tail length and locomotor performance probably arose from their correlated responses to some other, unmeasured variable.

THE ECOLOGICAL QUESTIONS

Determining the ecological consequences of variation in performance is the second step in an ecomorphological analysis (fig. 3.1). Behavioral performance may underlie patterns of resource use, survival, and reproductive success of individuals. For all of these, the use of performance to explain patterns is a potentially powerful and generally unexploited approach. In this section, I consider the ecological issues that can be addressed within this framework, discuss how those studies can be accomplished, and present some examples.

My aim here is to emphasize the use of functional morphology in answering questions about patterns of resource use. In so doing I will mostly bypass an important third part of the paradigm illustrated in figure 3.1, relating functional design to individual fitness. Interest in studies of natural selection within functional morphology and physiology has recently expanded (Jayne and Bennett, 1990b; Schluter, 1989). These works illustrate the connection between morphology and fitness emphasized by Arnold (1983) and Lande (1979) and have emerged as a powerful way of studying the evolution of functional characters at the population level, but although fitness has become a major focus of ecomorphological research, the use of physiology and functional morphology to explain resource, population, and community dynamics has been almost completely neglected. This not only represents a fertile source of explanations for classical ecological questions but should become an integral part of selection studies in the future. Except in cases where selection acts directly on performance variation, the influence of performance on fitness is mediated by the influence of performance on resource use. In most cases, understanding how performance affects fitness requires understanding how it affects resource use (see, e.g., Boag and Grant, 1981; Grant and Grant, 1989).

Resource Use

The ecological identity of an individual is determined by the suite of environmental resources it uses. It is not surprising, then, that identifying the mechanistic basis of resource use patterns is one of the fundamental problems in ecology (Began et al., 1986; Schoener, 1986). The major environmental resources used by individuals are generally considered to be space, time, food, and reproductive mates. In the discussion that follows, I mostly consider the food resource for convenience, but the comments apply broadly to other resource types as well.

Many factors interact to determine patterns of prey use by predators, but two are fundamental. First, only prey that are physically available can be eaten. The density and relative abundance of different prey ultimately determine their relative desirability, but here I focus on presence or absence in the environment. Hence, the availability of particular prey is the first factor that determines which are consumed by a predator. The second factor is the predator's effectiveness. Only prey that the predator is able to locate, capture, handle, and digest can be included in the diet. Relative proficiency in feeding on different prey is also significant, but here I simply distinguish those prey that can be eaten from those that cannot. Thus, two factors interact to determine what a predator eats, one an environmental factor and the other an intrinsic property of the predator that is a consequence of the design of the feeding system. This principle will apply generally to the use of other environmental resources; patterns of resource use develop as a function of the *interaction* between availability of the resource and the individual's ability to exploit the resource. Because the individual's ability to exploit the resource may have a strong basis in the design of relevant morphological systems, functional morphology holds promise as a source of mechanistic explanations for resource-use patterns.

Other factors may further shape the pattern of prey use; but these factors must exert their effects within the limits determined by prey presence and the predator's capability. Among the most common and important additional factors are the relative energy return involved in foraging on various prey, competitive interactions with other individuals and other species, and the threat of predation associated with foraging in specific habitats. With each of these three forces— energetics, competition, and predation—the behavioral capabilities of the individual play a significant role. Thus, individual performance affects resource use directly by setting absolute limits and more subtly through its role in determining prey choice. For example, a central consideration in optimal foraging models is the relative energy return involved in locating and consuming various prey types (Stephens and Krebs, 1986). For a given prey type, this cost-benefit curve can differ among species or among individuals within species because the capabilities of predators differ (Emerson et al., chap. 6, this volume).

Linking Performance to Resource Use

The ability of an organism to perform some ecologically relevant task can be used to predict the limits of potential resource use. For example, the size of potential prey can be predicted from laboratory feeding experiments (e.g., Werner, 1977). These predictions are then compared with documented patterns of resource use with three possible outcomes. First, actual resource use may fall well below the maximal capabilities of the organism, indicating that some factor other than the constraint of functional capacity is limiting resource use. Second, the realized resource use may approach the potential resource use indicated by the maximal capabilities of the organism. In this instance performance and its functional basis provide a mechanistic explanation for the pattern of resource use. The third possible outcome is that real resource use may exceed that predicted on the basis of the performance experiments. In this seemingly impossible case, the explanation is that either the performance measures were inaccurate or the individuals are able to invoke some behavioral shift or other strategy that permits them to operate beyond the expected range. An example is the use of protected microhabitats within rocky intertidal shores to avoid the waveforce rigors of immediately adjacent environs (Denny, 1988). Sessile invertebrates and algae exposing themselves to the full force of incoming waves may face certain dislodgement, but by using specific microhabitats, they can avoid the high wave forces and face a more benign environment.

Remarkably, the role of maximal performance in shaping resource use has been tested in few cases (Grant, 1986; Hertz et al., 1988; Kiltie, 1982; Wainwright, 1987). How often do individuals use their maximal capabilities when capturing prey, escaping predators, or obtaining mates? Are maximal capabilities important on a day-to-day basis, or do they only take on prominence during ecological bottlenecks? Little is known of these critical ecomorphological issues.

A much more common approach, and one that has met with considerable success, has been to explain differences in resource-use patterns among species or among size classes within species by identifying key performance differences that allow one species or size class to exploit a resource that another cannot (Osenberg and Mittelbach, 1989; Wainwright, 1988). This approach has been a mainstay in physiological ecology since the 1950s (Bennett, 1987; Feder, 1987). Physiologists have directed considerable attention toward the functional capacities of organisms that live in extreme environments. In this context, behavioral capacity is viewed as a permissive feature for one group and a constraint for the other. One well-studied example within functional morphology involves two species of North American sunfishes (Centrarchidae) that commonly co-occur in midwestern lakes, the bluegill (*Lepomis macrochirus*) and the pumpkinseed (*L. gibbosus*). Adult pumpkinseed diets are dominated by snails, whereas bluegill virtually never eat snails, feeding more broadly on zooplankton and benthic invertebrates (Mittelbach, 1984). Pumpkinseeds are able to crush snails with their pharyngeal jaws, but bluegill cannot crush snails (Lauder, 1983; Mittelbach, 1984). The capacity for predation on snails by pumpkinseeds has a clear functional basis in the hypertrophied muscles and bones in the pharyngeal-jaw crushing apparatus and the presence of a phylogenetically derived neuromuscular pattern (Lauder, 1983). This novel muscle activity pattern permits pumpkinseeds to use their pharyngeal jaws to exert a lethal crushing force against snail shells. Even when bluegill eat snails they do not crush them. Thus, adult pumpkinseeds gain a competitive refuge from bluegill by feeding on a prey resource that bluegill are incapable of eating. In this case the difference between the co-occurring fish species in prey-use patterns can be explained by a difference in snail-crushing ability, which has a functional basis in the design of the feeding system.

Beyond Individual Resource Use to Populations and Communities

Schoener (1986) argued for a hierarchy within ecology where patterns and processes observed at each level have a mechanistic basis in lower levels. He emphasized a three-tiered hierarchy of individual ecology, population ecology, and community ecology. Thus, understanding patterns at the community level ultimately requires the identification of mechanisms determining patterns at the individual-ecological level. I have argued above that individual patterns of resource use are shaped by the interaction between resource availability and behavioral capabilities. The utility of functional morphology in establishing mechanisms that forge community level patterns will depend on the future success of work with individuals.

Case studies of the impact of organismal design on population and community dynamics have been presented by the Grants and their coworkers (summarized by Grant, 1986; Grant and Grant, 1989), and Kingsolver (1989). Theoretical work has also begun to incorporate concepts of organismal function into models of population dynamics (Nisbet et al., 1989; Tilman, 1982). One particularly well-studied example of the link between functional morphology and population and community ecology comes, again, from work with the North American sunfishes (Lauder, 1983; Mittelbach, 1981, 1984, 1986, 1988; Osenberg and Mittelbach, 1989; Osenberg et al., 1988; Werner, 1977; Werner and Hall, 1976, 1977, 1979; Werner et al., 1983). This body of work has shown that the limits of feeding capability of the bluegill and pumpkinseed sunfishes and their primary predator, the largemouth bass (*Micropterus salmoides*) interact with environmental variables to shape patterns of habitat use and the relative abundance of these species.

As discussed above, pumpkinseed sunfish possess phylogenetically derived morphological and neuromuscular specializations that permit adult individuals to feed on snails, a prey resource that bluegill are ineffective at utilizing (Lauder, 1983; Mittelbach, 1984). Correlated with the difference in diet between these species is a difference in habitat use by the adults (Mittelbach, 1984). Pumpkin-seeds forage in vegetated areas where snails are found, whereas bluegill feed on open-water zooplankton.

Diet and habitat use by small individuals (<75 mm) of both species contrast with the adult patterns. Small fish of both species feed on soft invertebrate prey associated with vegetation, even though they might more profitably forage on open-water zooplankton (Mittelbach, 1981). Two explanations for this pattern of resource use by the smaller size classes have been advanced and both have clear bases in functional morphology of the feeding mechanism. First, small pumpkin-seeds are unable to use gastropod prey effectively because their crushing muscles are not sufficiently developed (Lauder, 1983; Wainwright et al., 1991). Second, predation pressure from largemouth bass constrains the ability of small fish to exploit open-water habitats. Piscivory by bass is limited by mouth diameter. Prey-handling time increases exponentially as prey diameter approaches that of the bass's mouth (Werner, 1977). Thus, the vulnerability of bluegill and pump-kinseeds drops sharply between 50 and 100 mm body size (Hall and Werner, 1977). At about this body size bluegill leave the vegetated habitat and begin feed-ing in open water (Hall and Werner, 1977; Werner and Hall, 1977). In a controlled field experiment, Werner et al. (1983) showed that bluegill of all sizes will forage in open-water habitats if they are more profitable but that, in the pres-ence of largemouth bass, smaller fish use the vegetated habitat more heavily. Thus, several lines of evidence indicate that the threat of predation by largemouth bass has a differential impact on patterns of habitat use by fish of different size. In this example, the functional constraint of bass mouth size plays a central role in shaping patterns of habitat use by bluegill and pumpkinseed sunfish.

The differences in diet between bluegill and pumpkinseed are related to varia-tion among lakes in population sizes. Mittelbach (1984) found that the abun-dance of bluegill relative to pumpkinseeds varied considerably among three lakes, from about 25:1 to 1:1. Relative abundance of the species was directly correlated with the relative abundance of the vegetated and open-water habitats. Habitat abundance is further correlated with abundance of the dominant prey of each species, vegetated habitat with snail abundance and open-water habitat with zooplankton abundance. The population sizes of bluegill and pumpkinseeds, thus, seem to be determined by the availability of the adult trophic resource of each species. Here, the interaction between adult fish feeding mechanics and prey availability shapes not only adult diet but also adult population sizes.

The sunfish example illustrates how the availability of resources (snail and zooplankton prey and their associated habitats) and the threat of predation (by largemouth bass) interact with the feeding capabilities of the bluegill, pumpkin-seed, and largemouth bass to provide a mechanistic explanation for at least three prominent features of the bluegill-pumpkinseed communities: (1) patterns of prey use by small and large bluegill and pumpkinseeds, (2) patterns of habitat use by the two major size classes of each species, and (3) the relative abundance of the two species. Functional morphological analyses have identified the basis of snail-eating performance in pumpkinseeds and the basis of size-limited predation in largemouth bass. In future work, these insights may provide the basis for making predictions about the community level consequences of such things as a change in the morphology of the primary predator species or a change in growth rates of the snail-crushing musculature in pumpkinseeds.

CONCLUSIONS

Functional morphology provides the crucial link between individual variation in morphology and fundamental ecological parameters, patterns of resource use, and survival. I have argued that functional morphology may provide mechanistic explanations for ecological patterns by describing individual variation in performance. The limits of performance capacity interact with environmental resources to shape actual patterns of resource use and fitness of individuals. Individual patterns interact and sum to produce population- and, ultimately, community-level characteristics. Functional morphology thus shows promise as a tool in ecological research, a potential explanation for ecological patterns at many levels. Understanding the functional basis of performance is a task for functional morphologists and physiologists, but relating performance to ecological questions will require integrative efforts by both functional biologists and ecologists. As research in this area accumulates, several questions will be crucial to a general understanding of the role of functional morphology in ecological systems. Some of these are listed below:

First, how important are the maximal capabilities of individuals in shaping ecological patterns? What is the relationship between the limits of performance and actual patterns of resource use? Are individuals frequently limited by their performance capacities, or do other factors (i.e., competition, threat of predation) constrain them first, causing them to operate away from these limits? Thus, how precise is the predictive power of functional morphology in ecology?

Second, how frequently do organisms mitigate the effects of functional constraints by means of behavioral adjustments? It is clear that behavioral responses are a fundamental component of the individual's repertoire, but under what conditions is the effectiveness of behavioral adjustment maximized and minimized?

Third, do functional systems that differ in their importance for resource use and fitness in a population show corresponding differences in their expression? Are the phenotypic components of less significant systems expressed more variably and with looser organization than systems experiencing strong directional or stabilizing selection?

Finally, how do functional systems respond to selection on individual performance? Do changes occur at all levels of design of the system, or are certain components more conservative while others change readily (Lauder, 1990)?

Future studies integrating ecology and functional morphology promise to answer these questions and to make unexpected contributions to both fields. Biologists have disputed the predictive power of morphology in ecology for years (Alexander, 1988; Bock, 1981; Bock and von Wahlert, 1965; Dullemeijer, 1972; Frazzetta, 1975; James, 1983; Lewontin, 1978; Morse, 1980), but no consensus has yet emerged. Integrative efforts that capitalize on the experimental and theoretical heritage of both functional morphology and ecology offer the greatest hope for a more holistic understanding of organismal design.

ACKNOWLEDGMENTS

I thank S. Emerson, Z. Eppley, M. McPeek, K. Nishikawa, B. Richard, and two anonymous reviewers for critical comments on an earlier draft of the manuscript. The Cocos Foundation is gratefully acknowledged for crucial financial support of the symposium.

REFERENCES

Alexander, R. McN. 1988. The scope and aims of functional morphology and ecological morphology. *Neth. J. Zool.* 38:3–22.
Arnold, S. J. 1983. Morphology, performance and fitness. *Amer. Zool.* 23:347–361.
Arnold, S. J., and A. F. Bennett. 1988. Behavioural variation in natural populations. V. Morphological correlates of locomotion in the garter snake (*Thamnophis radix*). *Biol. J. Linnean Soc.* 34:175–190.
Bartholomew, G. A. 1987. Interspecific comparison as a tool for ecological physiologists. In *New Directions in Ecological Physiology,* ed. M. E. Feder, A. F. Bennett, W. W. Burggren, and R. B. Huey, 11–34. Cambridge: Cambridge University Press.
Begon, M., J. L. Harper, and C. R. Townsend. 1986. *Ecology.* Sunderland, MA: Blackwell Scientific.
Bennett, A. F. 1980. The thermal dependence of lizard behavior. *Anim. Behav.* 28:752–762.
Bennett, A. F. 1984. Thermal dependence of muscle function. *Amer. J. Physiol.* 247 (Reg. Integr. Comp. Physiol. 16):R217–R229.
Bennett, A. F. 1987. The accomplishments of ecological physiology. In *New Directions in Ecological Physiology,* ed. M. A. Feder, A. F. Bennett, W. W. Burggren, and R. B. Huey, 1–8. Cambridge: Cambridge University Press.
Bennett, A. F. 1989. Integrated studies of locomotor performance. In *Complex Organismal Functions: Integration and Evolution in Vertebrates,* ed. D. B. Wake and G. Roth, 191–202. New York: Wiley.
Bennett, A. F. 1990. Thermal dependence of locomotor capacity. *Amer. J. Physiol.* 259 (Reg. Integr. Comp. Physiol. 14):R253–R258.
Boag, P. T., and P. R. Grant. 1981. Intense natural selection in a population of Darwin's finches (Geospizinae) in the Galapagos. *Science* 214:82–85.
Bock, W. J. 1981. Functional-adaptive analysis in evolutionary classification. *Amer. Zool.* 21:5–20.
Bock, W. J., and G. von Wahlert. 1965. Adaptation and the form-function complex. *Evolution* 19:269–299.
Davies, K. J. A., J. J. Maguire, G. A. Brooks, P. R. Dallman, and L. Packer. 1982. Muscle mitochondrial bioenergetics, oxygen supply, and work capacity during dietary iron deficiency and repletion. *Amer. J. Physiol.* 242 (Endocrinol. Metab. 5):E418–E427.
Denny, M. W. 1988. *Biology and the Mechanics of the Wave-swept Environment.* Princeton: Princeton University Press.

Dullemeijer, P. 1972. Explanation in morphology. *Acta Biotheor.* 21:260–273.

Ehlinger, T. J., and D. S. Wilson. 1988. Complex foraging polymorphism in bluegill sunfish. *Proc. Natl. Acad. Sci. USA* 85:1878–1882.

Emerson, S. B., and S. J. Arnold. 1989. Intra- and interspecific relationships between morphology, performance, and fitness. In *Complex Organismal Functions: Integration and Evolution in Vertebrates,* ed. D. B. Wake and G. Roth, 295–314. New York: Wiley.

Emerson, S. B., and D. Diehl. 1980. Toe pad morphology and mechanisms of sticking in frogs. *Biol. J. Linnean Soc.* 13:199–216.

Emerson, S. B., and M. A. R. Koehl. 1990. The interaction of behavioral and morphological change in the evolution of a novel locomotor type: "Flying" frogs. *Evolution* 44:1931–1946.

Emerson, S. B., J. Travis, and M. A. R. Koehl. 1990. Functional complexes and additivity in performance: A test case with "flying" frogs. *Evolution* 44:2153–2157.

Feder, M. E. 1987. The analysis of physiological diversity: The prospects for pattern documentation and general questions in ecological physiology. In *New Directions in Ecological Physiology,* ed. M. A. Feder, A. F. Bennett, W. W. Burggren, and R. B. Huey, 38–70. Cambridge: Cambridge University Press.

Frazzetta, T. H. 1975. *Complex Adaptations in Evolving Populations.* Sunderland, MA: Sinauer Associates.

Garland, T., Jr. 1983. The relation between maximal running speed and body mass in terrestrial mammals. *J. Zool.* (London) 99:157–170.

Garland, T., Jr. 1984. Physiological correlates of locomotory performance in a lizard: An allometric approach. *Amer. J. Physiol.* 247 (Reg. Integr. Comp. Physiol. 16):R806–R815.

Grant, B. R., and P. R. Grant. 1989. *Evoutionary Dynamics of a Natural Population.* Chicago: University of Chicago Press.

Grant, P. R. 1986. *Ecology and Evolution of Darwin's Finches.* Princeton: Princeton University Press.

Hall, D. J., and E. E. Werner. 1977. Seasonal distribution and abundance of fishes in the littoral zone of a Michigan lake. *Trans. Amer. Fish. Soc.* 106:545–555.

Hertz, P. E., R. B. Huey, and T. Garland. 1988. Time budgets, thermoregulation, and maximal locomotor performance: Are reptile Olympians or Boy Scouts? *Amer. Zool.* 28:927–938.

Huey, R. B., and R. D. Stevenson. 1979. Integrating thermal physiology and ecology of ectotherms: A discussion of approaches. *Amer. Zool.* 19:357–366.

James, F. C. 1983. Environmental component of morphological differentiation in birds. *Science* 221:184–186.

Jayne, B. C., and A. F. Bennett. 1989. The effect of tail morphology on locomotor performance of snakes: A comparison of experimental and correlative methods. *J. Exp. Zool.* 252:126–133.

Jayne, B. C., and A. F. Bennett. 1990a. Scaling of speed and endurance in garter snakes: A comparison of cross-sectional and longitudinal allometries. *J. Zool.* (London) 220:257–277.

Jayne, B. C., and A. F. Bennett. 1990b. Selection on locomotor performance capacity in a natural population of snakes. *Evolution* 44:1204–1219.

Jayne, B. C., A. F. Bennett, and G. V. Lauder. 1990. Muscle recruitment during terrestrial locomotion: How speed and temperature affect fibre type use in a lizard. *J. Exp. Biol.* 152:101–128.

Josephson, R. K. 1981. Temperature and the mechanical performance of insect muscle. In *Insect Thermoregulation,* ed. B. Heinrich, 20–44. New York: Wiley.

Kiltie, R. A. 1982. Bite force as a basis for niche differentiation between rain forest peccaries (*Tayassu tajacu* and *T. pecari*). *Biotropica* 14:188–195.

Kingsolver, J. G. 1989. Weather and population dynamics of insects: Integrating physiological and population ecology. *Physiol. Zool.* 62:314–334.

Lande, R. 1979. Quantitative genetic analysis of multivariate evolution, applied to brain:body size allometry. *Evolution* 33:402–416.

Lauder, G. V. 1983. Functional and morphological bases of trophic specialization in sunfishes (Teleostei, Centrarchidae). *J. Morphol.* 178:1–21.

Lauder, G. V. 1990. Functional morphology and systematics: Studying functional patterns in an historical context. *Ann. Rev. Ecol. Syst.* 21:317–340.

Lauder, G. V., and S. M. Reilly. 1988. Functional design of the feeding mechanism in salamanders: Causal basis of ontogenetic changes in function. *J. Exp. Biol.* 134:219–233.

Lauder, G. V., P. C. Wainwright, and E. Findeis. 1986. Physiological mechanisms of aquatic prey capture in sunfishes: Functional determinants of buccal pressure changes. *Comp. Biochem. Physiol.* 84A:729–734.

Lawrence, J. M. 1957. Estimated sizes of various forage fishes largemouth bass can swallow. *Proc. S. E. Asso. Game Fish Comm.* 11:220–226.

Lewontin, R. C. 1978. Adaptation. *Sci. Amer.* 239:212–230.

Losos, J. B. 1990a. Ecomorphology, performance capability, and scaling of West Indian *Anolis* lizards: An evolutionary analysis. *Ecol. Mon.* 60:369–388.

Losos, J. B. 1990b. The evolution of form and function: Morphology and locomotor performance in West Indian *Anolis* lizards. *Evolution* 44:1189–1203.

McPeek, M. A. 1990. Behavioral differences between *Enalagma* species (Odonata) influencing differential vulnerability to predators. *Ecology* 71:1714–1726.

Mittelbach, G. G. 1981. Foraging efficiency and body size: A study of optimal diet and habitat use by bluegills. *Ecology* 62:1370–1386.

Mittelbach, G. G. 1984. Predation and resource use in two sunfishes (Centrarchidae). *Ecology* 65:499–513.

Mittelbach, G. G. 1986. Predator-mediated habitat use: Some consequences for species interactions. *Envir. Biol. Fish.* 16:159–169.

Mittelbach, G. G. 1988. Competition between refuging sunfishes and effects of fish density on littoral zone invertebrates. *Ecology* 69:614–623.

Morse, D. H. 1980. *Behavioral Mechanisms in Ecology.* Cambridge: Harvard University Press.

Nisbet, R. M., W. S. C. Gurney, W. W. Murdoch, and E. McCauley. 1989. Structured population models: A tool for linking individual and population levels. *Biol. J. Linnean Soc.* 37:79–99.

Nishikawa, K., and G. Roth. 1991. The mechanisms of tongue protraction in *Discoglossus pictus*. *J. Exp. Biol.* 159:217–234.

Norberg, U. M., and J. M. V. Rayner. 1987. Ecological morphology and flight in bats (Mammalia; Chiroptera): Wing adaptations, flight performance, foraging strategy and echolocation. *Phil. Trans. R. Soc.* (London) B316:335–427.

Osenberg, C. W., and G. G. Mittelbach. 1989. Effects of body size on the predator-prey interaction between pumpkinseed sunfish and gastropods. *Ecol. Mon.* 59:405–432.

Osenberg, C. W., E. E. Werner, G. G. Mittelbach, and D. J. Hall. 1988. Growth patterns in bluegill (*Lepomis macrochirus*) and pumpkinseed (*L. gibbosus*) sunfish: Environmental variation and the importance of ontogenetic niche shifts. *Can. J. Fish. Aq. Sci.* 45:17–26.

Prosser, C. L. 1973. *Comparative animal physiology.* Philadelphia: Saunders.

Reilly, S. M., and G. V. Lauder. 1988. Ontogeny of aquatic feeding performance in the eastern newt, *Notophthalmus viridescens* (Salamandridae). *Copeia* 1988:87–91.

Reilly, S. M., and G. V. Lauder. 1991. Experimental morphology of the feeding mechanism in the salamander *Ambystoma tigrinum*. *J. Morphol.* 209:1–12.

Schluter, D. 1989. Bridging population and phylogenetic approaches to the evolution of complex traits. In *Complex Organismal Functions: Integration and Evolution in Vertebrates*, ed. D. B. Wake and G. Roth, 79–96. New York: Wiley.

Schoener, T. W. 1986. Mechanistic approaches to community ecology: A new reductionism? *Amer. Zool.* 26:81–106.

Sinervo, B. 1990. The evolution of maternal investment in lizards: An experimental and comparative analysis of egg size and its effects on offspring performance. *Evolution* 44:279–294.

Sinervo, B., and R. B. Huey. 1990. Allometric engineering: An experimental test of the causes of interpopulational differences in performance. *Science* 248:1106–1109.

Stephens, D. W., and J. R. Krebs. 1986. *Foraging Theory.* Princeton: Princeton University Press.

Tilman, D. 1982. *Resource Competition and Community Structure*. Princeton: Princeton University Press.

Wainwright, P. C. 1987. Biomechanical limits to ecological performance: Mollusc-crushing by the Caribbean hogfish, *Lachnolaimus maximus* (Labridae). *J. Zool.* (London) 213:283–297.

Wainwright, P. C. 1988. Morphology and ecology: Functional basis of feeding constraints in Caribbean labrid fishes. *Ecology* 69:635–645.

Wainwright, P. C. 1991. Ecomorphology: Experimental functional anatomy for ecological problems. *Amer. Zool.* 31:167–194.

Wainwright, P. C., C. W. Osenberg, and G. G. Mittelbach. 1991. Trophic polymorphism in the pumpkinseed sunfish (*Lepomis gibbosus*): Effects of environment on ontogeny. *Func. Ecol.* 5:40–55.

Wankowski, J. W. J. 1979. Morphological limitations, prey size selectivity, and growth response of juvenile Atlantic salmon, *Salmo salar*. *J. Fish Biol.* 14:89–100.

Wardle, C. S. 1975. Limit of fish swimming speed. *Nature* 255:725–727.

Webb, P. W. 1977. Effects of median-fin amputation on fast-start performance of rainbow trout (*Salmo gairdneri*). *Can. J. Zool.* 68:123–135.

Werner, E. E. 1974. The fish size, prey size, handling time relation in several sunfishes and some implications. *J. Fish. Res. Board Can.* 31:1531–1536.

Werner, E. E. 1977. Species packing and niche complementarity in three sunfishes. *Amer. Nat.* 111:553–578.

Werner, E. E., and D. J. Hall. 1976. Niche shifts in sunfishes: Experimental evidence and significance. *Science* 191:404–406.

Werner, E. E., and D. J. Hall. 1977. Competition and habitat shift in two sunfishes (Centrarchidae). *Ecology* 58:869–876.

Werner, E. E., and D. J. Hall. 1979. Foraging efficiency and habitat switching in competing sunfishes. *Ecology* 60:256–264.

Werner, E. E., J. F. Gilliam, D. J. Hall, and G. G. Mittelbach. 1983. An experimental test of the effects of predation risk on habitat use in fish. *Ecology* 64:1540–1548.

Zweers, G. 1979. Explanation of structure by optimization and systemization. *Neth. J. Zool.* 29:418–440.

4

Adaptation, Constraint, and the Comparative Method: Phylogenetic Issues and Methods

Jonathan B. Losos and Donald B. Miles

INTRODUCTION

Ecomorphological analyses generally seek to link the structure and function of organisms with relevant features of the environment. The scope of ecomorphological studies is far reaching and includes the adaptive significance of morphological design, convergence, evolution of function, morphological evolution, structure-function correlations, and community organization. Two schools of ecological morphology have independently flourished (see Ricklefs and Miles, chap. 2, this volume). One approach, characterized by functional morphologists, focuses on the effects of design and function on ecology (e.g., habitat choice or diet breadth). Another, championed by ecologists, uses morphology as an index to ecology and, through the analysis of morphological variation within an assemblage of species, derives inferences about ecological processes. Both schools share a common assumption that variation in morphology bears a strong relationship to variation in a suite of ecological characteristics.

Regardless of the goals of a specific project, many studies assume that the patterns revealed in an ecomorphological analysis reflect adaptation to prevailing selective pressures. In the absence of historical information, however, such interpretations are problematic. For example, suppose that one postulated a relationship between the presence of lamella-bearing subdigital pads and invariant clutch size in lizards. At least 89% of all lizard species with invariant clutch size have pads (from Shine and Greer, 1991)—seemingly a strong relationship. But, when one takes a phylogenetic perspective, a different picture emerges. The relationship exists because two clades, the geckos (>850 species) and the anoles (>300 species), evolved both pads and invariant clutch sizes. However, invariant clutch size has evolved at least twenty times in taxa without pads (Shine and Greer, 1991). Thus, the historical view indicates that the evolution of invariant clutch size is not linked to the presence of subdigital pads, contrary to what the distribution of the traits among extant taxa might suggest.

As this example illustrates, ecomorphological studies are beset with problems attributable to the shared ancestry of species. Taxa which share a trait due to common ancestry cannot be viewed as independent points for statistical analysis; further, in the absence of historical information, evolutionary rate and direction cannot be deduced (Felsenstein, 1985; Harvey and Pagel, 1991). Conversely, inclusion of phylogenetic information in ecomorphological and other studies presents several advantages. First, hypotheses regarding the rates and direction of evolutionary change are possible. When conducted with ecological data simultaneously, tests of the correlated evolution of ecomorphological relationships are also possible. Second, limits to the expression of morphological variation in design or function may be recognized and evaluated. Third, the statistical dilemma of non-independence of species as data points is resolved. Fourth, phylogenetic analyses may suggest hypotheses that can be experimentally tested.

In this chapter, we discuss the importance of including phylogenetic information in ecomorphological analyses. Although many researchers recognize that the form of a trait is an admixture of genealogy and recent adaptive variation (e.g., Darwin, 1859), until recently the analytical methods necessary to account for phylogeny in comparative analyses were not available. In particular, we discuss the various aspects of ecomorphology that have benefited from the inclusion of a phylogenetic approach; how knowledge of phylogenies can refine an analysis; the methods for including phylogenetic information into ecomorphological analyses; and the new comparative methods that explicitly include phylogenetic information in the analyses. The first part of the paper deals with various areas to which a phylogenetic perspective is important. The second part, by reviewing and critiquing the rapidly burgeoning field of statistical comparative methods, highlights the advantages and assumptions of each method and indicates under which circumstances each method is appropriate.

Pertinence of Phylogeny to Ecological Morphology

Reconstruction of Evolutionary Pathways. The fossil record can often tell us a great deal about evolutionary pattern and process (e.g., Van Valkenburgh, chap. 7, this volume). In the absence of a sufficient fossil record (and even in its presence), the derivation of a phylogenetic hypothesis allows inferences to be drawn about the character states of hypothetical ancestral taxa. Comparisons between extant and inferred ancestral taxa enable tests of the number of times a trait has evolved independently and the existence of directional trends of evolution within a clade. Phylogenetic information is crucial for studying a variety of questions ranging from patterns of morphological evolution to the adaptive significance of functional trait complexes. Below we discuss how phylogenetic approaches can refine ecomorphological analyses.

The presence of a given character in more than one species in a clade may be attributable to either shared ancestry or independent evolution of the trait. By mapping the character states on a phylogeny, these possibilities can often be distinguished. Phylogenetic studies can be critical in assessing how plastic a trait is evolutionarily (i.e., how frequently evolutionary change in the trait may have occurred) and can be used in testing scenarios about the evolution of a trait. For example, Losos (1990a) concluded that body size has evolved infrequently among northern Lesser Antillean *Anolis* lizards (see below), whereas Luke (1986) concluded that toe fringes had evolved in lizards from seven families no fewer than twenty-six times!

One initial procedure in an evolutionary analysis is the determination of character polarity: Does character *a* resemble an ancestral condition or has it changed since divergence? Character transformations can be determined by referring to a cladogram and using cladistic techniques (see Wiley et al., 1991; Maddison and Maddison, 1992, and references therein). Only with such information can one meaningfully attempt to understand the evolution of a character and test scenarios concerning its evolution. For example, a number of authors have questioned the selective features leading to the evolution of leaf retention in the common oak, *Quercus robur,* and the beech, *Fagus sylvatica.* Wanntorp (1983), however, noted that leaf retention, rather than shedding, appears to be the ancestral condition of the Fagaceae. Thus, one should look at characteristics of the family as a whole, rather than at attributes of particular species, to understand why leaf retention evolved (unless leaf retention is characteristic of an even larger group). More proximate explanations would be appropriate in understanding those taxa that have evolved leaf-shedding (Wanntorp, 1983; for another example see Coddington, 1986).

Further, one can ask why taxa are similar. Taxa may share the same condition of a trait either through common ancestry or because they have evolved the same condition independently. These possibilities often require radically different scenarios. In the absence of a detailed fossil chronology, the determination of character transformations can only be accomplished with reference to the phylogenetic relationships of the constituent taxa (Mickevich and Johnson, 1976; Wake, 1991). However, caution must be exercised because different methodologies for reconstructing ancestral states may tend to under- or overestimate the occurrence of parallel evolution (see below). Russell's (1976, 1979) studies of the evolution of pad structure in geckos elegantly demonstrates the utility of a phylogenetic perspective, in this case distinguishing when taxa are padless due to the inheritance of the primitive padless condition from when padlessness has resulted from convergent secondary loss of the pads (fig. 4.1).

Phylogenetic analyses can also be used to investigate proposed evolutionary sequences and scenarios. For example, plethodontid salamanders display a bewildering array of shapes of the premaxilla bone. Mapping variation in premaxillary shape along a phylogeny revealed that a paired premaxillary structure in adults represents the primitive condition, but a single, fused bone has evolved repeatedly. Further, consideration of ontogenetic data in a phylogenetic context suggested that the fusion has resulted from paedomorphosis in several cases, but peramorphosis in others (Larson, 1984; Lombard and Wake, 1986; Wake and

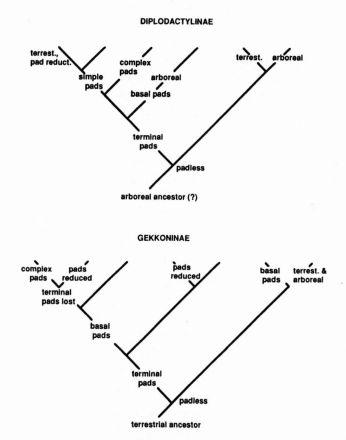

FIGURE 4.1 The evolution of subdigital pad structure in two gecko subfamilies. Comparison of trends for the two subfamilies indicates parallel patterns of pad evolution and subsequent secondary reduction (from Russell, 1979, with permission).

Larson, 1987; Wake, 1991). In a similar manner, Liebherr and Hajek (1990), testing predictions of the taxon pulse and taxon cycle theories concerning the evolution of habitat use in New World carabid beetles, found no trends in evolutionary direction relative to a null model of random change in habitat use. Dorit (1990) examined a proposed trend in the evolution of neurocranial shape in a clade of cichlid fishes. Rather than the proposed series of incremental changes in which each morphotype is derived from the next less specialized form, Dorit found that the most and least generalized species were sister taxa, and the intermediate species formed a separate monophyletic group.

Understanding the evolution of structurally complex phenotypic features, such as the avian wing or the vertebrate eye, has been a difficult problem for evolutionary biologists. But by examining the nested pattern of appearance of features in a phylogenetic hierarchy, it is possible to reconstruct how a complex structure has evolved through time. The recognition of intermediate states can facilitate investigation concerning the long-standing question about how such forms may have arisen (e.g., Mivart, 1871). For example, many authors have discussed under what selective circumstances the flying apparatus of modern birds evolved. However, the flight apparatus did not evolve in one step, but rather step-by-step, involving the sequential evolution of many characters over a fifty million-year period (fig. 4.2; Cracraft, 1990). Thus, one cannot simplistically talk about the evolution of flight capacity as if it were a single element. Similarly, King (1991) suggested that the cardia organ in the digestive tract of muscoid flies evolved gradually through a series of intermediate forms.

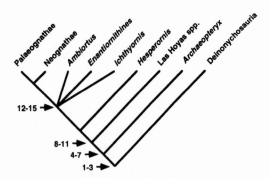

FIGURE 4.2 Evolution of the flight apparatus in birds. The fifteen characters of the flight apparatus did not evolve simultaneously, but, rather, constitute derived characters of four hypothetical ancestral taxa spanning more than fifty million years. Characters 1–3: elongated forelimbs, bowed metacarpal III, sternum; 4–7: feathers, loss of scapulocoracoid fusion, rotation of the forelimb, enlarged brain; 8–11: pygostyle, strutlike coracoid, procoracoid process, shift in scapulocoracoid articulation; 12–15: keel, carpometacarpus, external condyle of the ulna developed as a semilunate ridge, and modified humerus (modified from Cracraft, 1990, with permission).

This phylogenetic approach can be expanded to simultaneously consider more than one trait. Evolutionary biologists often wonder whether the evolution of various organismal traits is related. By examining the evolution of traits in the context of a phylogeny, one can determine whether the traits have evolved independently of each other. For example, Huey and Bennett (1987; Garland et al., 1991) used phylogenetic methods to investigate whether preferred body temperature and optimal temperature for maximal sprinting speed evolved in a correlated manner among scincid lizards. As a second example, Sillén-Tullberg (1988) found that gregariousness evolved fifteen times in butterflies subsequent to the evolution of aposematism, but never before it (though several lineages which evolved gregariousness never evolved aposematism). However, demonstration of an evolutionary correlation is only the first step in an adaptive analysis. Inferences concerning a causal explanation, be it adaptive or developmental, requires mechanistic analysis in a phylogenetic context (see below).

Phylogenetic reconstructions of character evolution can also test hypotheses suggested by studies of extant taxa. For example, by analyzing functional data phylogenetically, one can test whether structural and functional evolution are necessarily coupled (Lauder 1989, 1990; Losos, 1990b; Schaefer and Lauder, 1986). Data on extant taxa, for example, might suggest that a particular structure is causally related to a particular functional capability. This hypothesis is testable using a phylogenetic hypothesis, for if structure and function are necessarily correlated, then they should evolve concordantly (e.g., Wainwright and Lauder, 1993). Conversely, phylogenetic studies might suggest that traits evolve concordantly; subsequent functional studies could look for a mechanistic basis for such a relationship.

Adaptation and the Comparative Method. Adaptation has been a contentious topic among evolutionary biologists for more than a century. The validity of current approaches to the study of adaptation, and in some cases the concept itself, has been questioned on a number of grounds—for example, that it is untestable, relies on plausibility criteria, fosters the atomization of organisms into component parts, and ignores nonadaptive alternatives (e.g., Gould and Lewontin, 1979; Gould, 1980; Ho and Saunders, 1984; Reif et al., 1985). These criticisms have been ably countered by Mayr (1983), Fisher (1985), and Baum and Larson (1991), among others, and will not be discussed here.

Here, we follow Gould and Vrba (1982, p. 6) and define an adaptation as "any feature that promotes fitness and was built by selection for its current role." Features that currently promote fitness (i.e., have "current utility"), but arose for some other reason (either as an adaptation for something else or for nonadaptive reasons), are termed "exaptations," and features which currently promote fitness, but whose historical genesis is unknown, are simply called "aptations." A historical approach is obviously important for the study of adaptation. By contrast, other methods for the study of aptation, such as the optimality approach, which judges a feature by how closely it conforms to the optimal state based on functional criteria (e.g., engineering or foraging economics theory), or the microevolutionary approach, which assesses whether natural selection currently favors a trait, usually focus on current character states and ongoing evolutionary processes. In some cases these ahistorical methods can make a strong case for adaptation even in the absence of historical information (e.g., Schluter et al., 1985), but in most cases these approaches can more profitably be combined with historical analyses to shed light on the origin and evolution of a trait (see Baum and Larson, 1991, for a recent discussion).

Because the action of natural selection rarely can be directly demonstrated in the past, adaptive hypotheses must be evaluated in the same manner as hypotheses in the other historical sciences (Cracraft, 1981). That is, historical studies can unequivocally falsify adaptive hypotheses, but, because experiments are not possible, support for an adaptive hypothesis requires building a convincing case by integrating disparate lines of evidence. Several examples illustrate how historical data can negate an adaptive hypothesis. The hypothesis that coleoid cephalopods never developed hearing ability because whales would be able to stun them with blasts of intense sound was refuted by the observation that earless coleoids evolved 160 million years prior to the evolution of whales (Reid et al., 1986). Even in the absence of chronological data, phylogenetic information can negate hypotheses. In considering scenarios concerning the evolution of flight, Gauthier and Padian (1985) demonstrated that most of the "avian" characters of *Archaeopteryx* are shared by other coelurosaurs (the clade of dinosaurs to which birds belong) and consequently did not arise to facilitate flying, because other coelurosaurs were flightless.

Historical studies can make a convincing case for adaptive hypotheses by considering organismal analyses of function and selection in a phylogenetic context. Lauder (1981), Ridley (1983), Wanntorp (1983), and Greene (1986) were among the first to recognize that one must investigate the cause of the evolution of a trait at the phylogenetic level at which it arose. If a trait evolved as an adaptation to some problem posed by the environment, then that feature must have arisen in a lineage experiencing that selective regime. Further, a trait can confer enhanced fitness only when it leads to an increase in functional capabilities, which, in turn, affect reproductive success and survival (Alexander, 1990; Arnold, 1983; Garland and Losos, chap. 10, this volume; Huey and Stevenson, 1979). Consequently, if a trait evolved as an adaptation for a given selective regime, then the trait and its associated "performance advantage" should arise coincidentally (fig. 4.3; Coddington, 1990; Greene, 1986; Losos, 1990c; Wainwright and Lauder, in

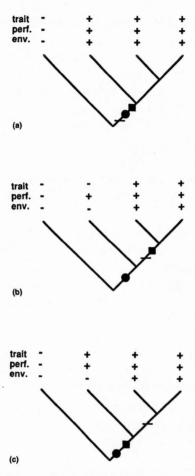

FIGURE 4.3 Evolution of a trait (■), a performance ability (●), and a selective regime (—). The scenario in *a* supports an hypothesis of adaptation; evolution of the trait and its associated performance advantage occur simultaneously and congruently with occupation of a new selective regime. In *b,* the trait evolves congruently with the new selective regime, but the performance capability is related neither to evolution of the trait nor to entrance to the new selective regime. If the trait is an adaptation to the regime, it must confer enhanced performance in some other manner. In *c,* the trait and the performance ability evolve simultaneously and prior to occupation of the new selective regime and thus may represent an exaptation. Other combinations are possible.

press). The historical approach to the study of adaptation thus has three components:

1. Identification of selectively important features in the environment and where, on a phylogeny, ancestral taxa experienced new selective conditions. Baum and Larson (1991) termed such features "selective regimes," which they defined as (p. 4) "the aggregate of . . . environmental and organismic factors

that combine to determine how natural selection will act upon character variation." Ideally, the biology of the organism would be well enough understood to predict how, in a selective regime, natural selection would direct the evolution of a trait (Baum and Larson, 1991).

2. Identification of which lineages in a phylogeny have experienced evolutionary change in the trait of interest to determine if evolution of the trait was concordant with entrance into the new selective regime.

3. Identification of which lineages have experienced change in functional capability and whether change in the trait of interest is associated with this change in capabilities. Performance advantages can be deduced by biomechanical theory or modelling as well as by direct measurement (Baum and Larson, 1991).

Few studies to date have fully employed this methodology. In a study of several taxa of sunfishes, Wainwright and Lauder (1993) found that evolutionary change in mouth size and gill raker structure evolved simultaneously with differences in foraging ability and diet. Similarly, limb proportions, jumping and running ability, and habitat use evolved concordantly among West Indian *Anolis* lizards (Losos, 1990c). Greene (1992) presented an example in which the evolution of a performance advantage was not concordant with evolutionary change in either morphology or selective regime. The thermally sensitive pit of pit-vipers has often been suggested to be an adaptation for detecting and capturing warm-blooded prey. However, no evidence to date indicates that the pit offers a performance advantage in prey-capture: non-pit-bearing vipers can detect thermal cues as well as pit-vipers. Further, there is no obvious shift in selective regime; no derived aspect of their dietary biology distinguishes pit-vipers from their near relatives (fig. 4.4; Greene, 1992).

Adaptive hypotheses can lead to the formulation of testable predictions by combining form-function-fitness studies on extant taxa (Arnold, 1983, 1986; Huey and Bennett, 1986; see Garland and Losos, chap. 10, this volume) with phylogenetic analyses (e.g., Schluter, 1989). For example, historical analysis might indicate that a particular trait evolved via natural selection in a given selective regime. If sufficient natural variation exists for the trait or can be created experimentally (e.g., Carothers, 1986; Sinervo, 1990), and if the selective regime in which a trait presumably arose still exists naturally or can be created artificially, then one can test whether the trait actually does increase fitness, relative to its antecedent condition, in that ecological context (Baum and Larson, 1991; Mitter and Brooks, 1983). Of course, this test could lead to false rejection of the adaptation hypothesis if subsequent evolutionary change has nullified the advantage of the trait or if the ecological context as it exists or is recreated is not the same as that experienced when the trait evolved.

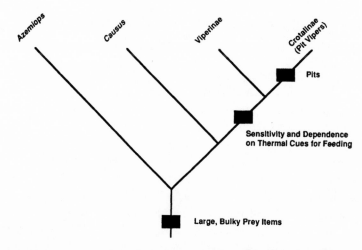

FIGURE 4.4 One possible phylogeny for the Viperidae (other possibilities do not alter these conclusions). The evolution of pits in the Crotalinae is not associated either with evolution of a new performance ability (the ability to use thermal cues for capturing prey) or use of a new selective regime (diet of large prey; based on Greene, 1992).

Two schools of thought exist on how to interpret a situation in which a trait is inferred to have evolved after a lineage enters a new selective regime (that is, trait evolution is reconstructed on a more terminal branch of the tree than is change in selective regime). Many authors (e.g., Baum and Larson, 1991; Ridley, 1983) consider that such a scenario would be consistent with a hypothesis of adaptation, but others indicate that one would expect adaptive change to occur rapidly relative to speciation. Thus, if trait evolution is adaptive, both changes (the trait and the new selective regime) should occur on the same branch of the phylogenetic tree (Björklund, 1991; Greene, 1986; Huey and Bennett, 1987; Losos, 1990c).

An adaptive scenario can be devised to explain any single evolutionary event (de Pinna and Salles, 1990). However, the morphology-performance-ecology approach to the study of adaptation also leads to the prediction that the same selective regime should lead to similar evolutionary responses in other lineages. This comparative approach to the study of adaptation has been a staple of biology for centuries (see Harvey and Pagel, 1991). As traditionally applied, the comparative method suggests that when distantly related taxa that experience the same selective regime evolve the same feature, then that feature must be an adaptation to that selective regime. Everyone is familiar with examples such as the convergent body shape of fast-moving aquatic organisms (e.g., sharks, tunas, dolphins, ichthyosaurs). Implicit in the notion of a comparative method is the idea that taxa independently evolve the same feature, but evolutionary independence is not always as obvious as in this example. In a phylogenetic framework, however, one can explicitly test whether the evolution of a trait has repeatedly been associated with a particular selective regime, as in Sillén-Tullberg's (1988) study of the evolution of gregariousness in butterflies (discussed above). Ideally, functional studies should be included in comparative analyses of adaptation rather than assumed. In the butterfly study, for example, the link between aposematism and unpalatability is assumed.

A potential difficulty with the comparative method is the assumption that there is a one-to-one match between selective regimes and organismal traits. To the extent that there are several adaptive "solutions" to a selective regime, or that one feature can evolve adaptively in several situations, then the comparative method may fail to identify adaptations (Bock, 1977, 1980; Bock and Miller, 1959; Harvey and Pagel, 1991). Several examples illustrate multiple adaptive solutions to a selective regime: some species of pocket gophers burrow with their forelimbs, others with their teeth (Lessa, 1989); diverse methods and structures for filter-feeding exist in taxa as different as whales, flamingos, and tadpoles (Sanderson and Wassersug, 1993). Conversely, some features can evolve in response to diverse selective regimes: for example, long legs can function to increase running ability in open habitats, as in antelopes, or jumping ability at habitat interfaces, as in frogs. Again, it is crucial to understand the functional basis for the relationship between morphology and environment (Bock and von Wahlert, 1965; Bock, 1977). For example, hawks, shrikes, and Australian shrike-tits have convergently evolved similar beaks, the first two for predation, but the shrike-tit for tearing bark off trees to capture insects. However, all three use their beaks for the same purpose, to seize and rend, even if the context is quite different (Simpson, 1978). In cases such as this, a detailed functional analysis can identify adaptation even when a less-detailed analysis fails to identify correlations between form and environment.

Quantitative traits can also be accommodated within the phylogenetic framework. If one were examining whether a quantitative trait evolved adaptively with respect to a quantitative selective regime (e.g., body size and prey size), then one would predict that evolutionary change in the two variables would be correlated. When considering quantitative characters and categorical selective regimes, one would predict that if the trait evolved adaptively, it would change more when a lineage enters a new selective regime than when the lineage remains in the same regime.

Constraints. Constraints on evolutionary change have been a popular topic for the past fifteen years. In response to the "adaptationist programme," Gould and Lewontin (1979) presented alternative explanations for evolutionary patterns. Certain traits may show a consistency of form within a clade for a variety of reasons, such as developmental canalization, intrinsic design, structural limitations, or stabilizing selection. These factors either prevent evolutionary change or channel it in particular directions. While a number of methods have been proposed to study constraints (e.g., Cheverud et al., 1985; Derrickson and Ricklefs, 1988; Stearns, 1983), explicitly phylogenetic methods allow direct examination of lineage specific effects on evolution of a trait (Carrier, 1991; Donoghue, 1989; Lauder, 1981; Lauder and Liem, 1989; Maddison, 1990; Miles and Dunham, 1992; Schluter, 1989; Wantorpp et al., 1990).

Hypotheses of constraint can be tested in a manner identical to hypotheses of adaptation: the evolution of a putatively constraining feature can be plotted on a phylogeny, allowing a comparison of the monophyletic clade containing that constraint with its sister taxon that does not have the constraint. If the putative constraint has actually constrained subsequent evolution, then the two clades should differ in their subsequent evolutionary history. As with studies of adaptation, mechanistic studies should be conducted as well to investigate whether the feature actually is the constraining element rather than a correlated trait (Carrier, 1991; Funk and Brooks, 1990).

The generality of the results could be tested by identifying multiple instances of the evolution of the constraint, with comparison in each case to sister clades. Emerson (1988) provided an example of this comparative method for studying constraints. She tested the hypothesis that evolutionary reduction in the number of elements in a system, in this case the pectoral girdle of anurans, leads to a reduction in morphological diversity among descendant species (following Lauder, 1981). Fusion of the epicoracoids has occurred in eight clades. The shape of the girdle in each clade was compared to the shape exhibited by the probable sister taxon. Emerson then tested the hypothesis that the evolution of epicoracoid fusion had constrained girdle shape to evolve in certain directions. As a generality, similar changes in shape occurred in the lineages experiencing fusion, which would seem unlikely as a null expectation given the morphological diversity and rampant evolution of girdle shape displayed by frogs (for another recent example, see Janson, 1992).

The recent emphasis on constraints and on historical influences has led to the notion that history can serve as a constraint. "Phylogenetic constraint" is often invoked to explain why a particular taxon has not evolved some feature. Along these lines, a number of authors have, either implicitly or explicitly, considered adaptation and history to be alternative explanations for why a particular taxon displays a particular feature (e.g., Derrickson, 1989; Lessios, 1990). However, a dichotomy between adaptation and history is fundamentally flawed. Whereas adaptation is a mechanistic process, as are developmental and other types of constraints, history is no more than a pattern. As such, history (or phylogeny) cannot constrain anything. The term "phylogenetic constraint" is shorthand for saying "the constraints that have previously evolved in and now characterize a lineage." For example, Gould and Lewontin (1979, p. 594) "invoke phyletic constraint in explaining why no molluscs fly in air and no insects are as large as elephants." Being a mollusc does not, in itself, preclude the evolution of flying ability. Rather, it is the developmental system and the materials of which molluscs are made that preclude the evolution of wings. Similarly, the surface area–dependent breathing mechanism of insects probably limits their maximum size to something less than elephantine. The point is, Gould and Lewontin do not really claim that phylogeny is a constraint; rather, the constraints exhibited by a particular clade, which were derived historically, are what limit evolutionary pathways.

This is more than mere semantics because adaptations, as well as constraints, may be inherited phylogenetically (Altaba, 1991; Clutton-Brock and Harvey, 1984; Fisher, 1985; Greene, 1986; Harvey and Pagel, 1991; Ridley, 1983)—historical and adaptational analyses are overlapping, rather than exclusive, perspectives. For example, enlarged and sharp canine teeth presumably arose at the base of the radiation of the order Carnivora as an adaptation for carnivory. Consequently, the presence of pointy canines in carnivores can be explained as a result of history. Nonetheless, they arose and, in most taxa, still function as adaptations—in this case adaptation and history are completely confounded.

Thus, a phylogenetic constraint simply refers to the observed pattern that a phenotypic trait does not evolve as expected within a given clade due to historical contingencies experienced by that clade. The term "phylogenetic constraint," then, should not be considered comparable to other types of constraints. Rather, a phylogenetic constraint results when a constraint (e.g., developmental, functional) evolves and affects subsequent evolution of the descendant clade. Phylogenetic constraints are thus the historical pattern, whereas other types of constraints refer to the actual mechanistic cause. Phylogenetic constraints may thus be recognized as "different evolutionary responses among taxa from different historical backgrounds when they are subjected to the same environmental selection pressure" (Schluter, 1989, p. 82; see also Maynard Smith et al., 1985). As such, a phylogenetic constraint is a relative concept investigated by comparing clades and seeing how they differ in evolutionary response to the same condition. Once a phylogenetic constraint is recognized, one can then determine, by plotting characters on a phylogeny, what features may be responsible for the constraint and what type of constraint it is (e.g., developmental).

Finally, it is important to distinguish between phylogenetic constraint and phylogenetic effect (Derrickson and Ricklefs, 1988; Miles and Dunham,

1992)—the latter simply indicates that taxa are similar, for whatever reason (including stabilizing selection), due to traits inherited from a common ancestor.

Community Structure. One of the major goals of ecology is the elucidation of the factors responsible for observed patterns of local diversity in biological communities. Characteristics such as species diversity and niche breadth, which species are present, how they interact, and what emergent properties characterize the community as a whole describe community structure. Ecologists attempt to explain differences in local diversity by invoking various mechanisms (e.g., competition, predation) that enhance or diminish the coexistence of species (MacArthur, 1972).

The development of ecological theory assumed that communities and their constituent species were at an equilibrium. Knowledge of contemporary species interactions were considered sufficient to describe processes responsible for maintaining diversity. The theories which were derived to explain the genesis and maintenance of diversity escaped the necessity of historical explanations by assuming that communities were saturated (Ricklefs, 1987). Consequently, the explanation for community-level phenomena is sought by examining conditions currently characterizing that community.

Species, however, are not interchangeable parts to be substituted into theoretical equations. Rather, they have specific adaptations and constraints that define how they can interact with and respond to other taxa and the environment. As with the study of adaptation and constraint, knowledge of the history of a community must be integrated with understanding of ongoing processes to understand patterns (Ricklefs, 1987; Brooks and McLennan, 1991).

Community ecology is contentious because the same pattern often can be produced by several processes, and experimental tests to distinguish alternatives are not always possible. Further, to the extent that observed patterns are the outcome of historical events, an historical perspective can be crucial in deciding between alternative possibilities. For example, in the northern Lesser Antilles, sympatric species of *Anolis* lizards are more dissimilar in size than would be expected by chance. This pattern could be produced either through *in situ* evolutionary change (e.g., character displacement) or differential colonization success (i.e., only dissimilar-sized species can successfully colonize the same island). Based on a phylogenetic analysis, Losos (1990a) argued that character displacement was responsible for the evolution of large and small taxa, but it may have occurred only once; the existence of size patterns across numerous islands consequently must be due to differential colonization success. The phylogenetic analysis thus indicates that both processes have operated in the northern Lesser Antilles—a conclusion not possible without phylogenetic information.

Ecological morphologists are often interested in comparing to what extent, if any, the structure of communities differs in diverse regions or habitats. Although in some cases differences may result purely from present-day causes, a historical perspective can often yield considerable insight. For example, carnivore communities in North America (Yellowstone) are more dispersed in morphological space than similar communities elsewhere, due primarily to the presence of bears (Van Valkenburgh, 1985, 1988). This difference in morphological diversity may exist because the environmental circumstances which have elicited the evolution of bear-like morphologies or allowed the immigration of ursids may never have existed in other communities, such as the Serengeti—differences in the use of "morphospace" would thus indicate ecological differences among sites. Alternatively, the difference in diversity may have a historical basis, resulting because the ursid lineage never colonized sub-Saharan Africa and the evolutionary flexibility of carnivore lineages present in Africa is constrained (Cadle and Greene, 1993)—perhaps bear-like morphologies are not possible in felid or viverrid lineages, for example. This latter perspective does not indicate that ecological considerations have been unimportant in the structuring of carnivore communities, but suggests that alternative lines of inquiry are appropriate. Why have bears never reached sub-Saharan Africa? What prevents other carnivore lineages from producing bear-like morphologies in the presence of bear-like ecological conditions? Similar phylogenetic effects are apparent in the differences in ecological structure of neotropical assemblages of colubrine snakes (Cadle and Greene, 1993), the abundance of nocturnal species in desert lizard assemblages around the world (Duellman and Pianka, 1990; Pianka, 1986), and the composition of helminth parasite assemblages of neotropical stingrays (Brooks and McLennan, 1991).

Theoretical analyses suggest that communities in similar environments should converge in structure. Convergence could result by the same ecological and/or morphological types occurring in each community, or it could result at the level of aggregate properties of communities, such as species richness and spacing (Blondel et al., 1984; Orians and Paine, 1983; Schluter, 1986, 1990; Strauss, 1987; Wiens, 1989). Although community convergence is clearly within the realm of historical approaches, its study will prove more difficult than studies involving single taxa and lineages because community structure results from differential colonization success, *in situ* evolutionary change of species within a community, and *in situ* production of new species. Schluter (1986) proposed a novel indirect method to study community convergence by comparing the variance between replicate communities in the same habitats on different continents to the variance between communities in different habitats within continents. To the extent that lineages are not shared between continents, replicate communities in the same habitats should be phylogenetically independent (see also Winemiller, 1991).

A more direct method of examining community structure would require phylogenetic information for each lineage in the communities, a forbidding task for all but the smallest communities. Even with such information, it will not be easy to relate to the evolution of aggregate properties of communities, such as niche packing, and few such approaches have appeared to date: for example, Nee et al. (1991) have shown how phylogenetic considerations can shed light on the relationship between abundance and body size in British birds (for another promising approach, see Winemiller, 1991). Perhaps by randomizing both evolutionary change within lineages and species co-occurrence, one could assess whether communities have converged relative to random expectations.

Brooks and McLennan (1991) and Gorman (1993) have devised an alternative approach, focusing more on species-level interactions, that incorporates aspects of vicariance biogeography and adaptational studies to partition the relative importance to community structure of colonization, interspecific interactions, and the interaction between the two. In contrast to Schluter's method, this approach requires that the members of the same lineages occur in the different communities so that changes in associations and traits can be compared and interpreted phylogenetically. In a particularly interesting example, Gorman (1993) found that interspecific interactions among stream fishes could be predicted with historical data—taxa that had a long history of association (as revealed by a combination of phylogenetic and biogeographical data) competed less strongly than species whose association was more recent.

The evolution of community (or assemblage) structure is simpler to comprehend when one is dealing with a monophyletic group, as is often the case on islands. The use of monophyletic groups also facilitates understanding of the sequence of addition and modification that occurs as a community assembles. For example, Williams (1972) used a phylogeny for the *Anolis* of Puerto Rico, which have diversified morphologically and ecologically, to formulate a hypothesis about the causal factors underlying the radiation (fig. 4.5). Based on the pattern of character evolution, he suggested that interspecific competition first leads to divergence in size followed by habitat partitioning. Subsequently, a phylogeny for Jamaican *Anolis* allowed factors governing the parallel radiations on the two islands to be compared (Losos, 1992; fig. 4.6). A comparison of the reconstructions in figures 4.5 and 4.6 reveals that the anole communities on the two islands are quite similar and that the sequence of ecomorph types (*sensu* Williams, 1972) evolved in identical order. For instance, the ecomorph type absent on Jamaica, the grass-bush anole, is the last to evolve on Puerto Rico, which offers a proximate explanation for its absence on the former island (Losos, 1992).

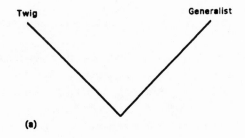

(a)

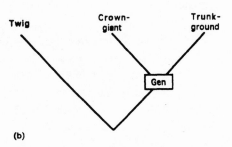

(b)

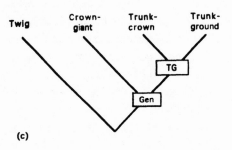

(c)

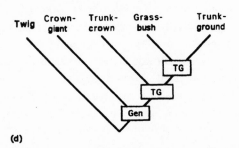

(d)

FIGURE 4.5 Diversification of anole fauna on Puerto Rico. Sequence in the evolution of ecomorph types (*a, b, c, d*).

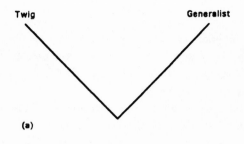

(a)

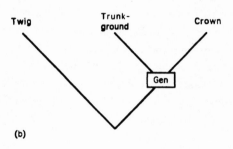

(b)

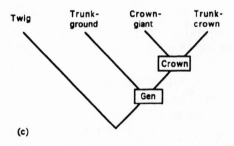

(c)

FIGURE 4.6 Diversification of anole fauna on Jamaica. Sequence in the evolution of ecomorph types (*a, b, c*).

METHODS FOR INCORPORATING PHYLOGENETIC INFORMATION INTO COMPARATIVE STUDIES OF ECOLOGICAL MORPHOLOGY

In this section, we first discuss problems posed by the phylogeny itself, and then detail the methodology, assumptions, and weaknesses of several of the most widely used methods for incorporating phylogenetic information into comparative analyses.

The Underlying Phylogenetic Hypothesis

Any historical analysis is only accurate if the phylogeny used is correct. Consequently, it is important to realize that phylogenies are hypotheses about patterns of evolutionary descent. Rarely, if ever, will it be possible to demonstrate that a phylogeny is correct. Rather, one can only assess to what extent the evidence corroborates a particular phylogenetic hypothesis. This is an important perspective, because historical analyses such as those discussed above tend to be based on a preferred phylogenetic hypothesis. However, in many systematic analyses, the preferred (or "best") phylogeny is only marginally better than competing hypotheses. Many cladistic analyses of morphological data, for example, simply choose as the preferred hypothesis the tree which requires the fewest evolutionary transitions in character states. The advent of statistical criteria for choosing between hypotheses will prove helpful in delimiting which trees to consider (e.g., DeBry, 1992; Cunningham and Buss, 1993), but these tests are only valid to the extent that their underlying assumptions are met (e.g., that parsimony criteria are appropriate or that transition state probabilities in maximum likelihood methods are accurate); these assumptions have generally proven difficult to verify.

Thus, future workers should not accept whatever phylogeny they can find or patch together, but instead should evaluate the evidence and include all appropriate phylogenetic hypotheses (Swofford, 1991). Eventually, it will be desirable to develop methods to weight the results of analyses using different phylogenies by the perceived likelihood that each phylogeny is correct. In a less thorough fashion, one could conduct an analysis using the preferred phylogeny, but perform sensitivity analyses to see how vulnerable the results are to changes in the phylogenetic structure (e.g., Richman and Price, 1992).

Recently, it has become common for comparative analyses to acknowledge the importance of phylogenetic considerations, but, because no phylogeny is available for the group in question, to conduct analyses ignoring phylogeny, or to use an available taxonomy. It is clear that ahistorical analyses have a high probability of being inaccurate (Felsenstein, 1985; Grafen, 1989; Martins and Garland, 1991); using taxonomies is perhaps less risky, but many available taxonomies probably poorly reflect phylogeny. An alternative approach (Losos, 1994) would be to conduct the analysis on many different phylogenies and to see if the results differ. Such an analysis might indicate, for example, that a significant relationship exists regardless of which phylogeny is used or that a significant result would obtain only when one used a phylogeny that seemed highly improbable. Rather than ignoring phylogeny, this approach would be more positive, laying a predictive groundwork for future systematic work.

Even when a comparative study focuses only on the preferred phylogenetic hypothesis, the analysis may be hampered by the inability to resolve the tree completely into a network of bifurcations. Polytomies (i.e., ancestral nodes, each producing more than two descendant lineages) may actually represent the history of diversification; population fragmentation could give rise to more than two descendant species. In many cases, however, polytomies reflect either lack of data or disagreement among data. Polytomies also will exist when taxonomies are recast as phylogenetic trees if any rank (e.g., genus) contains more than two taxa (such attempts also make the dangerous assumption that each taxon is monophyletic). Grafen (1989), Maddison (1989), and Harvey and Pagel (1991), among others, have developed methods for interpreting the evolution of a character when faced with a polytomy, but these methods involve assumptions that the state of the character can provide information useful in resolving uncertain phylogenetic relationships; Harvey and Pagel's (1991; see also Pagel, 1992) method, for instance, assumes that phenotypically more similar species are more closely related. Conducting separate analyses using each possible resolution of the polytomy would be less biased (Maddison, 1989). Similarly, when faced with multiple, equally supported phylogenetic hypotheses the most conservative course would be to conduct the analysis using each one (Harvey and Pagel, 1991; Harvey and Purvis, 1991). Rather than attempting to circumvent uncertainty in our knowledge with assumptions whose accuracy may be questionable, this approach would again put matters in a positive light, indicating to what extent different phylogenetic resolutions alter the outcome of historical analyses and encouraging further studies (Losos, 1994).

Variance Apportioning Methods

In some cases, investigators may wish to determine the correlation of a trait with phylogeny, which would provide an index of a trait's evolutionary lability. A number of quite distinct methods have been developed that partition variation between recently evolved versus inherited values (we will not discuss hierarchical taxonomic methods, which are not strictly phylogenetic and suffer from a number of problems; Felsenstein, 1988a; Harvey and Pagel, 1991; Losos, 1990c; Maddison and Maddison, 1992; Miles and Dunham, 1992). These techniques determine how much of the variation in a trait is attributable to phylogenetic inheritance and how much is due to evolution since divergence from a common ancestor.

Phylogenetic Autocorrelation Analysis. The elucidation of a phylogenetic constraint first requires information about the covariation between a character and phylogeny (Miles and Dunham, 1992). Indeed, if phenotypic values cannot be predicted as a function of phylogenetic relationships, then incorporating phylogenetic information into statistical analyses may be unnecessary (Gittleman and Luh, in press). Consequently, investigating whether variation in a trait is related to phylogenetic relationships of the group in question may be important. The difficulty with this approach is determining what one means by a correlation between phylogeny and trait variation.

Using an autocorrelation model, Cheverud et al. (1985) showed that correlation between the phenotype and phylogeny may be described by an autocorrelation coefficient which is the product of a phylogenetic connectivity matrix and a vector of standardized trait values. The phylogenetic connectivity matrix represents the pairwise phylogenetic similarity (or relatedness) of all the extant taxa included in the analysis. For each species, the value of a trait can be partitioned into two components: the phylogenetic value is estimated by the product of the autocorrelation and the matrix of phylogenetic weights and describes the degree of invariance of the trait with tree topology (i.e., the phylogenetic effect); the specific value is estimated by the variation that is independent of phylogenetic effects (Cheverud et al., 1985; Gittleman and Kot, 1990; Miles and Dunham, 1992; see fig. 4.7 for a hypothetical example). In other words, the expected value for each species is predicted by summing the phenotypic values for all other species, with the value for each of the other species weighted by some predetermined amount. The specific value, then, represents the difference between the observed and expected values for a species.

The weighting scheme is thus of crucial importance in determining the spe-

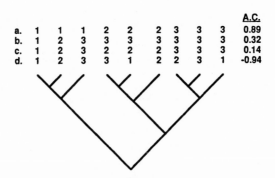

FIGURE 4.7 Sample autocorrelation values obtained using the method of Cheverud et al. (1985). *a–d*, different distributions of trait values. The autocorrelation value (A.C.) ranges from 1.0 (trait values completely predicted by phylogenetic topology) to 0 (no relationship) to −1.0 (closely related taxa dissimilar in trait value). In this example, the distance between any two taxa was assigned based on the number of speciation events since their common ancestor, with the provision that all members of a clade are equally distant from taxa outside that clade.

cific value. Weights are determined by the phylogenetic connectivity matrix, which in the past has been based on taxonomic rank (e.g., congeners weighted more highly than confamilial species, which are weighted more highly than members of the same order, and so on [Cheverud et al., 1985; Gittleman and Kot, 1990]). Alternatively, cladistic methods can also be used to determine the weighting scheme (Miles and Dunham, 1992), either by using the patristic distance (i.e., the number of nodes separating two taxa) or by other cladistically based distances. In an extension of the autocorrelation approach, Gittleman and Kot (1990) included a "variable weighting index," α, which was estimated by maximum likelihood procedures. The rationale for adding this exponent to the weighting matrix was the finding that the autocorrelation coefficient did not decay in monotonic fashion with taxonomic distance; by including α, pairwise distances among taxa are adjusted in the connectivity matrix to remove phylogenetic correlations at all hierarchical levels (see example in Gittleman and Kot, 1990).

Thus, the autocorrelation statistic is simply a measure of how well species' values can be predicted based on the phylogeny. Several authors have used the specific values to test adaptational hypotheses, arguing that the specific values, being free of phylogenetic influences, represent change due to environmental conditions. These analyses look for a correlation between specific values for two characters, such as clutch size and age of maturity (Miles and Dunham, 1992). Obviously, the key to this method revolves around the accuracy of the phylogenetic matrix and how successfully the values for other taxa can be used to predict a species' value. In addition, it remains to be determined what exactly the specific value measures in evolutionary terms. To some extent, the specific value can be seen as an amalgam of the evolutionary history of a taxon, with recent divergent events weighted more highly than more ancient events. The question, then, is whether a particular specific value could result from several, quite distinct, historical scenarios. If, for example, a small specific value could result either from slight change from a sister taxon or great change during a more distant cladogenetic event, then correlations between specific values for two variables might be suspect. In simulations based on taxonomic hierarchies, however, Gittleman and Luh (1992) found that autocorrelational studies of this sort work reasonably well, at least under certain circumstances.

Two cautionary comments need to be made concerning autocorrelation methods. First, Cheverud et al. (1985) argued that correlations based on the specific values may be used in assessing adaptive change independent of change experienced in other lineages. Because many factors may influence this variance component, such as genetic drift, pleiotropic effects, and sampling error, Lynch (1991) suggested caution in interpretation of such correlations and suggested that patterns of variation attributed to phylogeny may be viewed as estimates of macroevolutionary effects (see the discussion of Lynch's mixed model below) on a trait, whereas specific values represent patterns of variation due to microevolutionary events.

Second, partitioning variation into a phylogenetic and a specific component is not equivalent to partitioning into an adaptive and a nonadaptive component. As discussed above, adaptations can contribute to phylogenetic effects and phylogenetic effects can result from the action of stabilizing selection.

Lynch's Mixed Model. Lynch (1991), drawing a parallel between microevolutionary studies of selection and heritability and macroevolutionary analyses of adaptation and radiation, suggested extending quantitative genetics methodology to macroevolutionary questions. He suggested that the use of mixed model methods developed in animal breeding may be appropriate for partitioning the variation in a trait. Given n taxa for which k characters have been measured, the model for simultaneously analyzing variation in all traits is given by

$$z_{ij} = u_i + a_{ij} + e_{ij}$$

where u is the grand mean of the ith character for the whole phylogeny, a is the "heritable additive component" of the character for the jth taxon (similar to breeding values used in quantitative genetics), and e is the residual from the predicted value $(u + a)$. Although similar in spirit to the autocorrelation method of Cheverud et al. (1985), this approach differs in that all the traits must be included in the analysis to examine patterns of evolutionary change.

The inclusion of phylogenetic effects is accomplished by a matrix of phylogenetic relationships, G, derived from a phylogenetic tree (including branch lengths). In Lynch's (1991) example, phylogenetic relationships were estimated by the proportion of evolutionary time, from the root of the tree, shared by any two species. However, as with all methods, other information that captures the phylogenetic relationships among taxa could be used.

The parameters in the mixed model must be estimated directly from data from extant species using an iterative expectation-maximization algorithm. This algorithm produces estimates of the character means (over the whole phylogeny), the additive and residual values (analogous to the phylogenetic and specific values of Cheverud et al., 1985), and the variance-covariance matrices of the additive and residual values.

The variance-covariance matrix for the additive values allows the estimation of phylogenetic heritabilities as well as their associated standard errors, tests of hypotheses regarding the existence of phylogenetic heritability, and the prediction of phenotypes of ancestral species. Unlike other methods for inferring ancestral states, residual deviations (due, for example, to environmental causes or measurement error) do not contribute to the predicted mean of ancestral phenotypes. Standard regression and correlation procedures on the residual values may be used to describe adaptive patterns among traits. In addition, patterns deduced from the variance-covariance matrix of additive effects may represent macroevolutionary or phylogenetic phenomena, whereas those based on the residual variance-covariance matrix may represent variation specific to taxa since divergence from a common ancestor. To date, this method has been used only on one limited data set (Lynch, 1991).

Ancestral State Reconstruction and Directional Methods

A number of methods have been proposed to reconstruct the evolution of a character, given a phylogeny. It is important to recognize that all methods make implicit assumptions and depend on implicit or explicit evolutionary models (Felsenstein, 1988a; Harvey and Pagel, 1991; Harvey and Purvis, 1991; Maddison and Slatkin, 1991; Martins and Garland, 1991; Maddison and Maddison, 1992). Indeed, this is true of phylogenetic statistical methods in general (Harvey and Pagel, 1991; Pagel and Harvey, 1992; Martins, 1993). Qualitative (or categorical) characters have generally been reconstructed using parsimony methods that minimize the number of evolutionary transitions required throughout the tree (Farris, 1970; Swofford and Maddison, 1987). A number of variants exist, depending on particular assumptions (e.g., ordered versus unordered character states, irreversibility of evolution; see Felsenstein, 1983; Maddison and Maddison, 1992). In general, parsimony methods will produce a maximum likelihood estimation of evolutionary pattern when rates of character change have been low relative to cladogenesis or not excessively unequal among lineages (Felsenstein, 1983). To the extent that probability of character change differs on different branches of tree (e.g., due to differences in branch lengths in units of time), then parsimony methods are likely to overestimate the amount of change on some branches and underestimate it on others. Few studies have assessed the degree of error thus introduced: Maddison (1990) used simulations to demonstrate that parsimony did not generally introduce large or systematic errors in his method (discussed below); Hillis et al. (1992) found that parsimony methods were more than 98% accurate in reconstructing ancestral phenotypes in an experimentally derived clade of bacteriophage T7.

A number of methods have been introduced to reconstruct hypothetical ancestral states for quantitative characters. Linear parsimony methods, mathematically essentially identical to those for qualitative traits, minimize the change in trait value summed over all branches (Swofford and Maddison, 1987; Losos, 1990c; Maddison and Maddison, 1992). The squared-change parsimony method (Huey and Bennett, 1987; Maddison, 1991) minimizes the square of the amount of change on each branch summed over all branches. In contrast to conventional parsimony approaches, whose philosophy is to minimize the absolute amount of evolutionary change required, the purpose of minimizing the square of changes is not obvious (Losos, 1990c). The weighted squared-change parsimony method minimizes the square of the amount of change divided by the inverse of branch length (in units of expected variance) summed over all branches (Martins and Garland, 1991). The reconstruction of ancestral character states produced by this method has maximum posterior probability under a Brownian motion model of evolution (Maddison, 1991).

Given the differences in their underlying models of evolutionary change, the methods for reconstructing continuous traits can produce quite different estimates of values for ancestral taxa (Losos, 1990c). In general, linear parsimony tends to concentrate evolutionary changes on relatively few branches of the tree and minimize parallelism and reversal. The other methods spread evolutionary change more homogeneously among branches and exhibit enhanced parallelism and reversal. In addition, linear parsimony methods often yield ambiguities in character reconstruction (e.g., whether a character is interpreted as evolving in parallel in two lineages, or evolving once and subsequently being lost), whereas other methods yield a single estimate for each ancestral taxon. Although the latter methods are thus analytically more tractable, adherents of parsimony would argue that these methods obscure inherent uncertainties in the data (Losos, 1990c). Because there will usually be no way to determine the appropriate model for character evolution, it is important to recognize that choice of methodology will affect the character reconstruction (Harvey and Pagel, 1991; Harvey and Purvis, 1991). To date, little work has been conducted in assessing the extent of error that results when the assumptions of a model are incorrect, though Martins and Garland's (1991) simulation study suggested that in many cases the effect may be surprisingly small.

Once ancestral character states have been inferred, "directional" comparative methods (*sensu* Harvey and Pagel, 1991) can be used to investigate whether the evolution of two characters (both organismal or one environmental) is related by comparing the reconstruction of their ancestral states. One of two hypotheses is tested: that the characters evolved simultaneously (in continuous characters, that change in the two is correlated among branches of the tree), or that evolution in one variable is concentrated in clades exhibiting a particular state of the second character. Examples of these approaches have been discussed in the first section of this paper.

Appropriate null models for these tests are more complicated than they initially appear. Maddison (1990) developed a null model to test whether changes in a trait are distributed randomly throughout a tree or are clustered on branches with a particular state of a second trait. For example, although most instances of the evolution of gregariousness in butterflies occurred in aposematic lineages (Sillén-Tullberg, 1988; see above), this might not be surprising because most lineages are aposematic. Indeed, Maddison (1990) found no suggestion that gregariousness had evolved in aposematic lineages more often than expected by chance (for another example, see Donoghue, 1989).

A further complication is that some reconstruction methods do not consider differences in branch lengths. To the extent that the probability of change along a branch is proportional to time since divergence, one would expect changes to be concentrated on longer branches by chance alone, which could potentially bias Maddison's (1990) test (Harvey and Pagel, 1991; Maddison, 1990). Harvey and Pagel (1991) have proposed a maximum-likelihood method to estimate the transitional probabilities of change in a categorical character along all branches of a tree. The efficacy of this method is uncertain, however, because the maximum-likelihood estimates are derived from reconstructions of ancestral traits based on parsimony and ignoring branch lengths. Additional problems may arise because parsimony methods usually reconstruct many branches with no observed change, yet maximum-likelihood estimates lead to nonzero expectations for change on all branches. Consequently, if many branches are reconstructed with no change for both characters, standardized scores for the two variables on these branches will be highly correlated (Maddison and Maddison, 1992). Consequently, Harvey and Pagel (1991) suggest considering only branches on which change is reconstructed to have occurred, which necessarily omits some of the available information.

Most analyses that examine the evolution of categorical characters use parsimony methods, but often the existence of alternative, equally parsimonious reconstructions is not mentioned. Such ambiguities can make statistical analysis difficult, but ignoring the uncertainty is not the answer, because it may lead to incorrect conclusions. For example, Altaba (1991) found that resolution of ambiguities affected outcome of an analysis concerning whether evolutionary change in chemical warfare in beetles is related to habitat (see also Donoghue, 1989). Maddison (1990) recommended that when ambiguities in reconstructions exist, one should examine the reconstructions most and least favorable to the hypothesis to suggest how sensitive the analysis is to different reconstructions. Alternatively, one could check all possible combinations of reconstructions for both characters, or a random subset of that universe. Perhaps, if most (95%?) reconstructions favored a hypothesis, one could accept it as well supported.

Further, studies should not be limited to the most parsimonious reconstruction of character evolution. As with the decision concerning which phylogenies to consider, one should also conduct sensitivity analyses which consider slightly less parsimonious character reconstructions to see how they affect evolutionary interpretation (Maddison and Maddison, 1992). Indeed, one underappreciated difficulty with reconstructing ancestral states, regardless of which method is used, is that the reconstructions of the phylogeny and of the ancestral states are conducted sequentially rather than simultaneously, but there is no guarantee that this method produces the overall most parsimonious solution for evolution of the trait in question. For example, trees slightly less parsimonious than the best phylogeny may require considerably fewer changes in the trait in question. An analogous problem occurs in phylogeny reconstruction based on DNA data, when sequence alignment and phylogenetic analysis are conducted sequentially (Felsenstein, 1988b, and references therein).

Independent Contrast Methods

Felsenstein (1985) proposed an alternative to directional methods for studies of the evolution of continuous characters which does not require the inference of ancestral traits. By reference to a phylogeny, one can construct judiciously chosen comparisons that resolve the statistical and evolutionary dependence of species as data points. This method uses the information on tree topology and branch lengths to construct "contrasts," that is, a set of sequential comparisons between pairs of taxa (or nodes). Here, "contrasts" are defined as the difference of a trait between one extant species (or node) and its sister species (or node)—evolutionary change in each contrast is independent of change occurring elsewhere in the phylogeny. One continues down the tree, making comparisons from the tips to the root. For a sample of N taxa there will be $N-1$ possible comparisons (fig. 4.8). Statistical analyses can then be conducted on contrasts generated for two or more characters (see Garland et al., 1992).

Branch lengths, in units of expected variance of evolutionary change, are important to the contrast method in two ways. First, contrasts are based on the difference in character value between sister taxa. When one (or both) of the taxa are internal nodes of the tree, these nodes must be assigned values (these values are not equivalent to reconstruction of ancestral states). The assigned value is an average of the values of its two descendants, inversely weighted by branch length; further, internal branch lengths are lengthened by the addition of an error term, which is itself a function of branch lengths (Felsenstein, 1985; and Garland et al., 1992, provide details for calculations). Second, to meet the assumptions of parametric statistics, contrasts must be standardized by their expected variance. If one assumes that character change can be modeled as a Brownian motion process, then the cumulative variance will be a function of time since divergence. Therefore, previous workers have recommended that each contrast be scaled to a common expected variance through division by the standard deviation. Of course,

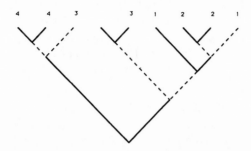

FIGURE 4.8 Hypothetical phylogeny indicating eight independent contrast comparisons. Each contrast is the difference in trait value between two sister taxa (each taxon being either an extant species or an internal node). With a phylogeny for nine species, it is possible to choose four unstandardized contrasts between extant species. The numbers above the phylogeny indicate one possible combination of four species-pairs. Note that in connecting the four sets of species through the phylogeny, no ancestral node lies on the path of more than one pair of species.

the Brownian motion model of evolutionary change was originally chosen more for its analytical tractability than for its biological realism (Felsenstein, 1985, 1988a; Harvey and Purvis, 1991). Other models are possible, the most obvious being punctuated equilibrium, in which expected variance on all branches would be equal (Felsenstein, 1988a; Harvey and Pagel, 1991; Martins and Garland, 1991). There are many sources that could provide branch lengths (e.g., genetic distance values or the number of systematic character changes inferred to have occurred along a branch; Garland, 1992; Garland et al., 1992); one only has to be willing to assume that expected variance in trait evolution should be proportional to branch length (one could also propose models in which the relationship between branch length and expected variance varied temporally). Harvey and Pagel (1991) and Garland et al. (1992) provide methods for verifying that contrasts have been adequately standardized.

The incorporation of branch length information is both the principal strength and weakness of the independent contrasts method. To the extent that ancestral nodes are correctly estimated, contrasts reflect nonoverlapping episodes of evolutionary divergence and are statistically independent, thus avoiding the problem of nonindependence inherent in ancestor reconstruction approaches (Felsenstein, 1985; Harvey and Pagel, 1991; Garland et al., 1992). But to the extent that the presumed model of change (e.g., Brownian motion) is inaccurate, then the contrasts lose their claim to statistical independence (Felsenstein, 1985, 1988a). The methods of Harvey and Pagel (1991) and Garland et al. (1992) can detect instances in which the underlying model for contrast standardization is inappropriate, but these methods probably will have low statistical power in many applications, particularly when the true evolutionary model is more complicated and nonlinear than the standardization model. This is currently an area of active research, with new developments appearing rapidly. Contrasts also may be non-independent because of correlated error (Harvey and Pagel, 1991). Because internal node values are calculated from the values of their descendants, an error in measurement of a character value for an extant taxon will affect all contrasts involving nodes ancestral to that taxon.

The need to standardize contrasts also imposes a cost: standardized contrasts no longer measure absolute change, but rather change per unit standardization (Garland, 1992; Losos, 1990b). This difference, though subtle, might be important in many contexts. For example, the exponent by which metabolic rate scales with body size has long been of interest to physiologists (reviewed in Harvey and Pagel, 1991). In a phylogenetic context, the question is: How does evolutionary change in metabolic rate scale relative to evolutionary change in body size? But when contrasts are scaled by branch length in units of time, the contrasts measure *rate* of change in metabolic rate relative to body size. It is not obvious that these two values should be the same. Matters become even more complicated when it becomes necessary to standardize by some quantity other than branch lengths in units of time. For example, the number of systematic characters inferred to change on a branch might be used for standardization (Garland, 1992; Garland et al., 1992). But if one is comparing rates of evolution in limb length between two clades (e.g., Garland, 1992), what would it mean that rates of change per unit cladistic character differ between clades? Certainly, this is not the same as saying that rates per unit time differ.

One solution to these difficulties in using independent contrasts, and a means of obviating the need for information on branch lengths, is to ignore branch lengths and not standardize the contrasts. By adopting this course, however, one cannot make comparisons involving internal nodes and is limited to comparisons among pairs of extant species, with no species used more than once and no species pair sharing a branch of the phylogeny (fig. 4.8; Felsenstein, 1985; Burt, 1989; for examples, see Voss, 1988; Møller and Birkhead, 1992). Thus, the numbers of degrees of freedom is cut in half, which may pose a problem for many studies. Further, unless one is willing to assume that the expected amount of change is equal in all comparisons (as would be the case if one used sister species and assumed that evolution was punctuational), then only a sign-test (not ranked) is valid. Although one might expect a sign-test to have low statistical power, re-analysis of two studies that used paired comparisons but failed to use the sign-test (Losos and Miles, unpubl.) indicates only a minor change in significance levels when the sign-test is employed (from $P = .026$ to $.049$ in Oakes, 1992, and from $P = .017$ to $.035$ in Balmford et al., 1993). The advantages of using such paired-species contrasts that ignore branch lengths and internal nodes is that they are unquestionably independent statistically, avoid the potential difficulties discussed above, and have the added benefit that information on branch lengths is unnecessary.

Most phylogenies will contain unresolved nodes, that is, polytomies, which prevent the calculation of independent comparisons. One method for resolving ambiguous nodes and constructing comparisons involves the calculation of contrast coefficients commonly used in ANOVA models (Harvey and Pagel, 1991; Pagel and Harvey, 1988). This entails the calculation of a weighted difference score that derives a single score from three or more taxa.

Grafen (1989) proposed an alternative method. The phylogenetic regression is a three-stage analysis for deriving contrasts from unresolved phylogenies. The first phase involves calculating an average value for each node. The average of each node is then expressed as a deviation from its parent node; this procedure results in a "long data set." Second, the values in the long data set are adjusted to account for differences in branch lengths. Grafen (1989, 1992) estimates a branch "height," which is described as one less than the number of taxa below a given node. Then, branch lengths are calculated as the difference between the heights of successive nodes. Grafen assumes that the expected amount of change will be proportional to the length of the branch (i.e., "Brownian motion"). The third step, after adjusting the nodal values by the branch lengths, is to construct a set of linear contrast vectors to remove (or "annihilate") the influence of phylogeny at the nodes with unresolved polytomies. In the phylogenetic regression, any number of independent variables may be annihilated (Grafen, 1992). Strengths and weaknesses of the phylogenetic regression have been discussed by Grafen (1992) and by Pagel and Harvey (1992).

Of course, as an alternative to these methods for dealing with polytomies, one could conduct analyses on all possible resolutions of the polytomy, as discussed above.

Phylogenetically independent contrasts have great potential in ecological morphology but have been used sparingly to date. Although we have elaborated several concerns, it is not clear how serious the biases they introduce may be, and all of these methods still are clearly superior to disregarding phylogenetic information entirely (Felsenstein, 1985), as Martins and Garland's (1991) simulation study clearly revealed.

Simulations

Randomization and phylogenetic simulation methods are other promising avenues to get around the statistical non-independence problem (Harvey and Pagel, 1991; Martins and Garland, 1991). Martins and Garland (1991) proposed another method for testing the significance of correlated evolution in two continuous characters given a phylogeny. They simulated the independent evolutionary change in each character through a phylogeny using either a punctuational or Brownian motion model of evolutionary change—the latter leads to greater expected change on longer branches. The correlations of change in two characters from 1000 simulations were then used as a null hypothesis in comparison to the observed correlation between two characters of interest. Garland et al. (1993) subsequently expanded this simulation method to consider questions normally investigated using ANOVA or ANCOVA approaches, such as whether home range size in carnivorous mammals is larger than in herbivorous mammals. In these simulations, the identity of the species (e.g., either ungulate or carnivore) remains the same, but character values for each species are assigned by simulating evolution of the character using the phylogeny for the group. A number of additional modes of character evolution were used in the simulations in addition to the two employed by Martins and Garland (1991). The surprising result of the Garland et al. (1993) study was that differences as extreme as those actually observed between ungulates and carnivores occur relatively often in the simulations.

Comparisons among Methods

Despite the development of numerous analytical procedures whose purpose is to incorporate phylogenetic information, few studies have compared the efficacy of these techniques (e.g., Grafen, 1989; Martins and Garland, 1991; Martins, 1993; Garland et al., 1993). Each of the statistical methods available for assessing the influence of phylogeny in ecomorphological analyses makes specific assumptions or requires varying details about the topology of the phylogenetic tree. For example, the independent contrasts method probably is more statistically justifiable than parsimony-based directional methods (but see concerns discussed above) but also requires more detailed phylogenetic information. Perhaps the best course of action would be to employ several methods; to the extent that they all produce similar results, one might be confident in reporting the outcome.

The applicability of the various methods also depends on the type of question involved. Some procedures compare ancestral and descendant taxa to estimate the direction and magnitude of character change, whereas nondirectional comparisons estimate the phylogenetic concordance between two traits, but do not provide information on direction. Other methods estimate how well knowledge of phylogenetic relationships predicts character values, that is, how frequently evolutionary change has occurred. Researchers should take care to choose the appropriate method for the question at hand.

A trend to date has been the unquestioned use of a particular technique without regard to the underlying evolutionary models and assumptions of each method—most methods (e.g., linear parsimony, autocorrelation), for example, fail to recognize parallel evolution in sister taxa. Rarely do studies evaluate the appropriateness of the models and analytical techniques employed, an empirical question which may prove difficult to get at. For example, Lynch (1990) found that rates of morphological evolution in mammals are generally too slow to match the expectations of neutral models of phenotypic evolution, which would indicate that Brownian motion models of change would not be appropriate in statistical analyses (see also Garland et al., 1992). The generality of these results remains to be determined. Similarly, we have little or no information about how robust the recent methods for apportioning variation are and little work has been conducted on the extent to which ancestral character states (i.e., nodal values) are estimated accurately. We see these questions as important foci in future comparative methods.

A few studies have evaluated the concordance of various statistical procedures. Martins and Garland (1991) evaluated the performance of three comparative methods—phylogenetic independent contrasts, minimum evolution reconstruction, and a nonphylogenetic correlation—for studying correlated evolutionary change. Two models of evolution were simulated, punctuational change (change only at speciation) and gradual evolution (Brownian motion). This simulation study demonstrated that either method was preferred over the nonphylogenetic approach; the tests yielded the best results when accurate branch length information was given. In another example, Gittleman (1991) controlled for the effects of phylogeny on olfactory bulb size in carnivores by using both the phylogenetic autocorrelation analysis (as modified in Gittleman and Kot, 1991) and the method of independent contrasts. Although Gittleman (1991) employed taxonomic levels as a measure of phylogenetic relationships, both methods yielded similar conclusions (for a more detailed study, still based on taxonomic structure, see Gittleman and Luh, in press). Other recent studies have found general concordance of results of analyses using various methods and models of evolutionary change, but with some important exceptions (Garland et al., 1993; Martins, 1993).

CONCLUSIONS

A historical perspective clearly has a great deal to offer to comparative biologists. Statistical considerations aside, the most important contribution of the phylogenetic approach is that it refocuses the questions being asked (Brooks and McLennan, 1991). Much of ecological morphology concerns how organisms have evolved in response to environmental circumstances. Although in some cases taxa evolve readily and rapidly enough that historical effects will be minor, in many (perhaps most?) cases, the key to understanding present character states will be an understanding of the sequence, timing, and cause of evolutionary events in the (sometimes distant) past. Consequently, explicitly directing investigations to this historical component is important.

The phylogenetic approach is also desirable because it can suggest research questions that otherwise might not have been obvious. For example, phylogenetic analyses might indicate that a structure and function evolve concordantly, a hypothesis that experimental or theoretical functional studies can directly test (Lauder, 1989, 1990), and can even suggest the importance of previously overlooked structures. For example, Lauder (1989) demonstrated that the hypochordal longitudinalis muscle arose prior to the evolution of an externally symmetrical tail in ray-finned fishes. Subsequent laboratory studies revealed that the presence of this muscle considerably alters the functional capabilities of the tail and may have played an important role in the evolution of external tail morphology (see also Futuyma and McCafferty, 1990; McLennan, 1991).

The new comparative methods hold great promise for application in ecomorphological studies. Although many studies recognize the importance of including phylogeny in ecomorphological analyses (e.g., James, 1982; Strauss, 1987; Losos, 1990c; Winemiller, 1991; Douglas and Matthews, 1992; Richman and Price, 1992), the use of these techniques is sparse. For the effects of phylogeny to be adequately recovered, future studies must include explicit phylogenetic hypotheses as a central analytical framework (Donoghue, 1989). On a positive note, reference to recent issues of most major journals indicates that the importance of phylogeny is being increasingly appreciated. With rapid developments in the field of comparative biology and its enhanced use, the next few years should prove exciting for systematists and the full spectrum of organismal biologists.

ACKNOWLEDGMENTS

We thank T. Garland, H. Greene, A. Larson, E. Martins, T. Price, S. Reilly, D. Stern, P. Wainwright, and two anonymous reviewers for helpful comments; B. Dial, T. Garland, J. Gittleman, O. Gorman, H. Greene, M. Lynch, W. Maddison, T. Price, L. Sanderson, and D. Wake for providing papers prior to publication; and the National Science Foundation (grant BSR 8617688 to DBM) for support. JBL was supported by a postdoctoral fellowship from the Center for Population Biology, University of California, Davis, while this paper was being written.

References

Alexander. R. McN. 1990. Apparent adaptation and actual performance. *Evol. Biol.* 25:357–373.
Altaba, C. R. 1991. The importance of ecological and historical factors in the production of benzaldehyde by tiger beetles. *Syst. Zool.* 40:101–105.

Arnold, S. J. 1983. Morphology, performance, and fitness. *Amer. Zool.* 23:347–361.

Arnold, S. J. 1986. Laboratory and field approaches to the study of adaptation. In *Predator-prey Relationships: Perspectives and Approaches from the Study of Lower Vertebrates,* ed. M. E. Feder and G. V. Lauder, 157–179. Chicago: University of Chicago Press.

Balmford, A., A. L. R. Thomas, and I. L. Jones. 1993. Aerodynamics and the evolution of long tails in birds. *Nature* 361:628–631.

Baum, D. A., and A. Larson. 1991. Adaptation reviewed: A phylogenetic methodology for studying character macroevolution. *Syst. Zool.* 40:1–18.

Björklund, M. 1991. Evolution, phylogeny, sexual dimorphism and mating system in grackles (*Quiscalus* spp.: Icterinae). *Evolution* 45:608–621.

Blondel, J., F. Vuilleumier, L. F. Marcus, and E. Terouanne. 1984. Is there ecomorphological convergence among Mediterranean bird communities of Chile, California, and France? *Evol. Biol.* 18:141–213.

Bock, W. J. 1977. Adaptation and the comparative method. In *Major Patterns in Vertebrate Evolution,* ed. M. K. Hecht, P. Goody, and B. Hecht, 57–82. New York: Plenum Press.

Bock, W. J. 1980. The definition and recognition of biological adaptation. *Amer. Zool.* 20:217–227.

Bock, W. J., and W. D. Miller. 1959. The scansorial foot of the woodpeckers, with comments on the evolution of perching and climbing feet in birds. *Amer. Mus. Nov.* 1931:1–45.

Bock, W. J., and G. von Wahlert. 1965. Adaptation and the form-function complex. *Evolution* 19:269–299.

Brooks, D. R., and D. A. McLennan. 1991. *Phylogeny, Ecology, and Behavior: A Research Program in Comparative Biology.* Chicago: University of Chicago Press.

Burt, A. 1989. Comparative methods using phylogenetically independent contrasts. In *Oxford Surveys in Evolutionary Biology,* vol. 6, ed. P. H. Harvey and L. Partridge, 33–63. Oxford: Oxford University Press.

Cadle, J. E., and H. W. Greene. 1993. Phylogenetic patterns, biogeography, and the ecological structure of neotropical snake assemblages. In *Species Diversity in Ecological Communities,* ed. R. E. Ricklefs and D. Schluter, 281–293. Chicago: University of Chicago Press.

Carothers, J. H. 1986. An experimental confirmation of morphological adaptation: Toe fringes in the sand-dwelling lizard *Uma scoparia. Evolution* 40:871–874.

Carrier, D. R. 1991. Conflict in the hypaxial musculo-skeletal system: Documenting an evolutionary constraint. *Amer. Zool.* 31:644–654.

Cheverud, J. M., M. M. Dow, and W. Leutenegger. 1985. The quantitative assessment of phylogenetic constraints in comparative analyses: Sexual dimorphism in body weight among primates. *Evolution* 39:1335–1351.

Clutton-Brock, T. H., and P. H. Harvey. 1984. Comparative approaches to investigating adaptation. In *Behavioural Ecology: An Evolutionary Approach,* 2d ed., ed. J. R. Krebs and N. B. Davies, 7–29. Oxford: Blackwell Scientific.

Coddington, J. 1986. The monophyletic origin of the orb web. In *Spider Webs and Spider Behavior,* ed. W. A. Shear, 319–363. Palo Alto: Stanford University Press.

Coddington, J. A. 1990. Bridges between evolutionary pattern and process. *Cladistics* 6:379–386.

Cracraft, J. 1981. The use of functional and adaptive criteria in phylogenetic systematics. *Amer. Zool.* 21:21–3.

Cracraft, J. 1990. The origin of evolutionary novelties: Pattern and process at different hierarchical levels. *Evolutionary Innovations,* ed. M. H. Nitecki, 21–44. Chicago: University of Chicago Press.

Cunningham, C. W., and L. W. Buss. 1993. Molecular evidence for multiple episodes of paedomorphosis in the family Hydractiniidae. *Biochem. Syst. Ecol.* 21:57–69.

Darwin, C. 1859. *On the Origin of Species by Means of Natural Selection, or the Preservation of Favoured Races in the Struggle for Life.* London: John Murray.

DeBry, R. W. 1992. Biogeography of new world taiga-dwelling *Microtus* (Mammalia: Arvicolidae): A hypothesis test that accounts for phylogenetic uncertainty. *Evolution* 46:1347–1357.

de Pinna, M. C. C., and L. O. Salles. 1990. Cladistic tests of adaptational hypotheses: A reply to Coddington. *Cladistics* 6:373–377.

Derrickson, E. M. 1989. The comparative method of Elgar and Harvey: Silent ammunition for McNab. *Func. Ecol.* 3:123–126.

Derrickson, E. M., and R. E. Ricklefs. 1988. Taxon-dependent diversification of life-history traits and the perception of phylogenetic constraints. *Func. Ecol.* 2:417–423.

Donoghue, M. J. 1989. Phylogenies and the analysis of evolutionary sequences, with examples from seed plants. *Evolution* 43:1137–1156.

Dorit, R. L. 1990. The correlates of high diversity in Lake Victoria haplochromine cichlids: A neontological perspective. *Causes of Evolution: A Paleontological Perspective*, ed. R. M. Ross and W. D. Allmon, 322–353. Chicago: University of Chicago Press.

Douglas, M. E., and W. J. Matthews. 1992. Does morphology predict ecology? Hypothesis testing within a freshwater stream fish assemblage. *Oikos* 65:213–224.

Duellman, W. E., and E. R. Pianka. 1990. Biogeography of nocturnal insectivores: Historical events and ecological filters. *Ann. Rev. Ecol. Syst.* 21:57–68.

Emerson, S. B. 1988. Testing for historical patterns of change: A case study with frog pectoral girdles. *Paleobiology* 14:174–186.

Farris, J. S. 1970. Methods for computing Wagner trees. *Syst. Zool.* 19:83–92.

Felsenstein, J. 1983. Parsimony in systematics: Biological and statistical issues. *Ann. Rev. Ecol. Syst.* 14:313–333.

Felsenstein, J. 1985. Phylogenies and the comparative method. *Amer. Nat.* 125:1–15.

Felsenstein, J. 1988a. Phylogenies and quantitative characters. *Ann. Rev. Ecol. Syst.* 19:445–472.

Felsenstein, J. 1988b. Phylogenies from molecular sequences: Inferences and reliability. *Ann. Rev. Genet.* 22:521–565.

Fisher, D. C. 1985. Evolutionary morphology: Beyond the analogous, the anecdotal, and the ad hoc. *Paleobiology* 11:120–138.

Funk, V. A., and D. R. Brooks. 1990. Phylogenetic systematics as the basis of comparative biology. *Smith. Cont. Bot.* 73:1–45.

Futuyma, D. J., and S. S. McCafferty. 1990. Phylogeny and the evolution of host plant associations in the leaf beetle genus *Ophraella* (Coleoptera, Chrysomelidae). *Evolution* 44:1885–1913.

Garland, T., Jr. 1992. Rate tests for phenotypic evolution using phylogenetically independent contrasts. *Amer. Nat.* 140:509–519.

Garland, T., Jr., R. B. Huey, and A. F. Bennett. 1991. Phylogeny and coadaptation of thermal physiology in lizards: A reanalysis. *Evolution* 45:1969–1975.

Garland, T., Jr., P. H. Harvey, and A. R. Ives. 1992. Procedures for the analysis of comparative data using phylogenetically independent contrasts. *Syst. Biol.* 41:18–32.

Garland, T., Jr., A. W. Dickerman, C. M. Janis, and J. A. Jones. 1993. Phylogenetic analysis of covariance by computer simulation. *Syst. Biol.* 42:265–292.

Gauthier, J., and K. Padian. 1985. Phylogenetic, functional, and aerodynamic analyses of the origin of birds and their flight. In *The Beginnings of Birds: Proceedings of the International Archaeopteryx Conference, Eichstätt 1984*, ed. M. K. Hecht, J. H. Ostrom, G. Viohl, and P. Wellnhofer, 185–197. Eichstätt: Freunde des Jura-Museums Eichstätt.

Gittleman, J. L. 1991. Carnivore olfactory bulb size: Allometry, phylogeny, and ecology. *J. Zool.* (London) 225:253–272.

Gittleman, J. L., and M. Kot. 1990. Adaptation: Statistics and a null model for estimating phylogenetic effects. *Syst. Zool.* 39:227–241.

Gittleman, J. L., and H.-K. Luh. 1992. On comparing comparative methods. *Ann. Rev. Ecol. Syst.* 23:383–404.

Gittleman, J. L., and H.-K. Luh. In press. Phylogeny, evolutionary models, and comparative methods. In *Pattern and Process: Phylogenetic Approaches to Ecological Problems*, ed. P. Eggleton and D. Vane-Wright. London: Academic Press.

Gorman, O. 1993. Evolutionary ecology and historical ecology: Assembly, structure, and organization of stream fish communities. In *Systematics, Historical Ecology, and North American Freshwater Fishes,* ed. R. L. Mayden. 659-688 Palo Alto: Stanford University Press.

Gould, S. J. 1980. The evolutionary biology of constraint. *Daedalus* 109:39–52.

Gould, S. J., and R. C. Lewontin. 1979. The spandrels of San Marcos and the panglossian paradigm: A critique of the adaptationist programme. *Proc. R. Soc.* (London) B205:581–598.

Gould, S. J., and E. S. Vrba. 1982. Exaptation—a missing term in the science of form. *Paleobiology* 8:4–15.

Grafen, A. 1989. The phylogenetic regression. *Phil. Trans. R. Soc.* (London) B326:119–157.

Grafen, A. 1992. The uniqueness of the phylogenetic regression. *J. Theor. Biol.* 156:405–423.

Greene, H. W. 1986. Diet and arboreality in the emerald monitor, *Varanus prasinus,* with comments on the study of adaptation. *Fieldiana Zool. New Series* 31:1–12.

Greene, H. W. 1992. The ecological and behavioral context for pitviper evolution. In *Biology of Pitvipers,* ed. J. A. Campbell and E. D. Brodie, Jr. 107–118. Tyler, TX: Selva.

Harvey, P. H., and M. D. Pagel. 1991. *The Comparative Method in Evolutionary Biology.* Oxford: Oxford University Press.

Harvey, P. H., and A. Purvis. 1991. Comparative methods for explaining adaptations. *Nature* 351:619–624.

Hillis, D. M., J. J. Bull, M. E. White, M. R. Badgett, and I. J. Molineux. 1992. Experimental phylogenetics: Generation of a known phylogeny. *Science* 255:589–592.

Ho, M.-W., and P. T. Saunders, eds. 1984. *Beyond Neo-darwinism: An Introduction to the New Evolutionary Paradigm.* London: Academic Press.

Huey, R. B., and A. F. Bennett. 1986. A comparative approach to field and laboratory studies in evolutionary biology. In *Predator-Prey Relationships: Perspectives and Approaches from the Study of Lower Vertebrates,* ed. M. E. Feder and G. V. Lauder, 82–98. Chicago: University of Chicago Press.

Huey, R. B., and A. F. Bennett. 1987. Phylogenetic studies of coadaptation: Preferred temperatures versus optimal performance temperatures of lizards. *Evolution* 41:1098–1115.

Huey, R. B., and R. D. Stevenson. 1979. Integrating thermal physiology and ecology of ectotherms: A discussion of approaches. *Amer. Zool.* 19:357–366.

James, F. C. 1982. The ecological morphology of birds: A review. *Ann. Zool. Fenn.* 19:265–275.

Janson, C. H. 1992. Measuring evolutionary constraints: A Markov model for phylogenetic transitions among seed dispersal syndromes. *Evolution* 46:136–158.

King, D. G. 1991. The origin of an organ: Phylogenetic analysis of evolutionary innovation in the digestive tract of flies (Insecta: Diptera). *Evolution* 45:568–588.

Larson, A. 1984. Neontological inferences of evolutionary pattern and process in the salamander family Plethodontidae. *Evol. Biol.* 17:119–217.

Lauder, G. V. 1981. Form and function: Structural analysis in evolutionary morphology. *Paleobiology* 7:430–442.

Lauder, G. V. 1989. Caudal fin locomotion in ray-finned fishes: Historical and functional analyses. *Amer. Zool.* 29:85–102.

Lauder, G. V. 1990. Functional morphology and systematics: Studying functional patterns in an historical context. *Ann. Rev. Ecol. Syst.* 21:317–340.

Lauder, G. V. and K. F. Liem. 1989. The role of historical factors in the evolution of complex organismal functions. In *Complex Organismal Functions: Integration and Evolution,* ed. D. B. Wake and G. Roth, 63–78. New York: John Wiley.

Lessa, E. P. 1989. Morphological evolution of subterranean mammals: Integrating structural, functional, and ecological perspectives. In *Evolution of Subterranean Mammals at the Organismal and Molecular Levels,* ed. E. Nevo and O. A. Reig, 211–230. New York: Wiley-Liss.

Lessios, H. A. 1990. Adaptation and phylogeny as determinants of egg size in echinoderms from the two sides of the isthmus of Panama. *Amer. Nat.* 135:1–13.

Liebherr, J. K., and A. E. Hajek. 1990. A cladistic test of the taxon cycle and taxon pulse hypotheses. *Cladistics* 6:39–59.

Losos, J. B. 1990a. A phylogenetic analysis of character displacement in Caribbean *Anolis* lizards. *Evolution* 44:558–569.

Losos, J. B. 1990b. The evolution of form and function: Morphology and locomotor performance in West Indian *Anolis* lizards. *Evolution* 44:1189–1203.

Losos, J. B. 1990c. Ecomorphology, performance capability, and scaling of West Indian *Anolis* lizards: An evolutionary analysis. *Ecol. Mon.* 60:369–388.

Losos, J. B. 1992. The evolution of convergent structure in Caribbean *Anolis* communities. *Syst. Biol.* 41:403–420.

Losos, J. B. 1994. An approach to the analysis of comparative data when a phylogeny is unavailable or incomplete. *Syst. Biol.*

Lombard, R. E., and D. B. Wake. 1986. Tongue evolution in the lungless salamanders, family Plethodontidae. IV. Phylogeny of plethodontid salamanders and the evolution of feeding dynamics. *Syst. Zool.* 35:532–551.

Luke, C. 1986. Convergent evolution of lizard toe fringes. *Biol. J. Linnean Soc.* 27:1–16.

Lynch, M. 1990. The rate of morphological evolution in mammals from the standpoint of the neutral expectation. *Amer. Nat.* 136:727–741.

Lynch, M. 1991. Methods for the analysis of comparative data in evolutionary biology. *Evolution* 45:1065–1080.

MacArthur, R. 1972. *Geographical Ecology.* Princeton: Princeton University Press.

Maddison, W. P. 1989. Reconstructing character evolution on polytomous cladograms. *Cladistics* 5:365–377.

Maddison, W. P. 1990. A method for testing the correlated evolution of two binary characters: Are gains or losses concentrated on certain branches of a phylogenetic tree? *Evolution* 44:539–557.

Maddison, W. P. 1991. Squared-change parsimony reconstructions of ancestral states for continuous-valued characters on a phylogenetic tree. *Syst. Zool.* 40:304–314.

Maddison, W., and D. Maddison. 1992. *MacClade Version 3: Analysis of Phylogeny and Character Evolution.* Sunderland, MA: Sinauer Associates.

Maddison, W. P., and M. Slatkin. 1991. Null models for the number of evolutionary steps in a character on a phylogenetic tree. *Evolution* 45:1184–1197.

Martins, E. P. 1993. A comparative study of the evolution of *Sceloporus* push-up displays. *Amer. Nat.,* in press.

Martins, E. P., and T. Garland, Jr. 1991. Phylogenetic analyses of the correlated evolution of continuous characters: A simulation study. *Evolution* 45:534–557.

Maynard Smith, J., R. Burian, S. Kauffman, P. Alberch, J. Campbell, B. Goodwin, R. Lande, D. Raup, and L. Wolpert. 1985. Developmental constraints and evolution. *Q. Rev. Biol.* 60:265–287.

Mayr, E. 1983. How to carry out the adaptationist programs? *Amer. Nat.* 121:324–334.

McLennan, D. A. 1991. Integrating phylogeny and experimental ethology: from pattern to process. *Evolution* 45:1773–1789.

Mickevich, M. F., and M. S. Johnson. 1976. Congruence between morphological and allozyme data in evolutionary inference and character evolution. *Syst. Zool.* 25:260–270.

Miles, D. B., and A. E. Dunham. 1992. Comparative analyses of phylogenetic effects in the life history patterns of iguanid reptiles. *Amer. Nat.* 139:848–869.

Mitter, C., and D. R. Brooks. 1983. Phylogenetic aspects of coevolution. In *Coevolution,* ed. D. J. Futuyma and M. Slatkin, 65–98. Sunderland, MA: Sinauer Associates.

Mivart, St. G. 1871. *On the Genesis of Species.* London: Macmillan.

Møller, A. P., and T. R. Birkhead. 1992. A pairwise comparative method as illustrated by copulation frequency in birds. *Amer. Nat.* 139:644–656.

Nee, S., A. F. Read, J. J. D. Greenwood, and P. H. Harvey. 1991. The relationship between abundance and body size in British birds. *Nature* 351:312–313.

Oakes, E. J. 1992. Lekking and the evolution of sexual dimorphism in birds: Comparative approaches. *Amer. Nat.* 140:665–684.

Orians, G. H., and R. T. Paine. 1983. Convergent evolution at the community level. In *Coevolution,* ed. D. J. Futuyma and M. Slatkin, 431–458. Sunderland, MA: Sinauer Associates.

Pagel, M. D. 1992. A method for the analysis of comparative data. *J. Theor. Biol.* 156:431–442.

Pagel, M. D., and P. H. Harvey. 1988. Recent developments in the analysis of comparative data. *Q. Rev. Biol.* 63:413–440.

Pagel, M. D., and P. H. Harvey. 1992. On solving the correct problem: Wishing does not make it so. *J. Theor. Biol.* 156:425–430.

Pianka, E. R. 1986. *Ecology and Natural History of Desert Lizards: Analyses of the Ecological Niche and Community Structure.* Princeton: Princeton University Press.

Reid, M. L., C. G. Eckert, and K. E. Muma. 1986. Booming odontocetes and deaf cephalopods: Putting the cart before the horse. *Amer. Nat.* 128:438–439.

Reif, W.-E., R. D. K. Thomas, and M. S. Fischer. 1985. Constructional morphology: The analysis of constraints in evolution dedicated to A. Seilacher in honour of his 60. birthday. *Acta Biotheor.* 34:233–248.

Richman, A. D., and T. Price. 1992. Evolution of ecological differences in the Old World leaf warblers. *Nature* 355:817–821.

Ricklefs, R. E. 1987. Community diversity: Relative roles of local and regional processes. *Science* 235:167–171.

Ridley, M. 1983. *The Explanation of Organic Diversity: The Comparative Method and Adaptations for Mating.* Oxford: Oxford University Press.

Russell, A. P. 1976. Some comments concerning interrelationships amongst gekkonine geckos. In *Morphology and Biology of Reptiles, Linnean Society Symposium Series 3,* ed. A. d'A. Bellairs and C. B. Cox, 217–244. London: Academic Press.

Russell, A. P. 1979. The origin of parachuting locomotion in gekkonid lizards (Reptilia: Gekkonidae). *Zool. J. Linnean Soc.* 65:233–249.

Sanderson, S. L., and R. Wassersug. 1993. Convergent and alternative designs for vertebrate suspension feeding. In *The Vertebrate Skull,* vol. 3, *Function and Evolutionary Mechanisms,* ed. J. Hanken and B. K. Hall. Chicago: University of Chicago Press.

Schaefer, S. A., and G. V. Lauder. 1986. Historical transformation of functional design: Evolutionary morphology of feeding mechanisms in loricarioid catfishes. *Syst. Zool.* 35:489–508.

Schluter, D. 1986. Tests for similarity and convergence of finch communities. *Ecology* 67:1073–1085.

Schluter, D. 1989. Bridging population and phylogenetic approaches to the evolution of complex traits. In *Complex Organismal Functions: Integration and Evolution,* ed. D. B. Wake and G. Roth, 79–95. New York: John Wiley.

Schluter, D. 1990. Species-for-species matching. *Amer. Nat.* 136:560–568.

Schluter, D., T. D. Price, and P. R. Grant. 1985. Ecological character displacement in Darwin's finches. *Science* 227:1056–1059.

Shine, R., and A. E. Greer. 1991. Why are clutch sizes more variable in some species than in others? *Evolution* 45:1696–1706.

Sillén-Tullberg, B. 1988. Evolution of gregariousness in aposematic butterfly larvae: A phylogenetic analysis. *Evolution* 42:293–305.

Simpson, G. G. 1978. Variation and details of macroevolution. *Paleobiology* 4:217–221.

Sinervo, B. 1990. The evolution of maternal investment in lizards: An experimental and comparative analysis of egg size and its effect on offspring performance. *Evolution* 44:279–294.

Stearns, S. C. 1983. The influence of size and phylogeny on patterns of covariation among life-history traits in mammals. *Oikos* 41:173–187.

Strauss, R. E. 1987. The importance of phylogenetic constraints in comparisons of morphological structure among fish assemblages. In *Community and Evolutionary Ecology of North American Stream Fishes*, ed. W. J. Matthews and D. C. Hein, 136–143. Norman: University of Oklahoma Press.

Swofford, D. L. 1991. When are phylogeny estimates from molecular and morphological data incongruent? In *Phylogenetic Analysis of DNA Sequences*, ed. M. M. Miyamota and J. Cracraft. Oxford: Oxford University Press.

Swofford, D. L., and W. P. Maddison. 1987. Reconstructing ancestral character states under Wagner parsimony. *Math. Biosci.* 87:199–229.

Van Valkenburgh, B. 1985. Locomotor diversity within past and present guilds of large predatory mammals. *Paleobiology* 11:406–428.

Van Valkenburgh, B. 1988. Trophic diversity in past and present guilds of large predatory mammals. *Paleobiology* 14:155–173.

Voss, R. S. 1988. Systematics and ecology of Ichthyomyine rodents (Muroidea): Patterns of morphological evolution in a small adaptive radiation. *Bull. Amer. Mus. Nat. Hist.* 188:259–493.

Wainwright, P. C., and G. V. Lauder. 1993. The evolution of feeding biology in sunfishes (Centrarchidae). In *Systematics, Historical Ecology, and North American Freshwater Fishes*, ed. R. Mayden, pp. 472–491. Palo Alto: Stanford University Press.

Wake, D. B. 1991. Homoplasy: The result of natural selection, or evidence of design limitations? *Amer. Nat.* 138:543–567.

Wake, D. B., and A. Larson. 1987. Multidimensional analysis of an evolving lineage. *Science* 238:42–48.

Wanntorp, H.-E. 1983. Historical constraints in adaptation theory: Traits and non-traits. *Oikos* 41:157–159.

Wanntorp, H.-E., D. R. Brooks, T. Nilsson, S. Nylin, F. Ronquist, S. C. Stearns, and N. Wedell. 1990. Phylogenetic approaches in ecology. *Oikos* 57:119–132.

Wiens, J. A. 1989. *The Ecology of Bird Communities*, vol. 1, *Foundations and Patterns*. Cambridge: Cambridge University Press.

Wiley, E. O., D. Siegel-Causey, D. R. Brooks, and V. A. Funk. 1991. *The Compleat Cladist: A Primer of Phylogenetic Procedures*. Lawrence: University of Kansas. Mus. Nat. Hist.

Williams, E. E. 1972. The origin of faunas. Evolution of lizard congeners in a complex island fauna: A trial analysis. *Evol. Biol.* 6:47–89.

Winemiller, K. O. 1991. Ecomorphological diversification in lowland freshwater fish assemblages from five biotic regions. *Ecol. Mon.* 61:343–365.

5

Evaluating the Adaptive Role of Morphological Plasticity

Joseph Travis

INTRODUCTION

The size and shape of morphological characters are rarely constant within a species. Although some characters typically exhibit low levels of phenotypic variation, others vary extensively among individuals from different populations or among individuals from different generations of the same population. Much of this variation is attributable to differences among such groups in the state of some environmental variable experienced during sensitive periods in development. A simple view of such environmental effects is that they represent the necessary and predictable results of biochemical processes that are carried out under different physical and chemical conditions.

In this chapter, I explore another interpretation of these environmental effects, that in many cases they represent an interaction between the genetic system and the environment that has been designed by natural selection to produce different average phenotypes under different conditions. The capacity of a single genotype to produce a variety of phenotypes is called *phenotypic plasticity,* and the thesis I explore is that some patterns of plasticity in morphological traits are adaptive ones.

This thesis is an attractive one from an empirical perspective and can be developed from two sets of empirical observations. First, the average values of morphological traits that are seen in a species or a population do appear to have been molded by natural selection, and different ecological demands appear to have produced different sizes and shapes in the same morphological traits. Several lines of argument lead to this conclusion: the matching of physical principles of design considerations with the observed organism performance (Denny, chap. 8, this volume); the observations and experiments on the ecological roles of morphological features (Wainwright, chap. 3, this volume); and comparisons and correlations across a range of morphological variation and ecological milieux (this volume: Ricklefs and Miles, chap. 2; Emerson et al., chap. 6; Norberg, chap. 9). Second, most organisms live in an environment that is heterogeneous is some fashion. For some, significant spatial variation occurs within the normal range of dispersal, such that an individual's offspring might encounter a range of ecological challenges as they are broadcast across the environment. For others, different ecological conditions may occur from one generation to the next even in the same place.

These two sets of observations lead to the obvious expectation that, if the ecological heterogeneity experienced by a species or a population is sufficiently great, natural selection ought to mold the pattern of phenotypic plasticity so that an organism can develop the most appropriate morphology for the ecological conditions that it is likely to experience. This notion becomes even more appealing when one realizes that there are some obvious examples in morphology of this very phenomenon (which will be discussed in subsequent sections).

However, even an intuitively appealing idea with some readily observable support need not reflect a general principle. Evolution via natural selection alters morphology and development to maximize fitness but can only do so from the set of variations available to it, and it is forced to do so in a historical context in which initial conditions might play an overwhelming role. These considerations mean that an assortment of constraints guide which evolutionary outcomes are realized, from ones associated with phylogenetic heritage (this volume: Losos and Miles, chap. 4; Reilly, chap. 12) to those associated with the interactions with other kinds of traits and multiple ecological demands (Bradley, chap. 11, this volume). The degree to which adaptive plasticity evolves, and how well such plasticity can adapt morphology to ecology across the relevant range of environments, involves these considerations as well, which are perhaps less appealing to the enthusiastic empiricist. As a result, to explore the thesis that some patterns of plasticity in morphological traits are adaptive is to explore a series of questions about the theoretical likelihood of the phenomenon as well as the accumulated empirical evidence for its reality.

In the first section of this essay, I delimit the types of morphological traits to which the thesis is likely to apply, define the contexts in which the evidence should be sought, and summarize the spectrum of examples of phenotypic plasticity that can be found within those guidelines. In the subsequent section I examine the arguments that such responses can be considered products of adaptive evolution, describe the conditions under which adaptive plasticity should be favored, and discuss how selection might create a pattern of adaptive plasticity. In the final section I examine two examples of adaptive plasticity that illustrate these themes. In some cases I have cited empirical studies that do not focus on morphological traits, but I have tried to limit those citations to cases in which the point cannot be illustrated by a morphological example. I have not considered "body size" a relevant trait in this review unless the variation in body size examined in an empirical study is based on demonstrable variation in structure and not simply variation in mass that might reflect variation in condition or fat levels.

AN OVERVIEW OF MORPHOLOGICAL PLASTICITY

Candidates for Hypotheses of Adaptive Plasticity

The thesis of "adaptive plasticity" indicates two criteria for morphological examples. First, the trait itself must be "ecologically significant": in any one environment, variation in the value of the trait must be causally associated with variation in fitness in that environment (Antonovics et al., 1988; Sober, 1984). This requirement constrains the possible examples to those in which at least an ecologically plausible connection exists between trait values and fitness. It eliminates characters such as the number, size, or shape of scales in fish or reptiles that are affected by environmental variation, that often appear to be under selection, but that are likely only to be pleitropic markers for other traits that are causally related to fitness variation (Travis, 1989).

Second, the best value of that trait, the value that confers the highest fitness on its bearers, must differ across environmental conditions for some ecological reason. This criterion indicates a focus on traits expressed in populations occupying environments that are variable in either space or time *with respect to the relationship between trait value and fitness.* This situation is distinguished from the circumstance in which there is variation in the environmental factors experienced during development but no variation in which trait values are best in each condition. In the latter circumstance, stabilizing selection (Schmalhausen, 1949) will mold the development system to insulate the formation of the trait from environmental effects (Gibson and Bradley, 1974; Kaufman et al., 1977; Scharloo et al., 1967; see Orzack, 1985, for theoretical discussion).

Contexts for Hypotheses of Adaptive Plasticity

For the thesis to be operational, a clear definition is needed for the contexts in which it is meant to apply. Phenotypic plasticity has been defined as the ability of a particular genotype to produce a range of phenotypic values of a trait in response to variation in some environmental variable (after West-Eberhard, 1989). I restrict this definition further to describe a repeatable, directional change in the phenotype that is produced by a particular genotype in response to a directional change in a value of a specific environmental variable. The direction and degree of such a response in phenotype elicited from a genotype is denoted as the genotype's *norm of reaction* to that variable. The particular genotype must be embedded in a natural genetic background, and the environmental variation must be within the range normally experienced by the population.

These restrictions distinguish the subject of this essay from related phenomena that are often called phenotypic plasticity but that are qualitatively distinct. The first of these phenomena might be called "destabilizing selection" (Belayaev, 1979), a process of selection for enhanced phenotypic variation. A set of genotypes that control the value of a trait may differ in their sensitivity to random environmental factors such that some produce a narrow range of phenotypic variation but others produce a wide range of variation around a common mean value. There is no net directional response. A large body of theory addresses how and when selection would favor those genotypes with enhanced environmental sensitivity so that a larger phenotypic variance is expressed in spatially or temporally variable selective environments (Bull, 1987; Bulmer, 1984; Capinera, 1979; Cohen, 1967; Ellner, 1985a, b; Garcia-Dorado, 1990; Kaplan and Cooper, 1984; Leon, 1985; McGinley et al., 1987; Slatkin and Lande, 1976; Venable and Brown, 1988). The high variability of certain life-history traits has often been interpreted in these terms (Crump, 1981, 1984; Kaplan, 1985; Meffe, 1987; Venable, 1989).

The remaining phenomena involve changes in the phenotypic mean and variance of a trait that result from the disruption of the developmental system caused by close inbreeding (Strauss, 1987), hybridization and introgression (Burton, 1990; Graham and Felley, 1985), or the exposure of individuals to novel environmental conditions (Harkey and Semlitsch, 1988; Heslop-Harrison, 1959). I do not imply that the enhanced phenotypic variation seen in these situations is uninteresting or uninformative but that a norm of reaction observed in such contexts is not a candidate for an evolved adaptation to a variable environment.

Because phenotypic plasticity as I have defined it can embrace both reversible and irreversible effects on the phenotype, precise hypotheses must be formed about the adaptive significance of the norm of reaction and the role of the trait through the life cycle. Two examples illustrate this point. *Daphnia pulex* in the presence of predatory *Chaoborus* spp. produce offspring with neck spines that are associated with reduced vulnerability to predation. The spines are lost in later instars, and the presumption is that individuals of sufficiently large body size do not require the spines to deter predators (Havel, 1985). The costs in reduced fitness of maintaining these spines when they are no longer needed is thought to have been the source of the selection pressure for the reversibility of their appearance within a single life cycle. In fishes that live in temporally variable environments, acclimatization to changes in temperature or salinity involves structural changes in cellular organization and membrane compositions that vary in their schedule and ease of reversibility among species with different life histories (Daikoku et al., 1982; Evans, 1984; Zwingelstein et al., 1980). In each case the reversibility of the effects must be accommodated in any testable hypothesis for their adaptive significance. The most general theories for phenotypic plasticity accommodate both reversible and irreversible effects but maintain clear distinctions between them when their evolutionary expectations are examined (Clark and Harvell, 1992; Gomulkeiwicz and Kirkpatrick, 1992).

The Spectrum of Morphological Plasticity

The simplest cases of environmentally induced phenotypic variation are discrete alternative phenotypes, or polyphenisms. These represent discrete variants that encompass one or a number of traits and that are induced by one or a few very specific factors in the environment. The processes producing such alternatives have been variously termed "autoregulatory morphogenesis" (Schmalhausen, 1949), "developmental conversion" (Smith-Gill, 1983), and "conditional choice" (Lloyd, 1984).

Several striking examples of polyphenisms are apparent in trophic structures. Cannibalistic larval morphs of *Ambystoma tigrinum* with distinct head widths and vomerine tooth morphologies (Reilly et al., in press) are induced by high larval densities (Collins and Cheek, 1983), and similarly striking morphs of cannibalistic tadpoles are induced in *Scaphiopus multiplicatus* by the presence of fairy shrimp in the pond (Pfennig, 1990). In a number of trophic polyphenisms in fishes (often called "trophic polymorphisms"), the specific cues are unknown, but the indirect evidence for environmental induction is strong (Grudzien and Turner, 1984; Kornfield et al., 1982; Sage and Selander, 1975; Turner and Grosse, 1980).

A multitude of polyphenisms are evident in nontrophic structures. Wing "polymorphisms" in four orders of insects (Harrison, 1980) and other discrete dispersal polyphenisms (e.g., Kennedy, 1956) are induced by environmental cues that indicate changing resource levels per individual and future habitat quality. Discrete worker caste polyphenisms in many ant species are determined by nutritional treatments set by the older workers as a function of caste ratios (Wheeler and Nijhout, 1983). The presence of predators can induce alternative shell morphologies in barnacles (Lively, 1986a) and defensive structures that deter predators in rotifers (Gilbert and Stemberger, 1984), cladocerans (Grant and Bayly, 1981; Krueger and Dodson, 1981), and bryozoans (Harvell, 1984; Yoshioka, 1982). The appearance of distinctive morphological structures in a variety of endoparasites can be induced by development in different host species (reviewed by Downes, 1990).

Continuous variation in size or shape of structures as a function of environmental variation is a more familiar example of phenotypic plasticity. This process has been termed "dependent development" by Schmalhausen (1949), "phenotypic modulation" by Smith-Gill (1983), and "continuous lability" by Lloyd

(1984). Perhaps the most striking examples involve the variation in trophic structure induced by variation in food type in insects (Bernays, 1986), fish (Meyer, 1987; Wainwright et al., 1991), and mammals (Moore, 1965; Watt and Williams, 1951). An intriguing example of an environmental effect on bill shape in nestling blackbirds was described by James (1983). Although the specific factors that induced the variation in the field remain unknown, variation in temperature and relative humidity in the nestling stage can induce variation in bill shapes even among birds fed a common diet (NeSmith, 1984).

Environmentally induced continuous variation in the size or shape (or both) of nontrophic structures that may covary with fitness has been studied since the turn of the century. Factors such as temperature and food level that can alter either overall rates of growth and differentiation or the timing of developmental onset and offset have been found to produce morphological variation in a plethora of traits in a diversity of animals (Arnold and Peterson, 1990; Arthur, 1982; Blouin and Loeb, 1991; Dentry and Lindsey, 1978; Emerson, 1986; Etter, 1988; Fowler, 1970; Fox et al., 1961; Gabriel, 1944; Kemp and Bertness, 1984; Lindsey, 1966; Martin, 1949; Osgood, 1978; Palmer, 1985; Perrin, 1988; Sumner, 1909, 1915; Warwick, 1944; Weaver and Ingram, 1969; Wilson, 1953; see also the review by Harvell, 1990).

Geographic variation in morphology offers additional potential examples of phenotypic plasticity. These examples involve the univariate or multivariate correlation of geographic variation in the value of a morphological trait with variation in the value of some environmental variable or the correlation of a matrix of differences among populations in the value of a morphological trait with a matrix of differences among the locations of those populations in the value of some environmental variable (Bamber and Henderson, 1985; Chernoff, 1982; James, 1970; Jameson et al., 1973). In one example (Jameson et al., 1973) a spatial correlation among populations corroborated a previously observed temporal correlation of morphology with environment within a single population (Vogt and Jameson, 1970). Geographic variation in the shell morphology of molluscs has been especially well documented; variation in shell morphology has been correlated with variation in wave exposure (O'Loughlin and Aldrich, 1987), substrate type (Edwards, 1988), predation regime (Thomas and Himmelman, 1988), and general water chemistry (Chambers, 1980).

Geographic variation can only suggest the extent of plasticity because of the complexity of interactions among environmental and genetic effects. Experimental studies usually reveal that both effects are important sources of variation (James, 1983; Sumner, 1932). However, discoveries of complex interactions between genetic and environmental effects (Berven et al., 1979; Conover and Present, 1990) indicate that observational studies alone cannot reveal the potential magnitude of phenotypic plasticity. In the extreme, the interplay among multiple environmental effects can completely mask the influence of individual environmental variables even when the responses of different populations are similar. Travis and Trexler (1987) could not uncover any correlations of body length of the sailfin molly, *Poecilia latipinna,* with any of a dozen environmental variables measured over an extensive geographic area, even though the effect of some of the variables on body length could be demonstrated experimentally (Trexler et al., 1990).

Certain cases of environmentally induced morphological variation in insects defy strict classification into "discrete" or "continuous." Many ant species have complex patterns of worker caste variation in which two or more groups of workers are distinguishable by size but wide variation in size and size-associated shape is apparent within each group. Patterns of allometric shape variation differ between the groups. These variations are generated by worker-controlled nutritional variation. Wilson (1953) described this pattern as "diphasic allometry" in species with two groups, and these patterns are reviewed by Oster and Wilson (1978) and by Brian (1980). Variation in horn size among individuals in some species of beetles appears to represent an analogous situation (Eberhard, 1980).

HOW AND WHEN MIGHT PLASTICITY ITSELF BE ADAPTIVE?

Empirical Evidence: Are Some Norms of Reaction Adaptive, and How Widespread Is the Raw Material for Such Adaptation?

There is direct evidence that the phenotypic plasticity in many morphological structures is adaptive. The most direct examples involve discrete variation in defensive structures in aquatic invertebrates. The phenotypic variations are expressed in response to the presence of a predator and provide effective deterrents. In the absence of predators a different phenotype is expressed and is more fit in that context than the predator-resistant one (see below for details). A less dramatic but equally persuasive line of evidence is provided by structural remodeling in response to physical stress in vertebrate bone (Biewener, 1990; Currey, 1984; Lanyon and Rubin, 1985), molluscan muscle (Etter, 1988), and sponge tissue (Palumbi, 1984) that enhances organism performance (and in the case of sponges, actual survivorship and growth). A third persuasive line of evidence is illustrated by the discrete trophic polyphenisms in larval amphibians. Alternate morphologies are induced that demonstrably facilitate growth, development, and survival under differing conditions of intraspecific density stress or food quality.

Several lines of indirect evidence suggest convincingly that certain patterns of phenotypic plasticity are adaptive. Morphological trait values often change in response to environmental stimuli in the direction that would enhance performance under the environmental conditions indicated by those stimuli. Changes in trophic structures in response to different textures or qualities of food are examples (Bernays, 1986; Meyer, 1987; Reilly et al., in press; Thompson, in press; Wainwright et al., 1991). Fitness differences are not demonstrated in these situations but are plausible from the arguments of functional morphology. Another example along these lines involves caste polyphenisms in ants: caste ratios in the imported fire ant (*Solenopsis invicta*) change with colony size and age in a fashion consistent with an adaptive scenario (Markin et al., 1973, but see Porter and Tschinkel, 1985). Analogous evidence is available for traits other than morphological structures, for example diapause responses to variations in temperature and photoperiod in insects (Bradshaw and Lounibos, 1977; Dingle et al., 1980).

Genetically based variation among conspecific populations in norms of reaction offers weak evidence for the plausibility of adaptive plasticity. Such variation is known in a variety of traits in plants (Cook and Johnson, 1968; Quinn and Hodgkinson, 1984; Scheiner and Goodnight, 1984; Schlichting and Levin, 1984; Taylor and Aarssen, 1988; Wilken, 1977; Winn and Evans, 1991; Zangerl and Berenbaum, 1990) and animals (Berven et al., 1979; Bradshaw, 1986; Bradshaw and Lounibos, 1977; Conover and Heins, 1987; Conover and Present, 1990; Coyne and Beecham, 1987; Dingle et al., 1980; Etter, 1988). These differences by themselves constitute weak evidence for a role for natural selection because they could have evolved through genetic drift, but they constitute *prima facie* evidence that norms of reaction do evolve and have done so frequently.

The contention that genetically based variations in reaction norms represent products of adaptive evolution is convincing when different norms of reaction necessarily produce different consequences for fitness in the different environments experienced by each population (Dingle et al., 1980). In many cases differences in norms of reaction reflect plausible expectations for adaptive evolution based on well-studied differences among populations in ecological conditions that would involve the traits in question (e.g., Berven et al., 1979; Conover and Present, 1990; Zangerl and Berenbaum, 1990). For example, Zangerl and Berenbaum (1990) show that individual plants from a population of wild parsnip that is subjected regularly to insect herbivory increase their production of costly toxic compounds when their leaves are damaged. Individuals from a population without a history of herbivory do not have a comparable response to leaf damage. Although fitness differences were not demonstrated, this population-level distinction is a plausible expectation for a product of adaptive evolution. To the skeptic, the weakness in this line of argument is that most studies of this type compare norms of reaction only among single, presumably, representative populations, chosen one from each distinct ecological scenario, and the studies thereby confound the "treatment" with the "sample."

Variation in reaction norms among related species that experience different ecological scenarios would offer analogously weak evidence for the plausibility of adaptive plasticity. There is some persuasive evidence of this nature for differences in the norm of reaction of diapause and dispersal to photoperiod variation among closely related species from different latitudes (Dingle et al., 1980).

Variation among congeneric species and among conspecific populations in norms of reaction demonstrates that norms of reaction are evolutionary products and are not ineluctable responses to physical and chemical stimuli. The raw material for such evolution, adaptive or not, is demonstrable. Genetic variation for norms of reaction has been found within many laboratory stocks of *Drosophila* species (Gebhardt and Stearns, 1988; Gupta and Lewontin, 1982; Scheiner and Lyman, 1989, 1991; Schnee and Thompson, 1984; Waddington and Robertson, 1966) and within many natural populations of plants and animals (Jain, 1979; Newman, 1988; Pomeroy, 1981; Service and Lenski, 1982; Spitze, 1992; Trexler and Travis, 1990; Via, 1984). In one case males and females exhibit different levels of plasticity (Trexler et al., 1990). To be sure, not all studies that have searched for such genetic variation have uncovered it (Groeters and Dingle, 1988; Travis, 1983). When it exists, how might natural selection act on that raw material to fashion the appropriately adaptive norm?

Theories for Adaptive Plasticity

The intriguing nature of this problem has inspired many reviews from a variety of perspectives and many verbal and mathematical models for the evolution of adaptive phenotypic plasticity and correlative phenomena (Baker, 1965; Bamber and Henderson, 1988; Berven and Gill, 1983; Bradshaw, 1965; Caswell, 1983; Cavalli-Sforza, 1974; Clark and Harvell, 1992; Gabriel and Lynch, 1992; Gavrilets and Scheiner, 1993; Gomulkiewicz and Kirkpatrick, 1992; Harvell, 1990; Levins, 1963, 1968; Lively, 1986b; Lloyd, 1984; Moran, 1992; Newman, in press; Schlichting, 1986; Schmalhausen, 1949; Shapiro, 1976; Slobodkin and Rapoport, 1974; Smith-Gill, 1983; Stearns, 1989; Stearns and Koella, 1986; Sultan, 1987; Thoday, 1953; Trexler, 1989; Van Tienderen, 1991; Via and Lande, 1985, 1987; West-Eberhard, 1989). There is no unified mathematically theoretical treatment, although the models of Gavrilets and Scheiner (1993) and Gomulkiewicz and Kirkpatrick (1992) are quite broad in applicability. Some models address discrete trait variation in discretely varying environments (Lively, 1986b; Moran, 1992), others address continuously varying traits in discretely varying environments (Via and Lande, 1985, 1987), and still others address continuous variation in both trait and environment (Clark and Harvell, 1992; Gomulkiewicz and Kirkpatrick, 1992; Stearns and Koella, 1986). Environmental variation is modeled as either spatial (Lively, 1986b; Via and

Lande, 1985, 1987) or temporal (Bamber and Henderson, 1988), although some treatments encompass either mode (Gomulkiewicz and Kirkpatrick, 1992; Moran, 1992). The approaches range across models of fitness sets (Levins, 1968), evolutionary stable strategies (Lively, 1986b), dynamic programming (Clark and Harvell, 1992), single-locus population genetics (Bamber and Henderson, 1988), and quantitative genetics (Gomulkiewicz and Kirkpatrick, 1992; Van Tienderen, 1991). Some treatments address whether the optimum norm of reaction can evolve from the polygenic variation that contributes to environmental sensitivity (Via and Lande, 1985; Gavrilets and Scheiner, 1993), some address when a "population" of plastic phenotypes outperforms a genetically polymorphic one that embraces the same range of variation (Bamber and Henderson, 1988; Lively, 1986b; see also Moran, 1992, and Van Tienderen, 1991), and some address which specific norms of reaction would be favored among a set of such norms (Stearns and Koella, 1986).

Two criteria are necessary for adaptive phenotypic plasticity to evolve. First, the environment must vary in either space or time with respect to the optimal value of a phenotypic trait such that significantly different phenotypic values are favored under different conditions. The minimum difference necessary for plasticity to evolve will increase as the level of environmental heterogeneity (measured by the variance in conditions) decreases (Moran, 1992). Second, some cue must signal the state of the environment so that a directional response can be induced. The cue need not be perfectly reliable in every patch; the requisite degree of reliability appears to depend on how many types of environments there are, how frequently each appears, and, in a situation of spatial heterogeneity, how individuals from each type of environment remix through mating or dispersal (Lively, 1986b; Moran, 1992). Without a cue, plasticity cannot be selected for, and the course of adaptive evolution in a selectively heterogeneous environment might lead to genetic polymorphism or increased levels of random phenotypic variance.

These criteria represent necessary but not sufficient conditions for adaptive plasticity to evolve. There are three additional considerations. First, when plasticity carries a significant cost in fitness, the requirements that permit its evolution are more stringent. Bamber and Henderson (1988) examined the conditions under which a phenotypically plastic population would outperform a population that is genetically polymorphic at one locus for a comparable range of phenotypic values in a selective environment that varied at random across generations. When plasticity entailed no cost, the plastic population always outperformed the polymorphic one, because the plastic population harbors no phenotypes with a fitness below the maximum possible fitness and the fluctuation of conditions ensures that the polymorphic population always does. They introduced a cost to plasticity by discounting the fitness of the plastic form by a constant increment below that for the best genotype of the polymorphic population in each environment. The plastic population prevailed only when environmental conditions were significantly autocorrelated. Temporal autocorrelation acts on the variance of long-term fitness; low autocorrelations lower the variance in long-term fitness of the genetically polymorphic population and allow its geometric mean fitness to exceed that of the plastic population, which allows the polymorphic population to prevail. The greater the cost, the larger is the autocorrelation necessary for plasticity to prevail (see also Orzack, 1985, and Real, 1980, for more precise mathematical arguments, and Moran, 1992, or Van Tienderen, 1991, for analogous arguments in other theoretical contexts).

Second, migration among populations or high mutation rates can inhibit the evolution of plasticity or the differentiation of populations for norms of reaction. In contrast to the situation for simple trait differences, even small amounts of gene flow can preclude polygenic differentiation for norms of reaction (Via and Lande, 1985). Bamber and Henderson (1988) found that, when their genetically polymorphic population experienced a constant influx of new alleles (which simulated either migration or mutation), the tolerable cost of plasticity had to decrease and the autocorrelation level of environments had to increase beyond that in the case without an influx in order for the plastic population to outperform the polymorphic one. This result requires further investigation because the model of Bamber and Henderson (1988) actually examined genotype-specific migration or mutation that always reintroduced whichever allele was rare. Several other models also show that the mode of mixing among patches and the cost of plasticity are important determinants of its likely evolution (Gomulkiewicz and Kirkpatrick, 1992; Lively, 1986b; Van Tienderen, 1991).

Third, "environmental heterogeneity" alone is insufficient because the relative frequencies of the different conditions matter. When one alternative condition occurs rarely, whether in space or in time, any cost to maintaining phenotypic plasticity can outweigh the rare gain from that maintenance and preclude its ultimate evolution. The lower the level of environmental heterogeneity, the greater the fitness gain must be in producing the appropriate phenotypes in the rarer environments for plasticity to be favored. This argument can be seen explicitly or implicitly in a variety of mathematical models (Bamber and Henderson, 1988; Moran, 1992; Orzack, 1985). Indeed the importance of temporal autocorrelation in favoring plasticity diminishes rapidly as the relative frequencies of different conditions become more uneven, that is, as the overall level of heterogeneity decreases (Orzack, 1985).

Even when the necessary and sufficient conditions for the evolution of plasticity are realized, a variety of constraints on the process exert significant influence on the outcome. For example, some types of natural selection with spatial heterogeneity produce more than one locally stable norm of reaction as an outcome, and initial conditions determine the norm toward which the population will evolve (Gomulkiewicz and Kirkpatrick, 1992, the hard selection model). From the practical point of view, different norms of reaction among conspecific populations, therefore, need not implicate distinct selective regimes (a deduction that was discussed previously in the context of genetic drift among populations for *neutral* norms of reaction) but may result from different initial conditions under a *common* selective regime. In the case of temporal variation, different sequences of environmental conditions will produce different trajectories of evolution toward the equilibrial norm of reaction (Gavrilets and Scheiner, 1993; Gomulkiewicz and Kirkpatrick, 1992). Finally, although it is obvious that genetic constraints can prevent the final norm of reaction from exactly matching the optimal norm, the mode of selection with spatial heterogeneity can interact with genetic constraints such that different modes of selection can produce very different norms of reaction, none of which matches the optimal norm, even when the *same* genetic constraints are operating (Gomulkiewicz and Kirkpatrick, 1992). These complications mean that it is virtually impossible to make any inferences about adaptive plasticity from observations of the norms of reaction alone or even from observing distinct norms in distinct taxa.

How Does Norm of Reaction Evolve?

Selection can produce an appropriately adaptive norm of reaction in two ways: First, selection could act directly on genetic variation in norms of reaction when replicates of individual genotypes experience the full range of relevant environmental heterogeneity every generation. The genotype that produces the best phenotypes across the range of environments (or that comes closest) is favored, and other norms of reaction are eliminated. However, this description applies strictly only to clonal organisms. Analogous arguments have been made for diploid sexual organisms, and methods for estimating selection intensities on entire norms of reaction in such circumstances have been proposed by Rausher and Simms (1989) and Weis and Gorman (1990).

The dynamics of a second possibility, in which individual genotypes do not experience every environment in every generation, are more difficult to envision without mathematical models. This situation describes temporal heterogeneity or combined spatial and temporal heterogeneity with extensive mixing every few generations. Selection never "sees" the full norm of reaction and therefore must work indirectly to favor those genetic elements that, across each realization of the variable environment, most consistently produce the best phenotype (Gupta and Lewontin, 1982). For diploid, sexual organisms, Via and Lande (1985) have shown that, given certain assumptions, a quantitative genetic model of discrete spatial heterogeneity predicts the evolution of the appropriate norm of reaction among the polygenic variants that contribute to environmental sensitivity. An assortment of subsequent treatments have described more general models and examined the operation of a number of constraints (Gavrilets and Scheiner, 1993; Gomulkiewicz and Kirkpatrick, 1992; Van Tienderen, 1991). Scheiner and Lyman (1991) review a large body of work on laboratory systems that indicate that this mode of evolution is more likely to produce an evolutionary response in a norm of reaction than is direct selection on the full norm.

EVALUATING SPECIFIC EXAMPLES OF ADAPTIVE PLASTICITY

The actual norms of reaction for a trait carry no information about its potential adaptive significance. If any pattern of plasticity is to be judged as adaptive, the phenotypes that are produced must be examined for their performance and, ultimately, their fitness under each of the appropriate ecological conditions. Two sets of examples illustrate how this approach might be taken profitably.

Defensive Structures in Aquatic Invertebrates

The defensive structures induced in aquatic invertebrates offer an excellent example of adaptive plasticity in many ways. The structures are all induced by the presence of a predator and are associated with reduced vulnerability to predation (Havel and Dodson, 1984; Krueger and Dodson, 1981; Spitze, 1989). In the absence of predation, the presence of the defensive structures is usually associated with a reduction in fitness below that of the equivalent phenotype without them. This reduction occurs through different fitness components in different species, from more expensive locomotion (Stenson, 1987) to decreased fecundity (Dodson, 1984; Kerfoot, 1977; Stemberger, 1988) to increased development time and slower population growth (Riessen and Sprules, 1990). The two phenotypic types have appropriately different fitnesses under different conditions, the conditions (the presence or absence of the predator) are highly variable in space and time, and the presence or absence of the predator provides an eminently reliable cue. The basic pattern has evolved repeatedly in a variety of taxa and thus represents a general phenomenon and not an isolated curiosity.

However, the evolution of adaptive plasticity, at least for cladocerans, is not completely understood. Some studies have not found any reduction in the fitness of the "toothed" morph in the absence of predation risk (e.g., Dodson and Havel, 1988), and indeed some have found the toothed morph to outperform its alternate (e.g., Gilbert, 1980; O'Brien et al., 1980). Spitze (1992) has shown that the relative expression of the neck tooth in *Daphnia pulex* is actually a poor predictor of the variation among clones in the gain in fitness associated with the presence of a predator. Several other traits vary in concert with the expression of the neck tooth, and one or more of them may actually counter the reduced vulnerability to predation (Spitze, 1992). Although it is clear that the plasticity in morphology is adaptive, it is unclear which traits have actually been selected to be plastic and which traits represent correlated responses generated by a common developmental pathway.

Anuran Developmental Patterns

The joint plasticity of development time and size at metamorphosis in anurans represents an example of apparently adaptive plasticity that requires evaluation of the contribution to fitness of two traits. In this case both traits are plastic; dual plasticity appears to maximize fitness across a range of conditions. It does so, however, because there is a change in which trait is more responsible for fitness variation as conditions change, and the developmental link between the traits enforces joint plasticity even though that link can produce a disadvantageous value for one of the traits in any single circumstance.

The starkest examples of the phenomenon occur in *Bufo americanus*. Tadpoles increase development rate in the presence of predatory odonate naiads and metamorphose earlier and at lower body sizes than in the absence of these predators (Skelly and Werner, 1990; van Buskirk, 1988; Wilbur and Fauth, 1990). This effect appears to be mediated by the reduced growth rates caused by the tadpoles' alteration of habitat use patterns and foraging efforts when the predators are present (Skelly and Werner, 1990). Corroborative evidence for growth rate mediation comes from studies of the effects of food quantity (Skelly and Werner, 1990) and variation in the conditions for growth in the field in a closely related species (Travis and Trexler, 1986). Toad tadpoles are remarkably sensitive to their "growth environment" throughout development (Alford and Harris, 1988) and thus may respond to alterations in predation risk at any point in their lives.

The shortened development time enhances survival by reducing the period during which larvae are exposed to predation. Predation on many tadpoles is size-limited (references in Richards and Bull, 1990; Travis et al., 1987), and escape occurs only on attainment of a threshold size or, for small species, metamorphosis. Metamorphosis is an especially sound option for toads because they become unpalatable to many predators as juveniles. Slow growth increases the duration of the larva's susceptibility to predators. Under these conditions, rapid development is an advantage. Accelerated development can also be advantageous if the aquatic environment of the tadpoles begins to disappear through drying; other anuran larvae can accelerate development in such circumstances (Crump, 1989; Newman, 1987, 1988; Sokol, 1984).

The consequent metamorphosis at small size carries a high cost in reduced fitness. Smaller size at metamorphosis in toads has a number of functional disadvantages through the allometry of many physiological performance-trait properties (Pough and Kamel, 1984). These disadvantages, among others, cause smaller metamorphic toads to have lower survival rates (Clarke, 1974), a result common to many amphibians (Berven, 1982; Ischenko and Schchupak, 1976; Semlitsch et al., 1988; Smith, 1987). Small size at metamorphosis confers no direct advantages.

The phenotypic plasticity in both traits appears to optimize fitness in the unpredictable larval environment (see Skelly and Werner, 1990). When larval conditions are poor, the larvae metamorphose more quickly; by doing so they achieve a higher fitness through the inverse relationship between development time and probability of survival but are forced to compromise on the value of size at metamorphosis. When larval conditions are favorable, the toads achieve higher fitness through the larger size at metamorphosis that results from slowing development rate. The plasticity in size at metamorphosis produces a poor trait value when the more important determinant of fitness is development time and a better trait value when development time is less critical.

The adaptive interpretation of these patterns is supported by the patterns of behavioral responses of amphibian larvae (including toads) to predators. All species do not respond alike to the same predators; larvae respond only to predators with which they normally occur and to which they are at risk (Kats et al., 1988). Thus the example does not illustrate a pattern of behavioral and somatic plasticity common to all anurans but an adaptive evolutionary response to a specific selective regime.

These examples indicate that the adaptive significance of phenotypic plasticity will usually involve several traits that interact to produce the optimal *multivariate* phenotype for each environment, or at least the best in each environment that the genetic variation available to selection can offer. In these cases the diagnosis of adaptive plasticity requires the diagnosis of fitness consequences of a suite of developmentally intertwined traits.

This point has been made either explicitly or implicitly by others (cf. Bradshaw, 1965; Schlichting, 1986). Stearns and Koella (1986) model a homologous situation through the example of delayed maturity. Several additional empirical examples exist, most notably that of Newman (1987, 1988). Trexler et al. (1990) hypothesize a compromise among traits to explain the patterns of plasticity in age and size at maturity they observed in a poeciliid fish. Berven and his colleagues (Berven, 1982; Berven and Gill, 1983; Berven et al., 1979) offer this type of argument to account for different norms of reaction in development rate and size at metamorphosis in different conspecific populations of anurans.

CONCLUSIONS

Environmental effects on the values of morphological traits are ubiquitous. However, the ubiquity of such effects is deceptive; the direction and magnitude of phenotypic plasticity is neither evenly nor randomly distributed among comparable species or populations of a single species. A considerable body of diverse evidence indicates that the response of the developmental system to environmental factors has been molded by natural selection. The ecological evidence is more persuasive for discrete variations than for continuous one, but it is not clear whether this situation reflects the relative likelihood of adaptive plasticity in each type or their relative tractability.

This contrast between discrete and continuous morphological plasticity is perhaps the most intriguing phenomenon. It seems likely that plastically determined discrete alternatives have been fashioned by disruptive selection on what was originally continuous variation in response to the environment. This is exactly the evolutionary history of caste polyphenisms in ants (Oster and Wilson, 1978). Examples of disruptive selection and consequent genetic differentiation usually involve some combination of dramatic environmental differences, reinforcement of fitness differences by habitat selection, and reduced genetic interchange, and mathematical models suggest that stringent requirements for mating and remixing may be necessary to achieve a stable pattern of genetic differentiation for the case of a continuous trait (Garcia-Dorado, 1990). The empirical circumstances surrounding evolution of plastically controlled discrete alternatives would seem to involve only dramatic environmental differences, but it is unclear whether models of disruptive selection can be modified to begin a theory for the evolution of plastic discrete alternative morphologies.

Evaluating the adaptive role of morphological plasticity may pose the most difficult challenge that an ecological morphology will face. It will require a combination of functional morphology, ecology, and developmental biology. But the number of potential examples and the persuasiveness of many of those that have been studied indicate that the challenge is not unanswerable. The evidence in hand indicates that Schmalhausen's (1949) enthusiasm for the thesis explored in this essay was not misplaced and that indeed the molding of developmental systems in heterogeneous environments represents selection's greatest achievement.

ACKNOWLEDGMENTS

I thank my colleague Dr. F. C. James for introducing me to Sumner's work thirteen years ago and thereby inspiring much of the thought that went into this essay. Discussions with Drs. D. Reznick and J. C. Trexler have helped clarify my thoughts on plasticity in the last few years. The perceptive comments of Drs. D. Harvell, F. C. James, R. A. Newman, S. M. Reilly, P. C. Wainwright, and A. Winn and an anonymous reviewer have improved this presentation enormously. The support of NSF grant BSR 88-18001 is gratefully acknowledged.

REFERENCES

Alford, R. A., and R. N. Harris. 1988. Effects of larval growth history on anuran metamorphosis. *Amer. Nat.* 131:91–106.

Antonovics, J., N. C. Ellstrand, and R. N. Brandon. 1988. Genetic and environmental variation: Expectations and experiments. In *Plant Evolutionary Biology*, ed. L. D. Gottlieb and S. K. Jain, 275–304. London: Chapman and Hall.

Arnold, S. J., and C. R. Peterson. 1990. A test for temperature effect on the ontogeny of shape in the garter snake *Thamnophis sirtalis. Physiol. Zool.* 62:1316–1333.

Arthur, W. 1982. Control of shell shape in *Lymnaea stagnalis. Heredity* 49:153–161.

Baker, H. G. 1965. Characteristics and modes of origin of weeds. In *The Genetics of Colonizing Species*, ed. H. G. Baker and G. L. Stebbins, 147–172. New York: Academic Press.

Bamber, R. N., and P. A. Henderson. 1985. Morphological variation in British atherinids, and the status of *Atherina presbyton* Cuvier (Pisces: Atherindae). *Biol. J. Linnean Soc.* 25:61–75.

Bamber, R. N., and P. A. Henderson. 1988. Pre-adaptive plasticity in atherinids and the estuarine seat of teleost evolution. *J. Fish. Biol.* 33:17–23.

Belyaev, D. K. 1979. Destabilizing selection as a factor in domestication. *J. Hered.* 70:301–308.

Bernays, E. A. 1986. Diet-induced head allometry among foliage-chewing insects and its importance for graminivores. *Science* 231:495–497.

Berven, K. A. 1982. The genetic basis of altitudinal variation in the wood frog *Rana sylvatica*. I. An experimental analysis of life history traits. *Evolution* 36:962–983.

Berven, K. A., and D. E. Gill. 1983. Interpreting geographic variation in life history traits. *Amer. Zool.* 23:85–97.

Berven, K. A., D. E. Gill, and S. J. Smith-Gill. 1979. Countergradient selection in the green frog, *Rana clamitans. Evolution* 33:609–623.

Biewener, A. A. 1990. Biomechanics of mammalian terrestrial locomotion. *Science* 250:1097–1103.

Blouin, M. S., and M. L. G. Loeb. 1991. Effects of environmentally induced development rate variation on head and limb morphology in the green treefrog, *Hyla cinerea. Amer. Nat.* 138:717–728.

Bradshaw, A. D. 1965. Evolutionary significance of phenotypic plasticity in plants. *Adv. Genet.* 13:115–155.

Bradshaw, W. E. 1986. Variable iteroparity as a life-history tactic in the pitcher-plant mosquito *Wyeomyia smithii. Evolution* 40:471–478.

Bradshaw, W. E., and L. P. Lounibos. 1977. Evolution of dormancy and its photoperiodic control in pitcher-plant mosquitos. *Evolution* 31:546–567.

Brian, M. V. 1980. Social control over sex and caste in bees, wasps, and ants. *Biol. Rev. Camb. Phil. Soc.* 55:379–415.

Bull, J. J. 1987. Evolution of phenotypic variance. *Evolution* 41:303–315.

Bulmer, M. G. 1984. Delayed germination of seeds: Cohen's model revisited. *Theor. Pop. Biol.* 26:367–377.

Burton, R. S. 1990. Hybrid breakdown in developmental time in the copepod *Tigriopus californicus. Evolution* 44:1814–1822.

Capinera, J. L. 1979. Qualitative variation in plant and insects: Effects of propagule size on ecological plasticity. *Amer. Nat.* 114:350–361.

Caswell, H. 1983. Phenotypic plasticity in life history traits: Demographic effects and evolutionary consequences. *Amer. Zool.* 23:35–46.

Cavalli-Sforza, L. L. 1974. The role of plasticity in biological and cultural evolution. *Ann. N. Y. Acad. Sci.* 231:43–59.

Chambers, S. T. 1980. Genetic divergence between populations of *Goniobasis* (Pleuroceridae) occupying different drainage systems. *Malacologia* 20:63–81.

Chernoff, B. 1982. Character variation among populations and the analysis of biogeography. *Amer. Zool.* 22:425–439.

Clark, C. W., and C. D. Harvell. 1992. Inducible defenses and the allocation of resources: A minimal model. *Amer. Nat.* 139:521–539.

Clarke, R. D. 1974. Postmetamorphic growth rates in a natural population of the Fowler's toad (*Bufo woodhousei fowleri*). *Can. J. Zool.* 52:1458–1498.

Cohen, D. 1967. Optimizing reproduction in a randomly varying environment when a correlation may exist between the conditions at the time a choice has to be made and the subsequent outcome. *J. Theor. Biol.* 16:1–14.

Collins, J. P., and J. E. Cheek. 1983. Effect of food and density on development of typical and cannibalistic salamander larvae in *Ambystoma tigrinum nebulosum. Amer. Zool.* 23:77–84.

Conover, D. O., and S. W. Heins. 1987. Adaptation variation in environmental and genetic sex determination in a fish. *Nature* 326:496–498.

Conover, D. O., and T. M. C. Present. 1990. Countergradient variation in growth rate: Compensation for length of the growing season among Atlantic silversides from different latitudes. *Oecologia* 83:316–324.

Cook, S. A., and M. P. Johnson. 1968. Adaptation to heterogeneous environments. I. Variation in heterophylly in *Ranunculus flammula* L. *Evolution* 22:496–516.

Coyne, J. A., and E. Beecham. 1987. Heritability of two morphological characters within and among natural populations of *Drosophila melanogaster. Genetics* 117:727–737.

Crump, M. L. 1981. Variation in propagule size as a function of environmental uncertainty for tree frogs. *Amer. Nat.* 117:724–737.

Crump, M. L. 1984. Intraclutch egg size variability in *Hyla crucifer* (Anura: Hylidae). *Copeia* 1984:302–308.

Crump, M. L. 1989. Effect of habitat drying on development time and size at metamorphosis in *Hyla pseudopuma. Copeia* 1989:794–797.

Currey, J. 1984. *The Mechanical Adaptations of Bones.* Princeton: Princeton University Press.

Daikoku, T., I. Yano, and M. Masui. 1982. Lipid and fatty acid compositions and their changes in the different organs and tissues of guppy, *Poecilia reticulata,* on sea water adaptation. *Comp. Biochem. Physiol.* 73A:167–174.

Dentry, W., and C. C. Lindsey. 1978. Vertebral variation in zebrafish (*Brachydanio rerio*) related to the prefertilization temperature history of their parents. *Can. J. Zool.* 56:280–283.

Dingle, H., B. M. Alden, N. R. Blakley, D. Kopec, and E. R. Miller. 1980. Variation in photoperiodic response within and among species of milkweed bugs (*Oncopeltus*). *Evolution* 34:356–370.

Dodson, S. I. 1984. Predation of *Heterocope septentrionalis* on two species of *Daphnia:* Morphological defenses and their cost. *Ecology* 65:1249–1257.

Dodson, S. I., and J. E. Havel. 1988. Indirect prey effects some morphological and life history responses of *Daphnia pulex* exposed to *Notonecta undulata. Limnol. Oceanogr.* 33:1274–1285.

Downes, B. J. 1990. Host-induced morphology in mites: Implications for host-parasite coevolution. *Syst. Zool.* 39:162–182.

Eberhard, W. G. 1980. Horned beetles. *Sci. Amer.* 242:166–182.

Edwards, A. L. 1988. Latudinal clines in shell morphologies of *Busycon carica* (Gmelin 1791). *J. Shellfish Res.* 7:461–466.

Ellner, S. 1985a. ESS germination strategies in randomly varying environments. I. Logistic-type models. *Theor. Pop. Biol.* 28:50–79.

Ellner, S. 1985b. ESS termination strategies in randomly varying environments. II. Reciprocal yield-law models. *Theor. Pop. Biol.* 28:116.

Emerson, S. 1986. Heterochrony and frogs: The relationship of a life history trait to morphological form. *Amer. Nat.* 127:167–183.

Etter, R. J. 1988. Asymmetrical developmental plasticity in an intertidal snail. *Evolution* 42:322–334.

Evans, D. H. 1984. The roles of gill permeability and transport mechanisms in euryhalinity. In *Fish Physiology,* vol. 10B, ed. W. S. Hoar and D. J. Randall, 239–283. New York: Academic Press.

Fowler, J. A. 1970. Control of vertebral number in teleosts: An embryological problem. *Q. Rev. Biol.* 45:148–167.

Fox, W. W., C. Gordon, and M. H. Fox. 1961. Morphological effects of low temperatures during the embryonic development of the garter snake, *Thamnophis elegans. Zoologica* 46:57–71.

Gabriel, M. L. 1944. Factors affecting the number and form of vertebrae in *Fundulus heteroclitus. J. Exp. Zool.* 95:105–147.

Gabriel, W., and M. Lynch. 1992. The selective advantage of reaction norms for environmental tolerance. *J. Evol. Biol.* 5:41–59.

Garcia-Dorado, A. 1990. The effect of soft selection on the variability of a quantitative trait. *Evolution* 44:168–179.

Gavrilets, S., and S. M. Scheiner. 1993. The genetics of phenotypic plasticity. V. Evolution of reaction norm shape. *J. Evol. Biol.* In press.

Gebhardt, M. D., and S. C. Stearns. 1988. Reaction norms for developmental time and weight at eclosion in *Drosophila mercatorum. J. Evol. Biol.* 1:335–354.

Gibson, J. B., and B. P. Bradley. 1974. Stabilising selection in constant and fluctuating environments. *Heredity* 33:293–302.

Gilbert, J. J. 1980. Further observations on developmental polymorphism and its evolution in the rotifer *Brachionus calyciflorus. Freshwater Biol.* 20:281–294.

Gilbert, J. J., and R. S. Stemberger. 1984. *Asplanchna*-induced polymorphism in the rotifer *Keratella slacki. Limnol. Oceanogr.* 29:1309–1316.

Gomulkiewicz, R., and M. Kirkpatrick. 1992. Quantitative genetics and the evolution of reaction norms. *Evolution* 46:390–411.

Graham, J. H., and J. D. Felley. 1985. Genomic coadaptation and developmental stability within introgressed populations of *Enneacanthus gloriosus* and *E. obesus* (Pisces, Centrarchidae). *Evolution* 39:104–114.

Grant, J. W. G., and I. A. E. Bayly. 1981. Predator induction of crests in morphs of the *Daphnia carinata* King complex. *Limnol. Oceanogr.* 26:201–218.

Groeters, F. R., and H. Dingle. 1988. Genetic and maternal influences on life history plasticity in milkweed bugs (*Oncopeltus*): Response to temperature. *J. Evol. Biol.* 1:317–333.

Grudzien, T. A., and B. J. Turner. 1984. Direct evidence that the *Ilyodon* morphs are a single biological species. *Evolution* 38:402–407.

Gupta, A. P., and R. C. Lewontin. 1982. A study of reaction norms in natural populations of *Drosophila pseudoobscura. Evolution* 36:934–948.

Harkey, G. A., and R. D. Semlitsch. 1988. Effects of temperature on the growth, development, and color polymorphism in the ornate chorus frog, *Pseudacris ornata. Copeia* 1988:1001–1007.

Harrison, R. G. 1980. Dispersal polymorphisms in insects. *Ann. Rev. Ecol. Syst.* 11:95–118.

Harvell, C. D. 1984. Predator-induced defense in a marine bryozoan. *Science* 224:1357–1359.

Harvell, C. D. 1990. The ecology and evolution of inducible defenses. *Q. Rev. Biol.* 65:323–340.

Havel, J. E. 1985. Cyclomorphosis of *Daphnia pulex* spined morphs. *Limnol. Oceanogr.* 30:853–861.

Havel, J. E., and S. I. Dodson. 1084. *Chaoborus* predation on typical and spined morphs of *Daphnia pulex:* Behavioral observations. *Limnol. Oceanogr.* 29:487–494.

Heslop-Harrison, J. 1959. Variability and environment. *Evolution* 13:145–147.

Ischenko, V. G., and E. L. Schchupak. 1976. Ecological regulation of the genetic constitution of the population of the small Asiatic frog (*Rana macrocnemis* Bigr.). *Sov. J. Ecol.* 6:130–137.

Jain, S. 1979. Adaptive strategies: Polymorphism, plasticity, and homeostasis. In *Topics in Plant*

Population Biology, ed. O. T. Solbrig, S. Jain, G. B. Johnson, and P. H. Raven, 160–187. New York: Columbia University Press.

James, F. C. 1970. Geographic size variation in birds and its relationship to climate. *Ecology* 51:365–390.

James, F. C. 1983. Environmental component of morphological differentiation in birds. *Science* 221:184–186.

Jameson, D. L., J. P. Mackey, and M. Anderson. 1973. Weather, climate, and the external morphology of Pacific tree toads. *Evolution* 27:285–302.

Kaplan, R. H. 1985. Maternal influences on offspring development in the California newt, *Taricha torosa. Copeia* 1985:1028–1035.

Kaplan, R. H., and W. S. Cooper. 1984. The evolution of developmental plasticity in reproductive characters: An application of the "adaptive coin-flipping" principle. *Amer. Nat.* 123:393–410.

Kats, L. B., J. W. Petranka, and A. Sih. 1988. Antipredator defenses and the persistence of amphibian larvae with fishes. *Ecology* 69:1865–1870.

Kaufman, P. K., F. D. Enfield, and R. E. Comstock. 1977. Stabilizing selection for pupa weight in *Tribolium castaneum. Genetics* 87:327–341.

Kemp, P., and M. D. Bertness. 1984. Snail shape and growth rates: Evidence for plastic shell allometry in *Littorina littorea. Proc. Natl. Sci. USA* 81:811–813.

Kennedy, J. S. 1956. Phase transformation in locust biology. *Biol. Rev.* 31:349–370.

Kerfoot, W. C. 1977. Competition in cladoceran communities: The cost of evolving defenses against copepod predation. *Ecology* 58:303–313.

Kornfield, I. K., D. C. Smith, P. S. Gagnon, and J. N. Taylor. 1982. The cichlid fish of Cuatro Ciénegas, Mexico: Direct evidence of conspecificity among distinct trophic morphs. *Evolution* 36:658–664.

Krueger, D. A., and S. I. Dodson. 1981. Embryological induction and predation ecology in *Daphnia pulex. Limnol. Oceanogr.* 26:219–223.

Lanyon, L. E., and C. T. Rubin. 1985. Functional adaptation in skeletal structures. In *Functional Vertebrate Morphology,* ed. M. Hildebrand, D. M. Bramble, K. Liem, and D. B. Wake, 1–25. Cambridge: Belknap Press.

Leon, J. 1985. Germination strategies. In *Evolution: Essays in Honour of John Maynard Smith,* ed. P. J. Greenwood, P. H. Harvey, and M. Slatkin, 129–142. Cambridge: Cambridge University Press.

Levins, R. 1963. Theory of fitness in a heterogeneous environment. II. Development flexibility and niche selection. *Amer. Nat.* 97:75–90.

Levins, R. 1968. *Evolution in Changing Environments.* Princeton: Princeton University Press.

Lindsey, C. C. 1966. Temperature-controlled meristic variation in the salamander *Ambystoma gracile. Nature* 209:1152.

Lively, C. M. 1986a. Predator-induced shell dimorphism in the acorn barnacle *Chthamalus anisopoma. Evolution* 40:232–242.

Lively, C. M. 1986b. Canalization versus developmental conversion in a spatially variable environment. *Amer. Nat.* 128:561–572.

Lloyd, D. G. 1984. Variation strategies of plants in heterogeneous environments. *Biol. J. Linnean Soc.* 21:357–385.

Markin, G. P., J. H. Dillier, and H. L. Collins. 1973. Growth and development of colonies of the red imported fire ant, *Solenopsis invicta. Ann. Entomol. Soc. Amer.* 66:803–808.

Martin, W. R. 1949. *The Mechanics of Environmental Control of Body Form in Fishes.* University Toronto Studies Biol. Ser., no. 58.

McGinley, M. A., D. H. Temme, and M. A. Geber. 1987. Parental investment in offspring in variable environments: Theoretical and empirical considerations. *Amer. Nat.* 130:370–398.

Meffe, G. K. 1987. Embryo size variation in mosquitofish: Optimality vs. plasticity in propagule size. *Copeia* 1987:762–768.

Meyer, A. 1987. Phenotypic plasticity and heterochrony in *Cichlasoma managuense* (Pisces, Cichlidae) and their implications for speciation in cichlid fishes. *Evolution* 41:1357–1369.

Moore, W. J. 1965. Masticatory function and skull growth. *J. Zool.* 146:123–131.

Moran, N. A. 1992. The evolutionary maintenance of alternative phenotypes. *Amer. Nat.* 139:971–989.

NeSmith, C. C. 1984. The effect of the physical environment on the development of red-winged blackbird nestlings: A laboratory experiment. M. S. thesis, Florida State University, Tallahassee.

Newman, R. A. 1987. Effects of density and predation on *Scaphiopus couchii* tadpoles in desert ponds. *Oecologia* 71:301–307.

Newman, R. A. 1988. Adaptive plasticity in development of *Scaphiopus couchii* tadpoles in desert ponds. *Evolution* 42:774–783.

Newman, R. A. In press. Coping with uncertainty: The role of phenotypic plasticity in variable environments. *Bioscience.*

O'Brien, W. J., D. Kettle, H. Riessen, D. Schmidt, and D. Wright. 1980. Dimorphic *Daphnia longiremis:* Predation and competition interactions between two morphs. In *Evolution and Ecology of Zooplankton Communities,* ed. W. C. Kerfoot, 497–505. Hanover: New England Press.

O'Loughlin, E. F. M., and J. C. Aldrich. 1987. An analysis of shell shape variation in the painted topshell *Calliostoma zizyphinum* (L.) (Prosobranchia: Trochidae). *J. Molluscan Studies* 53:62–68.

Orzack, S. H. 1985. Population dynamics in variable environments. V. *Amer. Nat.* 125:550–572.

Osgood, D. W. 1978. Effect of temperature on the development of meristic characters in *Natrix fasciata. Copeia* 1978:33–37.

Oster, G. F., and E. O. Wilson. 1978. *Caste and Ecology in the Social Insects.* Monographs in Population Biology, no. 12. Princeton: Princeton University Press.

Palmer, A. R. 1985. Quantum changes in gastropod shell morphology need not reflect speciation. *Evolution* 39:699–705.

Palumbi, S. R. 1984. Tactics of acclimation: morphological changes of sponges in an unpredictable environment. *Science* 225:1478–1480.

Perrin, N. 1988. Why are offspring born larger when it is colder? Phenotypic plasticity for offspring size in the cladoceran *Simocephalus vetulus* (Müller). *Func. Ecol.* 2:283–288.

Pfennig, D. 1990. The adaptive significance of an environmentally-cued developmental switch in an anuran tadpole. *Oecologia* 85:101–107.

Pomeroy, L. V. 1981. Developmental polymorphism in the tadpoles of the spadefoot toad *Scaphiopus multiplicatus.* Ph.D. diss., University of California, Riverside.

Porter, S. D., and W. R. Tschinkel. 1985. Fire ant polymorphism: The ergonomics of brood production. *Behav. Ecol. Sociobiol.* 16:323–336.

Pough, F. H., and S. Kamel. 1984. Post-metamorphic change in activity metabolism of anurans in relation to life history. *Oecologia* 65:138–144.

Quinn, J. A., and K. C. Hodgkinson. 1984. Population variability in *Danthonia caespitosa* (Gramineae) in response to increasing density under three temperature regimes. *Amer. J. Bot.* 70:1425–1431.

Rausher, M. D., and E. L. Simms. 1989. The evolution of resistance to herbivory in *Ipomoea purpurea.* I. Attempts to detect selection. *Evolution* 43:563–572.

Real, L. A. 1980. Fitness, uncertainty, and the role of diversification in evolution and behavior. *Amer. Nat.* 115:623–638.

Reilly, S. M., G. V. Lauder, and J. P. Collins. In press. Performance consequences of a trophic polymorphism: Feeding behavior in typical and cannibal phenotypes of *Ambystoma tigrinum. Copeia.*

Richards, S. J., and C. M. Bull. 1990. Size-limited predation on tadpoles of three Australian frogs. *Copeia* 1990:1041–1046.

Riessen, H. P., and W. G. Sprules. 1990. Demographic costs of antipredation defenses in *Daphnia pulex. Ecology* 71:1536–1546.

Sage, R. D., and R. K. Selander. 1975. Trophic radiation through polymorphism in cichlid fishes. *Proc. Natl. Acad. Sci. USA* 72:4669–4673.

Scharloo, W., M. S. Hoogmoed, and A. Ter Kuile. 1967. Stabilizing and disruptive selection on a mutant character in *Drosophila*. I. The phenotypic variance and its components. *Genetics* 56:709–726.

Scheiner, S., and C. J. Goodnight. 1984. The comparison of phenotypic plasticity and genetic variation in populations of the grass *Denathonia spicata*. *Evolution* 38:845–855.

Scheiner, S., and R. F. Lyman. 1989. The genetics of phenotypic plasticity. I. Heritability. *J. Evol. Biol.* 2:95–108.

Scheiner, S. M., and R. F. Lyman. 1991. The genetics of phenotypic plasticity. II. Response to selection. *J. Evol. Biol.* 4:23–50.

Schlichting, C. D. 1986. The evolution of phenotypic plasticity in plants. *Ann. Rev. Ecol. Syst.* 17:667–693.

Schlichting, C. D., and D. A. Levin. 1984. Phenotypic plasticity of annual phlox: Tests of some hypotheses. *Amer. J. Bot.* 7:252–260.

Schmalhausen, I. I. 1949. *Factors of Evolution*. Philadelphia: Blakiston. Reprint. Chicago: University of Chicago Press, 1986.

Schnee, F. B., and J. N. Thompson, Jr. 1984. Conditional polygenic effects in the sternopleural bristle system of *Drosophila melanogaster*. *Genetics* 108:409–424.

Semlitsch, R. D., D. E. Scott, and J. H. K. Pechmann. 1988. Time and size at metamorphosis related to adult fitness in *Ambystoma talpoideum*. *Ecology* 69:184–192.

Service, P. M., and R. E. Lenski. 1982. Aphid phenotypes, plant phenotypes, and genetic diversity: A demographic analysis of experimental data. *Evolution* 36:1276–1282.

Shapiro, A. M. 1976. Seasonal polyphenism. *Evol. Biol.* 9:259–333.

Skelly, D. K., and E. E. Werner. 1990. Behavioral and life-historical responses of larval American toads to an odonate predator. *Ecology* 71:2313–2322.

Slatkin, M., and R. Lande. 1976. Niche width in a fluctuating environment: Density independent model. *Amer. Nat.* 100:31–55.

Slobodkin, L. B., and A. Rapoport. 1974. An optimal strategy of evolution. *Q. Rev. Biol.* 49:181–200.

Smith, D. C. 1987. Adult recruitment in chorus frogs: Effects of size and date at metamorphosis. *Ecology* 68:344–350.

Smith-Gill, S. J. 1983. Developmental plasticity: Developmental conversion versus phenotypic modulation. *Amer. Zool.* 23:47–55.

Sober, E. 1984. *The Nature of Selection*. Cambridge: MIT Press.

Sokol, A. 1984. Plasticity in the fine timing of metamorphosis in tadpoles of the hylid frog *Litoria ewingi*. *Copeia* 1984:868–873.

Spitze, K. 1989. The consequences of predation by *Chaoborus americanus* on *Daphnia pulex*. Ph.D. diss. University of Illinois, Urbana-Champaign.

Spitze, K. 1992. Predator-mediated plasticity of prey life history and morphology: *Chaoborus americanus* predation on *Daphnia pulex*. *Amer. Nat.* 140:229–247.

Stearns, S. C. 1989. The evolutionary significance of phenotypic plasticity. *Bioscience* 39:436–445.

Stearns, S. C., and J. C. Koella. 1986. The evolution of phenotypic plasticity in life-history traits: Predictions of reaction norms for age and size at maturity. *Evolution* 40:893–913.

Stemberger, R. S. 1988. Reproductive costs and hydrodynamic benefits of chemically induced defenses in *Keratella testudo*. *Limnol. Oceanogr.* 33:593–606.

Stenson, J. A. E. 1987. Variation in capsule size of *Holopedium gibberum* (Zaddach): A response to invertebrate predation. *Ecology* 68:928–934.

Strauss, S. H. 1987. Heterozygosity and developmental stability under inbreeding and crossbreeding in *Pinus attenuata*. *Evolution* 41:331–339.

Sultan, S. E. 1987. Evolutionary implications of phenotypic plasticity in plants. *Evol. Biol.* 20:127–178.

Sumner, F. B. 1909. Some effects of external conditions on the white mouse. *J. Exp. Zool.* 7:97–155.

Sumner, F. B. 1915. Some studies on environmental influence, heredity, correlation, and growth in the white mouse. *J. Exp. Zool.* 18:325–432.

Sumner, F. B. 1932. Genetic, distributional, and evolutionary studies of the subspecies of deer mice (*Peromyscus*). *Bibliogr. Genet.* 9:1–106.

Taylor, D. R., and L. W. Aarssen. 1988. An interpretation of phenotypic plasticity in *Agropyron repens* (Gramineae). *Amer. J. Bot.* 75:401–413.

Thoday, J. M. 1953. Components of fitness. *Symp. Soc. Exp. Biol.* 7:96–113.

Thomas, M. L. H., and J. H. Himmelman. 1988. Influence of predation on shell morphology of *Buccinum undatum* L. on Atlantic coast of Canada. *J. Exp. Mar. Biol. Ecol.* 115:221–236.

Thompson, D. B. In press. Consumption rates and the evolution of diet-induced plasticity in the head morphology of *Melanoplus femurrubrum* (Orthoptera: Acrididae). *Oecologia.*

Travis, J. 1983. Variation in developmental patterns of larval anurans in temporary ponds. I. Persistent variation within a *Hyla gratiosa* population. *Evolution* 37:496–512.

Travis, J. 1989. The role of optimizing selection in natural populations. *Ann. Rev. Ecol. Syst.* 20:279–296.

Travis, J., and J. C. Trexler. 1986. Interactions among factors affecting growth, development and survival in experimental populations of *Bufo terrestris* (Anura: Bufonidae). *Oecologia* 69:110–116.

Travis, J., and J. C. Trexler. 1987. *Regional Variation in Habitat Requirements of the Sailfin Molly, with Special Reference to the Florida Keys.* Florida Game and Fresh Water Fish Comm. Nongame Wildl. Program Tech. Rep. no. 3.

Travis, J., S. Emerson, and M. Blouin. 1987. A quantitative genetic analysis of larval life history traits in *Hyla crucifer. Evolution* 41:145–156.

Trexler, J. C. 1989. Phenotypic plasticity in Poeciliid life histories. In *Ecology & Evolution of Livebearing Fishes (Poeciliidae),* ed. G. K. Meffe and F. F. Snelson, Jr., 201–216. Englewood Cliffs, NJ: Prentice Hall.

Trexler, J. C, and J. Travis. 1990. Phenotypic plasticity in the sailfin molly, *Poecilia latipinna* (Pisces: Poeciliidae). I. Field experiment. *Evolution* 44:143–156.

Trexler, J. C., J. Travis, and M. Trexler, 1990. Phenotypic plasticity in the sailfin molly, *Poecilia latipinna* (Pisces: Poeciliidae). II. Laboratory experiment. *Evolution* 44:157–167.

Turner, B. J., and D. J. Grosse. 1980. Trophic differentiation in *Ilyodon,* a genus of stream-dwelling goodeid fishes: Speciation vs. ecological polymorphism. *Evolution* 34:259–270.

van Buskirk, J. 1988. Interactive effects of dragonfly predation in experimental pond communities. *Ecology* 69:857–867.

Van Tienderen, H. P. 1991. Evolution of generalists and specialists in spatially heterogeneous environments. *Evolution* 45:1317–1331.

Venable, D. L. 1989. Modelling the evolutionary ecology of seed banks. In *Ecology of Soil Seed Banks,* ed. V. T. Parker, M. A. Leck, and R. L. Simpson, 67–87. Orlando, FL: Academic Press.

Venable, D. L., and J. S. Brown. 1988. The selective interactions of dispersal, dormancy, and seed size as adaptations for reducing risk in variable environments. *Amer. Nat.* 131:360–384.

Via, S. 1984. The quantitative genetics of polyphagy in an insect herbivore. I. Genotype environment interaction in larval performance on different host species. *Evolution* 38:881–895.

Via, S., and R. Lande. 1985. Genotype-environment interaction and the evolution of phenotypic plasticity. *Evolution* 39:505–522.

Via, S., and R. Lande. 1987. Evolution of genetic variability in a spatially heterogenous environment: Effects of genotype-environment interaction. *Genet. Res.* 49:147–156.

Vogt, T., and D. L. Jameson. 1970. Chronological correlation between change in weather and change in morphology of the Pacific treefrog in southern California. *Copeia* 1970:135–144.

Waddington, C. H., and E. Robertson. 1966. Selection for developmental canalisation. *Genet. Res.* 7:302–312.

Wainwright, P. C., C. W. Osenberg, and G. G. Mittelbach. 1991. Trophic polymorphism in the pumpkinseed sunfish (*Lepomis gibbosus*): Environmental effects on ontogeny. *Func. Ecol.* 5:40–55.

Warwick, T. 1944. Inheritance of the keel in *Pomatopyrgus jenkensi. Nature* 154:798–799.

Watt, D. G., and H. M. Williams. 1951. The effects of the physical consistency of food on the growth and development of the mandible and maxilla of the rat. *Amer. J. Orthod.* 37:895–928.

Weaver, M. E., and D. L. Ingram. 1969. Morphological changes in swine associated with environmental temperature. *Ecology* 50:710–713.

Weis, A. E., and W. L. Gorman. 1990. Measuring selection on reaction norms: An exploration of the *Eurosta-Solidago* system. *Evolution* 44:820–831.

West-Eberhard, M. J. 1989. Phenotypic plasticity and the origins of diversity. *Ann. Rev. Ecol. Syst.* 20:249–278.

Wheeler, D. E., and H. F. Nijhout. 1983. Soldier determination in the ant *Pheidole bicarinata:* Hormonal control of caste and size within castes. *J. Insect Physiol.* 29:847–854.

Wilbur, H. M., and J. Fauth. 1990. Experimental aquatic food webs: Interactions between two predators and two prey. *Amer. Nat.* 135:176–204.

Wilken, D. H. 1977. Local differentiation for phenotypic plasticity in the wild California annual *Collomia linearis* (Polemoniaceae). *Syst. Bot.* 2:99–108.

Wilson, E. O. 1953. The origin and evolution of polymorphism in ants. *Q. Rev. Biol.* 28:136–156.

Winn, A. A., and A. S. Evans. 1991. Variation among populations of *Prunella vulgaris* in plastic responses to light. *Func. Ecol.* 5:563–571.

Yoshioka, P. M. 1982. Predator-induced polymorphism in the bryozoan *Membranipora membranacea* (L.). *J. Exp. Mar. Biol. Ecol.* 61:233–242.

Zangerl, A. R., and M. R. Berenbaum. 1990. Furanocoumarin induction in wild parsnip: Genetics and populational variation. *Ecology* 71:1933–1940.

Zwingelstein, G., J. Portoukalian, G. Rubei, and G. Brichon. 1980. Gill sulpholipid synthesis and seawater adaptation in euryhaline fish, *Anguilla. Comp. Biochem. Physiol.* 65B:555–558.

6

Allometric Aspects of Predator-Prey Interactions

Sharon B. Emerson, Harry W. Greene, and Eric L. Charnov

INTRODUCTION

A correlative relationship between prey body size and predator body size has now been demonstrated for most major groups of vertebrates (Peters, 1983; Hespenheide, 1973, and references therein). However, the allometry or scaling of that interaction remains largely unexplored (but see Hespenheide, 1971, 1973; Peters, 1983; Vezina, 1985). Few studies have actually calculated the slope of the log relationship between prey body size and predator body size, but the available empirical data show that both intra- and interspecifically, prey body size and predator body size scale with ordinary least squares (OLS) slopes of between 0.70 and 1.20 (Inger and Marx, 1961; Toft, 1980; Vezina, 1985; Voris and Moffett, 1981). This scaling relationship holds true whether mean or maximum prey body size is considered and is suggestive of a general pattern. In this chapter, we will examine the effect of body size on predator-prey interactions. Our general approach incorporates aspects of population dynamics and foraging theory from ecology as well as biomechanical principles from functional morphology. (See chap. 3 for more on the relationship between ecology and functional morphology). Our goal is to explore several specific approaches toward assessing why there might be a regularity to the scaling of prey-predator body size interactions.

Optimal foraging and biomechanical models share the property that they provide a priori predictions of lower and upper boundaries to performance. As such they can be used to generate testable hypotheses regarding the scaling of prey-predator interactions with changes in body size, particularly when we consider maximum prey body size for a given-sized predator. Three somewhat different approaches to deriving the predicted scaling relationship will be examined. The first is an energy balance approach where we use a general equation from ecology which describes the capture rate of prey. In the second example, the predicted slope for a prey-predator interaction will be derived from a biomechanical model of performance limitation. The last approach will employ a model from foraging theory (Charnov, 1976) to predict the scaling relation.

BACKGROUND

The Predatory-Prey Interaction

Predator-prey interactions (PPIs) are composed of a number of steps, from searching for to digestion of a prey item. At every point in the sequence, both the predator and prey might have behavioral, physiological, and morphological specializations for increasing capture or escape success (Endler, 1986; Greene, 1988). By identifying the types of variables in each step it is possible to examine the consequences of changes in body size on these different components of the PPI.

Predation events can be divided into several activities: search, pursue, capture, immobilize, ingest, and process. During search and pursuit the locomotor morphology and sensory modalities of the predator are of paramount importance. Capture, immobilization, and ingestion involve the predator's trophic morphology and handling behavior. Processing food is dependent on physiological processes related in part to the length and surface area of the gut. In subsequent sections, we will review the empirical data on PPIs and then explore the importance of relative body size for different subsections of the PPI sequence.

In this paper, we focus primarily on the capture, immobilization, and ingestion events in a PPI, for three reasons. Those later stages of a PPI are more likely to involve high risk, energy expenditure, and prey-specific aspects of the predator and, therefore, show relationships with measurable variables (see Endler, 1991). Aspects of size obviously can play a role in detection and pursuit but are dealt with in more detail elsewhere (see Bakker, 1983; Endler, 1991). Finally, to date, much attention in ecomorphology has focused on trophic morphology, usually of the head, and thus, there are more data for consideration dealing with that topic.

The end results of a successful PPI from a predator's standpoint are the ingestion of the prey and the assimilation of nutrients. Clearly, inclusion of a prey item in the diet reflects a number of factors, including traits intrinsic to predator and prey, relative and absolute abundances of potential prey, and the extent to which a predator is selective (e.g., Bence and Murdoch, 1986; Osenberg and Mittlebach, 1989; Schmitt and Holbrook, 1984). In this analysis, we focus not on these details for particular predator-prey systems but rather on the general output of the interacting factors: intra- and interspecific scaling of prey body size to predator body size.

SCALING, ALLOMETRY, AND FUNCTIONAL EQUIVALENCE

For many biological variables the equation

$$Y = aM^b$$

is a good mathematical descriptor of how a particular feature, Y, changes with body size, M. For ease of interpretation, this equation can be transformed into a logarithmic version

$$\log Y = \log a + b \log M,$$

which has the form of a straight line with intercept $\log a$ and slope b. It is slope b which is often referred to as the scaling coefficient or exponent.

Allometry can be defined as the study of changes in shape with size. A model of geometric similarity is often used as the null hypothesis against which empirically determined coefficients of scaling, or regression slopes from log-transformed data, are compared. This model is used because it is based on the rules and predictions of Euclidean geometry and entails no a priori biological assumptions (Emerson and Bramble, 1993).

The geometric similarity model predicts that as an object scales up, its linear dimensions are increased by n, area dimensions by n^2, and volume (or mass) by n^3. These relationships define isometry, or maintenance of shape, with increases in size. Allometry, or change in shape with size, occurs when the scaling exponent is greater (positive allometry) or less (negative allometry) than that predicted on the basis of geometric similarity (Gould, 1966). The scaling expectations for some important trophic variables when maintaining geometric similarity are as follows: eye orbit area and toothrow surface area will scale as $(\text{mass})^{2/3}$; muscle force, which is a function of the cross-sectional area of a muscle, should scale as $(\text{mass})^{2/3}$; and linear measurements such as jaw length, jaw muscle moment arm length, and toothrow length should all scale as $(\text{mass})^{1/3}$.

The predicted geometric relationships indicate that there can be important functional consequences of maintaining shape with increasing or decreasing size. For example, larger animals will have relatively less muscle force, relatively smaller eyes, and relatively less tooth surface area than geometrically similar smaller animals; while shape is being maintained with increases in size, function is not. Functional equivalence in different-sized organisms occurs only when performance capability is maintained with changes in size. Geometrically similar organisms are not necessarily functional equivalents.

The maximum prey body size of an organism is a measure of function, specifically, feeding performance (*sensu* Wainwright, chap. 3). The allometry of predator-prey body sizes will, therefore, be examined from the perspective of scaling in relation to the maintenance of functional equivalence.

Predator-prey body size data generally come from cross-sectional samples. They are obtained by examining the stomach contents of a predator at a single point in time. Food items are retained in the stomach of an organism for some finite time period depending on the digestive rate of the particular type of predator. For this reason, prey body size can be conceptualized as a type of ingestion rate function. A plot of maximum prey body size to predator body size is, in actuality, a measure of maximum obtained prey body size over some unit of time in relation to body size of the predator.

Generally, ingestion rates scale to body size with slopes close to 0.75 (Peters, 1985; Schmidt-Nielsen, 1984). Therefore, functional equivalence in a predator-prey relationship might be maintained over increases in predator body size if the slope between prey body size and predator body size were also 0.75. This value will be used throughout the chapter to define functional equivalence.

PREDICTIONS FOR THE SCALING OF PREDATOR-PREY INTERACTIONS

Balance Approach

It seems intuitively reasonable that larger predators might eat bigger prey—after all, larger animals require more energy. It is less obvious what the slope of the relationship between maximum prey body size and predator body size might be. However, functional response theory from ecology (Holling, 1966) provides a general equation from which a prediction can be derived if we rewrite the variables as a function of predator body size and assume that all perceived prey are captured.

Capture rate of $prey_i$ (in mass) per predator $= V \cdot D_i \cdot W_i \cdot P_i$

where

V = velocity of the predator
D_i = density of the prey
W_i = mass of prey
P_i = perception distance of prey

and therefore,

$$\text{Total capture rate (TCR)} = \int_{W_{min}}^{W_{max}} V \cdot D_i \cdot W_i \cdot P_i dW.$$

Let

M = predator mass

and assume

$W_{min} \cong 0$
Predator energy needs $\propto M^{.75} \propto \text{TCR}$
$V \propto M^{.25}$ (Peters, 1983)
$D_i \propto 1/W_i$ (LaBarbera, 1989)
$P_i \propto M^\delta$.

$$\text{TCR} \propto M^{.75} \propto \int_0^{W_{max}} M^{.25} \cdot (1/W_i) \cdot W_i \cdot M^{\delta}dW$$

$$M^{.75} \propto \int_0^{W_{max}} M^{.25+\delta}dW$$

$$M^{.75-.25-\delta} \propto W_{max}$$
$$W_{max} \propto M^{.5-\delta}$$

Thus, the model predicts that the maximum mass of the prey (W_{max}) will scale to predator mass with a slope of 0.5 minus the slope of the relationship between the perception distance of prey and predator mass.

This represents the simplest possible model (we balance the predator's energy needs simply by increasing the range of prey it attacks), but nonetheless, many of its general aspects are taken from data in the literature. Generally, vertebrates have velocities that scale to body mass with slopes of less than 0.35 (Garland, 1985; Peters, 1983) so the use of the 0.25 exponent is not unreasonable. Similarly, density generally scales to mass^{-1} for vertebrates (LaBarbera, 1989).

The predicted slope from the model ($M^{.5-\delta}$) will be considerably lower than the $0.70 - 1.20$ range actually observed in prey-predator body size relationships, unless the slope of the relationship between perception distance and predator body size is quite small or negative. Reaction distance (P_i in the model) of predators to prey has been examined in fish (e.g., Breck and Gitter, 1983; Wright and O'Brien, 1984), and in the one study where predator body size was varied, reaction distance scaled to predator body size with a slope of 0.48 (Breck and Gitter, 1983). Substituting this value into the equation above gives a predicted slope of 0.02 between maximum prey body size and predator body size—clearly much lower than actually observed.

One unrealistic assumption of this simple model is that all perceived prey are captured. A more complicated version might add a probability of capture term dependent on handling time as well as the respective weights of the prey and of the predator (e.g., probability of capture $= e^{-H \cdot W_i/M_i}$). The main point of this exercise is to show how we might modify an energy balance equation from ecology to make a priori predictions for the scaling relationship of maximum prey body size to predator body size for a wide diversity of biological systems.

Performance Limitation Approach

In the last section we generated a predicted scaling relationship for maximum prey body size to predator body size. Here we will take as a given that the scaling relationship is one that maintains functional equivalence (i.e., slope = 0.75) and, instead, derive predictions for how (1) predator trophic morphology should scale with predator body size and (2) maximum prey body size should scale with predator trophic morphology for particular ecological situations.

It is common for morphological features of related function to be phenotypically correlated (e.g., Lande, 1979). Additionally, many aspects of morphology are correlated with body size. Correlation is, therefore, not the best assay for determining whether a particular trophic structure is directly limiting prey size. For example, bill shape in birds is often correlated with seed size. From this, it has been assumed that bill shape limits ingested seed size. In fact, it may be the correlated muscle size (and generated force of the muscle) that is actually limiting the size of seed that can be cracked and, therefore, eaten (Abbott et al., 1975). An allometric model allows for direct testing of which aspect of trophic morphology is limiting maximum prey body size because it makes specific predictions of what the scaling relationship must be between predator trophic morphology and body size for the trophic structure to be limiting.

Let the trophic morphology of a predator scale to some measure of predator body size with the following general relationship:

Predator trophic morphology $\propto$ predator body size ^{x_1}.

Similarly, maximum prey body size scales to predator trophic morphology as

Maximum prey body size $\propto$ predator trophic morphology ^{x_2}.

By substitution, therefore

Maximum prey body size $\propto$ (predator body size ^{x_1}) ^{x_2}.

Using this allometric model, assuming selection for maintenance of functional equivalence with increases in predator body size, and considering body size as mass, we present two examples that illustrate the use of this approach to test two common hypotheses in the ecological literature on PPIs.

Example 1: Suppose that maximum prey body size is determined by predator body size rather than size-specific trophic specialization, and functional equivalence is maintained over a large size range. This is the situation implicitly assumed by ecological studies where predator trophic structure and predator body size are used interchangeably (e.g., Schoener, 1968), and species in a community are thought to rely on body size alone for food niche separation (e.g., Brown and Maurer, 1989; Wilson, 1975). This includes the extensive literature on Hutchinson ratios (see Simberloff, 1983). Unfortunately, few of these studies have tested for possible allometry (i.e., changes of shape with size) between predator trophic structure and body size before calculating ratios. Allometry could explain some of the contradictory results (Hespenheide, 1973).

For maximum prey body size to be strictly a function of predator body size, there must be no size-specific trophic specialization or change of shape in predator trophic morphology with increases in predator body size. Predator trophic morphology must scale with geometric similarity to predator mass. For a linear trophic variable, $X_1 = 0.33$, $X_1X_2 = 0.75$, and therefore $X_2 = 2.27$. An area dimension trophic morphology would produce a slope for X_2 of 1.1 between prey mass and predator trophic morphology.

Example 2: Suppose that trophic morphology directly limits maximum prey body size. This is one variant of the idea that a correlation between prey body size and predator trophic morphology indicates that the trophic structure itself is causally limiting the body size of the prey (e.g., Boag and Grant, 1981; Janson, 1983). From this assumption of causality from correlation, particular morphological differences in predator trophic structure also sometimes are used as niche descriptors in multivariate analyses of community structure (e.g., Ricklefs and Travis, 1980; see also chap. 2, this volume, for more on inferring ecology from morphology).

Most morphological aspects of the trophic apparatus of vertebrate predators have linear or area dimensions. Jaw length, jaw width, mouth gape, and force of the adductor musculature have all been shown to be critical aspects of the feeding morphology of vertebrates (Emerson, 1985; Rieppel and Labhardt, 1979; Schoener, 1974; Toft, 1980; Wainwright, 1987, 1988; Wheelwright, 1985). If a linear aspect of predator trophic morphology is the only factor limiting maximum prey body size, prey length or diameter should scale to jaw length, mouth width, or gape with a slope of 1.0 (X_2). In this case, the slope of the predator-prey body size relationship (X_1X_2) must be the same as the slope of the relationship between predator trophic structure and predator body size (X_1).

In an interesting application of this "allometric" approach, Wainwright (1987) used expected scaling relationships for linear and area trophic variables to predator mass and live animal performance tests to examine whether a mollusc-eating hogfish was gape or force limited in its prey diameter. Animals were found to be force limited in the laboratory trials. These findings were extended to the natural situation by comparing the prey diameter–predator mass relationship found in the laboratory to the maximum-diameter snails found in the stomachs of a size series of field-collected fish.

Foraging Decision/Trophic Morphology Approach

The mollusc-eating fish represents a system where some aspect of predator trophic morphology is directly limiting prey body size. Often prey body size tracks predator trophic morphology but is not directly limited by it. For example, among species in an anuran forest litter community, mouth width was significantly correlated with prey length, and ant specialists had smaller mouths than generalists and non-ant specialists (Toft, 1980). An interspecific plot of mouth width, prey length, and predator length for the six ant specialist species shows that the slopes of prey length and predator mouth width to predator length are very similar (fig. 6.1). However, the prey are absolutely much smaller than the predator mouth width, indicating that mouth width or gape is not directly limiting the size of the prey.

A classic diet-breadth model from ecological foraging theory (Charnov, 1976; Stephens and Krebs, 1986) may provide insight into the tracking of predator trophic morphology by prey body size and why prey-predator body size curves show a limited range of slopes. From optimal foraging theory (Charnov, 1976), when net energy intake is to be maximized by a predator, prey types can be ranked in desirability by their ratio of handling time to energy content (or prey body size). Theoretically, the optimal prey set for a predator includes those prey types below some threshold value of handling time to prey body size. From the optimal diet model we know that the threshold ratio of handling time to prey body size will be equal to one divided by the inverse of the rate of food intake for the predator of interest. As noted earlier, this rate scales with predator mass to the 0.75 power (Peters, 1983). As prey body size increases, at some point handling time steeply escalates (Andrews et al., 1987; Dickman, 1988; Kislalioglu and Gibson, 1976; Pastorok, 1981; Sherry and McDade, 1982; Werner, 1974).

Figure 6.2 shows a hypothetical curve of handling time (H) as a function of prey body size. The dotted line is the ingestion rate^{-1} (or handling time/prey body size = constant) for the model predator. For a predator of a fixed size and with the illustrated prey body size versus handling time curve, the threshold (the

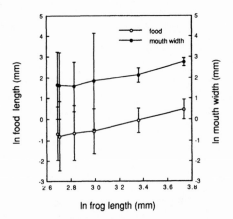

FIGURE 6.1 Natural logarithms of mean food length and mean mouth width in relationship to mean frog length for six species of ant-eating frogs (data from Toft, 1980). Vertical lines represent 95% confidence intervals.

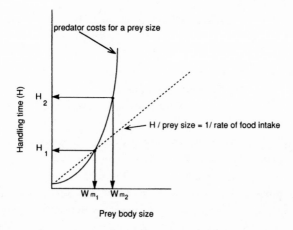

FIGURE 6.2 The theoretical relationship between handling time (H) and prey body size (W_m) for a given prey type (H/prey body size). Predator costs for a prey body size are shown as a sharply rising power function. Dotted line represents 1/rate of food intake.

largest prey eaten) is where the ingestion rate line crosses the handling time curve. To incorporate trophic morphology in relation to handling time curves and prey selection, we consider the following situation.

W_{m_2} represents near the largest prey body size a small predator can take, based on the physical limitations of its trophic morphology (fig. 6.2). But this prey body size has a very high handling time (H_2). W_{m_1}, lying on the same curve, has a slightly smaller size than W_{m_2} but takes much less handling time (H_1). For this reason, the predator chooses a smaller prey (W_{m_1}) than the absolutely largest prey (W_{m_2}) it can physically eat. The selection of the smaller prey (W_{m_1}) is still influenced by the trophic morphology of the predator but not directly limited by it.

If the handling time curve and the predator ingestion rate both change with predator size (see fig. 6.3 below), the argument may be used to generate a relationship between predator size and maximum prey body size. This will be illustrated later in the paper.

HANDLING COSTS AND THE SLOPE OF PREY-PREDATOR BODY SIZE RELATIONSHIPS

The rise of handling time with increased prey body size suggests that the shape of the handling time curve could generally set the upper boundary for the slopes of maximum prey-predator body size relationships. As mentioned earlier, the slopes of intra- and interspecific prey body size to predator body size relationships are quite consistent across vertebrates, hovering close to 1.0 (Peters, 1983; Toft, 1980; Vezina, 1985; Voris and Moffett, 1981). Published handling time curves do not depict the scaling of prey-predator body size relationships directly (i.e., what the slope of the curve of maximum prey body size to predator body size would be on a log-log graph) with regard to handling time because in most instances only prey or predator body size is changing (e.g., Sherry and McDade, 1982). Nonetheless, the general consequences of allometric versus isometric scaling of prey body size to predator body size relationships in relation to handling time can still be inferred from these data. If prey body size were scaling isometrically to predator body size with a slope of 1.0, the prey/predator body size ratio would remain the same, and handling time would be constant despite increases in prey and predator body sizes. The rapidly increasing handling time with a larger prey/predator body size ratio is analogous to what would happen to handling time if there were a strong positive allometry in the prey-predator body size relationship. The shape of the handling time curves also suggests it is to a prey's advantage to reach the point where handling costs sharply ascend as quickly as possible. Any prey feature that might increase handling time of the predator (e.g., shape, spines, toxicity) should show strong positive allometry.

TRACKING OF PREY-PREDATOR AND TROPHIC MORPHOLOGY-PREDATOR SLOPES

Figure 6.3 shows predator cost and handling time versus prey body size for both a small and a large predator. In this case, the curves have been given the same form. Both predators are shown taking the largest threshold prey (W_m) under the diet-breadth model, but this prey is proportionally smaller than that possi-

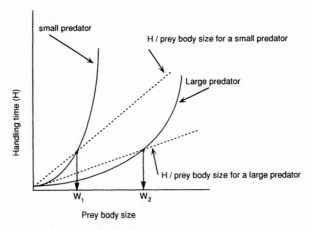

FIGURE 6.3 The theoretical relationship between handling time (H) and prey body size (W_m) for a small and large predator with similar shaped cost functions. Dotted lines represent 1/rate of food intake.

ble based on the physical limitations of the trophic morphology (see fig. 6.2). From this situation we can derive a predicted scaling relationship for maximum prey body size (W_m) to predator body size (M).

The threshold (H_m/W_m) is equal to 1/mean rate of food intake (R) and, in equilibrium (Charnov, 1976)

$$R = W_m/H_m \text{ and } R \propto M^{.75} \tag{eq. 6.1}$$

Suppose the handling curve can be described as a power function where

$$H = AW^\lambda$$

where $\lambda > 1$ gives the handling time curve its characteristic faster acceleration with larger prey sizes.

Now let

$$A \propto M^{-C}$$

to make handling time less for a given prey body size as predator body size increases. C describes the lateral displacement of the curve as a function of predator body size. We thus have, by substitution

$$H_m \propto M^{-C} W_m^\lambda$$

and from equation 6.1

$$M^{.75} \propto \frac{W_m}{M^{-C}W_m^\lambda}$$

$$M^{.75-C} \propto W_m^{1-\lambda}$$
$$W_m \propto M^{(.75-C)/(1-\lambda)} \text{ with } \lambda > 1 \text{ and } C > 0.75$$

Thus under optimal foraging theory (Charnov, 1976) and the model developed from it, maximum prey body size (W_m) will scale with predator body size (M) to the $(.75 - C)/(1-\lambda)$ power. This derivation suggests a possible link between the handling cost curves of ecologists (e.g., Werner, 1974) and the trophic morphology–predator body size relationships studied by functional morphologists (e.g., Emerson, 1985).

Interspecific comparisons of handling cost functions and prey/predator body size ratios show two general patterns. In some cases the curves for different species have the same form (e.g., Werner, 1974). In other situations the shapes of the curves are quite different (e.g., Andrews et al., 1987). We suggest that similar-shaped cost function curves between species indicate that trophic morphology has a similar shape or is scaling to predator body size similarly between the two species. Different shaped curves reflect nonidentical scaling relationships between trophic morphology and predator body size. That is, there is a unique trophic structure/prey body size interaction for each species when the curves differ in shape. Extrapolating this relationship to prey-predator body size curves, we predict that species with similar-shaped cost function curves and trophic morphology/predator body size slopes will have parallel slopes between prey body size and predator body size. Conversely, the slopes of the prey-predator body size curves should vary between species with different cost function curves and trophic morphology/predator body size relationships.

In most interspecific studies, trophic morphology has varied between predator species, but prey type has been kept constant (e.g., Andrews et al., 1987). Theoretically, prey differ in handling time to a predator depending not only on the trophic morphology of the particular predator and body size of its prey but also on prey characteristics such as shape, hardness, toxicity, and stickiness. Given the relationship between handling time curves and the scaling of prey-predator body size described above, intraspecific comparisons of prey body size to predator body size for a series of known prey items could (1) identify which prey items constitute a particular prey type for a predator, and, (2) predict the degree of similarity in cost functions for different prey types. The diet of a voiceless Bornean frog provides a possible intraspecific example.

Rana blythi is a large rain forest stream frog reaching more than 125 mm snout-vent length. The species has a broad diet (Inger and Greenberg, 1966), eating prey that range in body length over two orders of magnitude, from ants to vertebrates. Plotting the log of the maximum body length of prey found in each frog as a function of the log of the body length of the frogs (fig. 6.4) yields a weak

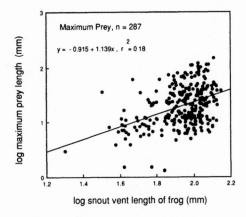

FIGURE 6.4 Log maximum prey length graphed in relationship to log snout vent length for *Rana blythi*.

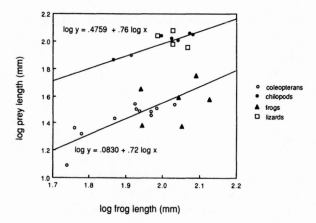

FIGURE 6.5 Log maximum prey length for four prey taxa graphed in relationship to log snout vent length for *Rana blythi*. Upper regression equation is for long, skinny prey (chilopods and lizards), and lower regression equation is for short, fat prey (coleopterans and frogs).

but significant correlation between prey body length and predator body length (r^2 $= 0.18, P < .001$). The slope of the least squares regression is 1.13 (± 0.29, 95% C.I.). *Rana blythi* shows a typical pattern of retaining small prey items while concurrently increasing maximum prey body length with increases in predator body length. This same pattern is repeated if one examines the PPI for different taxonomic groups of prey separately. The retention of small prey in the diet results in a large variance in maximum prey body length for a given predator body length (fig. 6.4), and the data in this form do not offer much insight into the PPIs of this species.

Figure 6.5 plots prey length to predator body length for four of the major food items for *Rana blythi:* frogs, centipedes, lizards, and beetles. To the extent that prey taxa lying along the same curve represent a single prey type to the predator, centipedes and lizards constitute a long, slender prey type; and coleopterans and frogs represent a short, fat prey type. If parallel slopes of prey body size to predator body size among prey types indicate that the two prey types have similar-shaped cost function curves (see argument above), long, slender prey and short, fat prey should have similarly-shaped handling time to prey/predator body size ratio curves. This prediction is somewhat counterintuitive and points to the need for observations and data on actual PPIs. The fact that taxonomically unrelated chilopods and lizards have similar body length relationships with predator body size presumably reflects the shape-mass relationships of these elongated prey and the general widespread correlation between shape and mass of prey and handling costs to the predator (Greene, 1983).

While it appears that steeply ascending handling time curves constrain the upper boundary to maximum prey body size for most predators, there are some instances where these constraints have been shifted. Viperine snakes offer an excellent example of a group where morphological, behavioral, and physiological modifications have changed the cost function, and, consequently, relative prey body size (Greene, 1983, 1992, unpubl.; Pough and Groves, 1983). In staged feeding trials, the number of maxillary protractions (and presumably handling time) is less for any prey/predator body size ratio in viperine versus nonviperine snakes, and the slopes of that relationship are different as well (Pough and Groves, 1983). In the feeding trials, these differences translated to nonvipers failing to eat prey that fell between 11 and 29% of their body mass while vipers failed to eat prey that were between 40 and 45% of their body mass. The largest prey taken by a nonviper was 18.4% of the snake's mass, while that of a viper was 36.4% (Pough and Groves, 1983). Stomach contents of field-caught snakes support the results of the feeding trials. Data from 262 individuals of eighty-two species in the families Aniliidae, Colubridae, Elapidae, and Viperidae (Greene, unpubl.) show that the largest prey taken by a highly venomous snake (Elapidae, Viperidae) was 157% of the snake's mass (Greene, 1992). In contrast, the largest prey taken by a nonvenomous snake (Aniliidae) was 82% of the snake's mass.

Context might play an important role in the allometry of PPIs. Data from the staged feeding trials show a linear relationship between number of maxillary protractions and relative prey-predator size, and Pough and Groves (1983) concluded that the snakes do not have disproportionate costs associated with eating larger prey. This is quite different from other species where there are disproportionately higher handling times for relatively larger prey. One reason for the discrepancy may be that the captive snakes were not eating large enough prey relative to their own body size to be in the sharply ascending part of the handling time curve. The snakes used in the feeding trials rejected prey of a smaller size than what was found in the stomachs of wild-caught individuals. While venomous snakes rejected prey that were between 40 and 45% of their body mass in feeding trials (Pough and Groves, 1983), field-caught venomous snakes (6/155 individuals) occasionally contained prey items that were equal to or greater than their own mass (Greene, unpubl.). Under natural conditions a viper bites from ambushes and releases bitten rodents, which then travel up to several meters before they are incapacitated by the venom. Under confined conditions, vipers cannot effectively ambush large prey, and even relatively small rodents sometimes elicit defensive behavior rather than post-strike searching from snakes as the bitten rodent jumps about a cage (Greene, pers. obs.). The possibility that very large prey are disproportionately costly to handle should be reexamined with vipers feeding in a more naturalistic context.

PROSPECTS AND CONCLUSIONS

We have presented three approaches toward better understanding the effect of body size on predator-prey interactions. Progress in explaining the geometry of the relationship of prey body size to predator body size can be served in several ways.

1. Scaling models from ecology and functional morphology can be used to generate a priori predictions that allow testing of ecomorphological hypotheses on the causal links among prey body size, predator body size, and trophic morphology.

2. We have emphasized the predator's half of PPIs because of the greater availability of quantitative examples. The scaling of prey defensive responses in relationship to predator body size has not been well studied and might explain, for example, why bluffs work against hungry, capable predators (Greene, 1988). Allometric approaches could provide significant insights on such problems by delineating patterns of ontogeny and magnitude for deceptive attributes across a variety of taxa.

3. With the availability of robust phylogenies it will be possible to explore the evolutionary history of shifts in allometric aspects of PPIs (see Brooks and McLennan, 1991, for overview).

4. Some of the most exciting questions to be explored in this area are currently unanswerable for lack of appropriate quantitative data. Students of diet and feeding behavior should seek to record functionally relevant body size variables. As the data on vipers illustrate, field measurements of performance in relation to size variables are particularly needed because they delimit the range of realistic conditions for laboratory studies.

ACKNOWLEDGMENTS

We appreciate support from the National Science Foundation (BSR-8822630 to SBE, BSR-8300346 to HWG). Discussions with J. Seger, D. M. Bramble, M. S. Foster, and S. J. Arnold were extremely helpful. R. F. Inger provided the stomach content data on *Rana blythi*. F. H. Pough let us reexamine his snake data. Comments by C. Osenberg, P. Motta, and an anonymous reviewer considerably improved this contribution.

REFERENCES

Abbott, I., L. K. Abbott, and P. R. Grant. 1975. Seed selection and handling ability of four species of Darwin's finches. *Condor* 77:332–335.

Andrews, R. M., F. H. Pough, A. Collazo, and A. de Queiroz 1987. The ecological cost of morphological specialization: Feeding by a fossorial lizard. *Oecologia* (Berlin) 73:139–145.

Bakker, R. T. 1983. The deer flees, the wolf pursues: Incongruencies in predator-prey coevolution. In *Coevolution*, ed. D. J. Futuyma and M. Slatkin, 350–382. Sunderland, MA: Sinauer Associates.

Bence, J. R., and W. W. Murdoch. 1986. Prey size selection by the mosquito fish: Relation to optimal diet theory. *Ecology* 67:324–336.

Boag, P. T., and P. R. Grant. 1981. Intense natural selection in a population of Darwin's finches (Geospizinae) in the Galapagos. *Science* 214:82–85.

Breck, J., and M. Gitter. 1983. Effect of fish size on the reactive distance of bluegill (*Lepomis macrochirus*) sunfish. Can. J. Fish. Aq. Sci. 40:162–167.

Brooks, D. R., and D. A. McLennan. 1991. *Phylogeny, Ecology, and Behavior: A Research Program in Comparative Biology.* Chicago: University of Chicago Press.

Brown, J. H., and B. A. Maurer. 1989. Macroecology: The division of food and space among species on continents. *Science* 243:1145–1150.

Charnov, E. L. 1976. Optimal foraging: Attack strategy of a mantid. *Amer. Nat.* 110:141–151.

Dickman, C. R. 1988. Body size, prey size, and community structure in insectivorous mammals. *Ecology* 69:569–580.

Emerson, S. B. 1985. Skull shape in frogs: Correlations with diet. *Herpetologica* 41:177–188.

Emerson, S. B., and D. M. Bramble. 1993. The vertebrate skull. In *Scaling, Allometry and Skull Design,* ed. J. Hanken and B. Hall, 384–421. Chicago: University of Chicago Press.

Endler, J. A. 1986. Defense against predators. In *Predator-prey Relationships: Perspectives and Approaches from the Study of Lower Vertebrates,* ed. M. E. Feder and G. V. Lauder, 109–134. Chicago: University of Chicago Press.

Endler, J. A. 1991. Interactions between predators and prey. In *Behavioural Ecology: An Evolutionary Approach,* ed. J. R. Krebs and N. B. Davies, 169–196. Oxford: Blackwell Scientific.

Garland, T. 1985. Ontogenetic and individual variation in size, shape and speed in the Australian agamid lizard *Amphibolurus nuchalis. J. Zool.* (London) A207:425–439.

Gould, S. J. 1966. Allometry and size in ontogeny and phylogeny. *Biol. Rev.* 41:587–640.

Greene, H. W. 1983. Dietary correlates of the origin and radiation of snakes. *Amer. Zool.* 23:431–441.

Greene, H. W. 1988. Antipredator mechanisms in reptiles. In *Biology of the Reptilia,* vol. 16, *Ecology B: Defense and Life History,* ed. C. Gans and R. B. Huey, 1–152. New York: Alan R. Liss.

Greene, H. W. 1992. The ecological and behavioral context for pitviper evolution. In *Biology of the Pitvipers,* ed. J. A. Campbell and E. D. Brodie, Jr., 107–117. Tyler, TX: Selva.

Hespenheide, H. A. 1971. Food preference and the extent of overlap in some insectivorous birds, with special reference to the Tyrannidae. *Ibis* 113:59–72.

Hespenheide, H. A. 1973. Ecological inferences from morphological data. *Ann. Rev. Ecol. Syst.* 4:213–229.

Holling, C. S. 1966. The functional response of invertebrate predators to prey density. *Mem. Entomol. Soc. Can.* 47:3–86.

Inger, R. F., and B. Greenberg. 1966. Ecological and competitive relations among three species of frogs (genus *Rana*). *Ecology* 47:746–759.

Inger, R. F., and H. Marx. 1961. The food of amphibians. In *Exploration du Parc National de l'Upemba,* Mission G. F. de Witte, Brussels: Imprimerie Haye.

Janson, C. H. 1983. Adaptation of fruit morphology to dispersal agents in a neotropical forest. *Science* 219:187–189.

Kislalioglu, M., and R. N. Gibson. 1976. Prey "handling time" and its importance in food selection by the 15-spined stickleback, *Spinachia spinachia* (L.) *J. Exp. Mar. Biol. Ecol.* 25:151–158.

LaBarbera, M. 1989. Analyzing body size as a factor in ecology and evolution. *Ann. Rev. Ecol. Syst.* 20:91–117.

Lande, R. 1979. Quantitative genetic analysis of multivariate evolution applied to brain:body size allometry. *Evolution* 33:402–416.

Osenberg, C., and G. G. Mittlebach. 1989. Effects of body size on the predator-prey interaction between pumpkinseed sunfish and gastropods. *Ecol. Mon.* 59:405–432.

Pastorok, R. A. 1981. Prey vulnerability and size selection by *Chaoborus* larvae. *Ecology* 62:1311–1324.

Peters, R. H. 1983. *The Ecological Implications of Body Size.* Cambridge: Cambridge University Press.

Pough, F. H., and J. D. Groves. 1983. Specializations of the body form and food habits in snakes. *Amer. Zool.* 23:443–454.

Rieppel, O., and L. Labhardt. 1979. Mandibular mechanics in *Varanus niloticus* (Reptilia: Lacertilia). *Herpetologica* 35:158–163.

Ricklefs, R. E., and J. Travis. 1980. A morphological approach to the study of avian community organization. *Auk* 97:321–338.

Schmidt-Nielsen, K. 1984. *Scaling: Why Is Animal Size so Important?* New York: Cambridge University Press.

Schmitt, R. S., and S. J. Holbrook. 1984. Gape-limitation, foraging tactics and prey size selectivity in two microcarnivorous species of fish. *Oecologia* 63:6–12.

Schoener, T. W. 1968. The *Anolis* lizards of Bimini: Resource partitioning in a complex fauna. *Ecology* 49:704–726.

Schoener, T. W. 1974. Resource partitioning in ecological communities. *Science* 185:27–39.

Sherry, T. W., and L. A. McDade. 1982. Prey selection and handling in two neotropical hover-gleaning birds. *Ecology* 63:1016–1028.

Simberloff, D. 1983. Sizes of coexisting species. In *Coevolution*, ed. D. J. Futuyuma and M. Slatkin, 404–430. Sunderland, MA: Sinauer Associates.

Stephens, D. W., and J. R. Krebs. 1986. *Foraging Theory.* Princeton: Princeton University Press.

Toft, C. A. 1980. Feeding ecology of thirteen syntopic species of anurans in a seasonal tropical environment. *Oecologia* (Berlin) 45:131–141.

Vezina, A. F. 1985. Empirical relationships between predator and prey size among terrestrial vertebrate predators. *Oecologia* (Berlin) 67:555–565.

Voris, H. K., and M. W. Moffett. 1981. Size and proportion relationship between the beaked sea snake and its prey. *Biotropica* 13:15–19.

Wainwright, P. C. 1987. Biomechanical limits to ecological performance: Mollusc crushing by the Caribbean hogfish, *Lachnolaimus maximus* (Labridae). *J. Zool.* (London) 213:283–297.

Wainwright, P. C. 1988. Morphology and ecology: Functional basis of feeding constraints in Caribbean labrid fishes. *Ecology* 69:635–645.

Werner, E. E. 1974. The fish size, prey size, handling time relation in several sunfishes and some implications. *J. Fish. Res. Board Can.* 31:1531–1536.

Wheelwright, N. T. 1985. Fruit size, gape-width, and the diets of fruit-eating birds. *Ecology* 66:808–819.

Wilson, D. S. 1975. The adequacy of body size as a niche difference. *Amer. Nat.* 109:769–784.

Wright, D. I., and W. J. O'Brien. 1984. The development and field test of a tactical model of the planktivorous feeding of white crappie (*Pomoxis annularis*). *Ecol. Mon.* 54:65–98.

7

Ecomorphological Analysis of Fossil Vertebrates and Their Paleocommunities

Blaire Van Valkenburgh

INTRODUCTION

Studies of ecomorphology form a cornerstone of paleontology. Fossils do not exhibit behavior, physiology, or ecology, but they do display morphology and it is this which must be interpreted when reconstructing the life of extinct organisms. The advantages of studying questions of form and function with fossils are evident. First, the fossil record documents the history of morphological adaptation. Second, the record frequently yields multiple examples of the evolution of a particular feature or suite of features under similar environmental conditions. Such examples of convergent evolution, if free from phylogenetic bias, are some of the strongest possible evidence for the adaptive nature of form (Fisher, 1985). Third, the history of extinct organisms expands our awareness of the possibilities of form; among the animals we might never have thought possible are dinosaurs, armored glyptodonts, the many sabertoothed catlike species, the chalicotheres with horselike heads and clawed limbs, and shovel-tusked gomphotheres. Although it is tempting to explain the limited range of body sizes and morphologies seen among present-day species as due to physiological and biomechanical constraints, the fossil record often reveals surprising and significant exceptions to present-day patterns.

In this chapter, I hope to give an overview of the goals, methods, and contributions of ecomorphological studies of extinct vertebrates. Although the examples and specifics will be from published investigations of vertebrates (mostly Cenozoic mammals), there are numerous similar studies of invertebrates which utilize analogous concepts and approaches. The emphasis on mammals is not simply a reflection of my own bias; fossil mammals lend themselves much more easily to ecomorphological analysis than do reptiles, amphibians, birds, and to a lesser extent fishes. Cenozoic mammals have a much more complete fossil record than the other tetrapods because of their relative recency and large body size. In addition and more importantly, our ability to infer function in extinct organisms depends heavily on comparisons with living analogs. In general, fossil mammals differ less from their living representatives than do many other vertebrates, such as dinosaurs and mammal-like reptiles. The analogies are closer, therefore, and functional inferences can be made with greater confidence. The relatively complex shapes of mammalian teeth and their close relationship with diet also allows for more refined interpretations of behavior in extinct mammals. Consequently, many more ecomorphological studies have been undertaken on mammals than other vertebrates.

Paleontologists have tended to use morphology to answer paleoecological questions concerning the association between particular morphologies and environments (e.g., hypsodonty [high-crowned teeth] and open grasslands), the possible causes of diversifications or declines within clades, and community evolution, structure, and stability. Often these are questions of long-term trends or effects that can only be answered with the fossil record. For example, is there a predictable relationship between speciation rate and some ecomorphology? Do herbivores speciate more readily than carnivores? Do small species diversify more rapidly than large species? Does diversification usually proceed from a relatively generalized ancestor to more specialized descendants? What is the impact of mass extinction on paleocommunities and their component species? How long do particular suites of coexisting ecomorphs persist as stable paleocommunities? What triggers their demise and how does the replacement paleocommunity develop? Are some ecomorphs more likely to persist despite severe environmental perturbations than others? All of these are problems that can best be studied with fossil data. Of course, the fossil record as a data base has numerous shortcomings, such as uneven and biased sampling, incomplete material, and uncertain temporal control, among others. The key to success in paleontological studies is asking an appropriate question that is answerable given the available data. Thus, a good understanding of the strengths and weaknesses of the fossil data base for the groups or communities under study is essential.

LIMITATIONS OF FOSSIL DATA

The primary limitations facing paleontologists concern sampling biases at all levels, from individual organisms to communities. Soft tissue anatomy is rarely preserved and some communities and organisms are much less likely to be represented than others. Due to their tough endoskeleton, vertebrates have been preserved much more frequently than terrestrial invertebrates (e.g., arthropods), and this makes studying potential interactions between these two groups over evolutionary time extremely difficult. Most terrestrial fossil deposits are the result of sediments laid down in basins or valleys in relatively arid environments that favor slow decomposition of carcasses; consequently, both upland and warm, humid environments are rarely preserved. It is often difficult to compare animals from a variety of environments or a large geographic area at the same point in time because of the vagaries of deposition and questions of contemporaneity between sites separated by distance. Although the record is often excellent for studying long-term phenomena among vertebrates, it is less amenable to studies of short-term variations within species or communities as these can be masked in the record due to the effects of fluctuating depositional rates and shifting source environments. Because of these shortcomings, most ecomorphological analyses of fossils concern organisms with hard endo- or exoskeletons that exist in relative abundance in depositional environments such as flood plains, deltas, and lake and ocean basins.

With few exceptions, the morphological record of extinct vertebrates is restricted to the hardest of tissues, bones and teeth. Occasionally, there have been fortuitous circumstances which resulted in the preservation of some soft tissue anatomy, such as the feathers of the ancient bird, *Archaeopteryx*, (Solnhofen limestones; Ostrom, 1976), impressions of the body outlines of Eocene mammals and birds (Grube Messel, Germany; Franzen, 1985), and frozen Pleistocene mammoths and woolly rhinoceros (Anderson, 1984). In addition, traces of extinct animal behavior, such as burrows, trackways, and cut- or bite-marks on bones, can occasionally be used to infer behavior (e.g., Haynes, 1982; Hunt et al., 1983; Lockley, 1986, 1987; Shipman, 1981). However, the bulk of the vertebrate record consists of bones and teeth, and consequently most ecomorphological studies must rely on a thorough understanding of the relationship between skeletal and dental form and function in living species. In the past, the study of dental and skeletal form was almost entirely one of outer shape and size, but this is no longer true. New approaches and technologies, such as the scanning electron microscope and radiography, have been used to examine subsurface and microscopic anatomical features such as trabeculae orientation (Thomason, 1985a, b), bone cortical thickness (Biknevicius and Ruff, 1992a, b; Ruff, 1989), bone histology (Reid, 1984; de Riqlés, 1980), and dental microwear (Teaford, 1988; Van Valkenburgh et al., 1990; Walker et al., 1978).

Sample sizes for most extinct vertebrate species are relatively small and it is difficult to answer questions at the intraspecific level, such as those that concern variability in morphology within species over time (but see Bell et al., 1985, 1987; Bell, 1988, Kurtén, 1958; Rose and Bown, 1986). Skeletons are rarely complete, and there are clear biases as to which elements are most likely to be preserved under differing conditions of death, burial, and exposure. In the last twenty years, there has been a great expansion of research into taphonomy (the study of the processes of burial and preservation of organisms) which has resulted in a substantial literature (for useful reviews see Behrensmeyer and Hill, 1980; Behrensmeyer and Kidwell, 1985; Behrensmeyer and Hook, 1992; Shipman, 1986; Wilson, 1988). Of obvious importance to the morphologist about to embark on a paleontological analysis is an understanding of which skeletal elements are most frequently preserved. For example, although the scapula appears to be an excellent indicator of locomotor function in primates (Oxnard, 1967) and some carnivores (Davis, 1949), it is a poor choice for most fossil studies of extinct mammals as it is rarely preserved in a complete, undistorted condition. Far more likely to be preserved are long bones and teeth.

The lack of complete skeletons creates other potential problems for the ecomorphologist. For example, intermembral limb proportions have proven to be useful indicators of locomotor behavior in a variety of mammals (Howell, 1944; primates, Napier and Walker, 1967, Rollinson and Martin, 1981; ungulates, Scott, 1985; carnivores, Taylor, 1974, 1976, 1989, Van Valkenburgh, 1987). Because the number of complete limbs available for fossil species is always much smaller than the number of isolated elements, the paleontologist may be forced to use average values for each element in calculating intermembral proportions. The accuracy of these ratios of means can be tested by comparing the values obtained from a sample of known, extant individuals with those obtained for the same sample by averaging the size of each element. Walker (1980) did just that for several Madagascan primates and found that the ratios of means were extremely close to the ratios based on individuals, based on samples sizes of eight to sixteen for each element. Thus, he could proceed with confidence in his analysis of fossil prosimian proportions.

Perhaps the overriding concern in working with fossils is the impossibility of testing our inferences of function or causality by observation. We rely on analogy with extant organisms and communities. Thus our conclusions tend to be based on a strong assumption of uniformitarianism, which may be false. For example, the presence of bladelike slicing molars in early felids is assumed to imply a highly carnivorous diet based on the correlation between the feature and the behavior in extant carnivores. Although this appears a reasonable conclusion that can be supported on biomechanical evidence, there are other examples, such as the inference of body mass from skeletal or dental dimensions, that may assume, incorrectly, an equivalency of proportions in ancient and modern species (cf. Damuth and MacFadden, 1990; Van Valkenburgh, 1990). These problems become increasingly acute as one moves further backwards in time and both the organisms and communities deviate more strongly from extant examples. In the reconstruction of communities and guilds, the assumption of uniformitarianism can be stifling; if the pattern observed in the fossil record does not conform to what is expected based on the recent, the fossil example may be discarded as unreasonable (see discussions in Kitchell, 1985; Olson, 1980; Van Couvering, 1980). If this procedure were never violated, there would be no possibility of gaining new insights concerning ecological processes from paleoecological studies, and all of paleoecology would be a matter of "me-too-ecology" (Olson, pers. comm.). Fortunately, the thinking of paleoecologists is not so rigid, and studies are now designed in part to test theory based on modern systems (e.g., Bambach, 1985; Janis, 1982; Kitchell and Carr, 1985; Van Valkenburgh, 1985, 1988; Wolfe and Schorn, 1989). In some of these studies, the unexpected result discovered in the fossil record can be supported through multiple examples. For example, comparisons of trophic structure between an ancient thirty-two-million-year-old paleoguild and a modern analog revealed a remarkable similarity, despite dramatic differences in the taxonomic composition of the two paleoguilds. This apparent stability of structure (array of ecomorphs) within predator paleoguilds was further supported by additional examples from communities widely spaced in time (Van Valkenburgh, 1988). A single observation is of little importance, but multiple, independent ones take on some significance.

THE INFERENCE OF ADAPTATION IN EXTINCT ORGANISMS

There are two common approaches to the problem of inferring function of features of extinct organisms: (1) the paradigm method (Rudwick, 1964) and (2) by analogy with living species. The paradigm method compares the feature in the fossil to one or more models or ideal structures designed to perform the presumed function. The determination of probable function is based on the degree of similarity between the feature and the paradigm. For example, a number of functions ranging from thermoregulation to sexual advertisement have been hypothesized for the enlarged crests on the skulls of some hadrosaurid dinosaurs. Using a paradigm approach, Weishampel (1981) examined the crests as potential acoustic resonators and found that their structure would have functioned well in producing vocalizations. Notably, Weishampel went further and analyzed hadrosaur inner ear structure, demonstrating that the vocal frequencies estimated to have been produced would have been audible to the vocalizer and its conspecifics. Although the paradigm approach has merits, especially when living analogs are not available (see discussion in Fisher, 1985), the alternative method of comparison with living species is simpler and more frequently employed. Excellent discussions of the principles of inference of adaptation in extinct species by analogy with living taxa are provided in Kay and Cartmill (1977) and Kay and Covert (1984) and are summarized here. They present four criteria that must be satisfied for a robust inference of the adaptive nature of a given feature: (1) a living analog exists; (2) all living species with the feature use it similarly; (3) there is some biomechanical justification for the alleged function of the feature; and (4) the first appearance of the feature in the fossil record should not precede evidence of the presumed object of the adaptation. For instance, specializations for arboreality, such as prehensile tails, should not appear in species before there is extensive tree cover.

While I agree in large part with the four criteria, they should not be viewed narrowly or treated as inviolable, with the possible exception of the biomechanical justification. If close living analogies had to exist before one could attempt functional analysis by analogy, then dinosaur behavior and ecology would be practically unworthy of study. Moreover, it would constrain examinations of past adaptations to those for which we have close living analogs or could construct plausible paradigms. Under such conditions, Weishampel (1983, 1984a) might not have discovered that hadrosaurs evolved an entirely different solution to the problem of transverse chewing than any extant vertebrate, nor could Tracy et al. (1986) have studied the thermoregulatory potential of pelycosaur sails. The second criterion, that a feature is used similarly in *all* species, should also not be taken as an absolute. Although this is certainly preferable, the multiple functions of many features and the possibility of co-opting a feature for a new function (without substantial or any modification) make it likely that functions may differ occasionally. For example, a mandibular tooth comb is present in some primates and other mammals and functions in grooming fur in almost all extant species, but also functions in feeding in a few species (Rose and Walker, 1981). This dual function complicated the interpretation of tooth combs in fossil mammals but was resolved through analysis of additional characters, that is, the microscopic wear patterns on the comb; the incisors of species which utilize the comb to groom exhibited distinct grooves that were not present on the incisors of non-grooming species (Rose and Walker, 1981). Thus, based on microwear they established that the oldest mammal (a Paleocene arctocyonid), known to have a tooth comb, utilized it for both grooming and food procurement. The fourth criterion, that the feature not appear prior to the object of its adaptation, depends on correct interpretation of the adaptation. For example, if high-crowned teeth are interpreted as an adaptation to feeding on grass but appears prior to the appearance of grasses in the fossil record, then the functional interpretation is not justified. However, if the function is redefined from the specific one, feeding on grass, to a more general one, feeding on relatively abrasive foods, then it may be reasonable if there is some independent evidence of an increase in the abrasive quality of the food, perhaps due to windblown grit or plants other than grass (cf. Fortelius, 1985).

In addition to the four criteria enumerated above, I would add the caveat that there should be some evidence against the presumed adaptation being a retained ancestral feature of greater phylogenetic than functional significance. For example, the small Brazilian bush dog, *Speothos venaticus*, shares a number of skull and teeth characteristics with the much larger, highly predacious dhole (*Cuon alpinus*) and the African hunting dog (*Lycaon pictus*) (Van Valkenburgh, 1991; Wayne and O'Brien, 1987). Because of its similarity to these larger canids, the bush dog is believed to be very carnivorous, but behavioral data are few. Alternatively, paleontological evidence suggests that *Speothos* descended from larger, wolf-sized taxa (Kurtén and Anderson, 1980), and thus the apparently carnivorous dentition might represent a retention of ancestral features rather than adaptation to present conditions. Without behavioral data it is difficult to distinguish between these two alternatives. Of course, with extinct species behavioral data are not available and the problem of discriminating between phylogenetic retention and adaptation is particularly difficult (see below).

In some cases, the presence of retained ancestral characteristics is a boon to the paleontologist and can be used to infer behavior in an extinct species. For example, Greene and Burghardt (1978) presented a convincing case for the assumption that Paleocene boid snakes subdued their prey by constriction based upon the shared distribution of a particular constriction mode in four living families (Boidae, Aniliidae, Acrochordidae, and Xenopeltidae). These families diverged in the Paleocene and the common constriction pattern is more parsimoniously viewed as a symplesiomorphy (shared primitive character) than as an example of convergence. Thus, in this example and in several others described in Brooks and McLennan (1991), phylogenetic "baggage" assisted in the interpretation of fossil behavior rather than confounding it.

The Ancestry/Function Dilemma

Recently, a number of papers have addressed the general problem of phylogenetic bias in comparative studies (Burghardt and Gittleman, 1990; Harvey and Purvis, 1991; Huey, 1987; Losos and Miles, chap. 4, this volume; Pagel and Harvey, 1988), but a clear solution has not emerged (but see Cheverud et al., 1985; Felsenstein, 1985; Gittleman and Kot, 1990). In many cases, the problem may be partially overcome by establishing a clear biomechanical basis for the inferred function and documenting convergent evolution of the feature in unrelated lineages (cf. Lauder, 1990). For example, all felids have a greatly reduced molar tooth row. Presumably, this represents a response to selection for increased bite force at the canines and a specialized diet of flesh (Emerson and Radinsky, 1980; Ewer, 1973; Van Valkenburgh and Ruff, 1987). However, because all felids share this feature it could represent a retained feature of little or no functional significance. This seems unlikely for two reasons: (1) the tooth row reduction clearly results in an increase in the mechanical advantage of the jaw muscles during a canine bite, and (2) a similar dental reduction appeared in several other carnivorous families, such as the Nimravidae, Viverridae, Mustelidae, and Hyaenidae (Butler, 1946; Van Valkenburgh, 1991). Of course, not all features are so clearly adaptive or exhibit such extensive convergence, and in these cases it may be difficult to estimate the role of ancestry.

Most of the discussions concerning the ancestry/function confusion have emphasized the potential for mistakenly assigning function to a feature that is better explained as a retained, ancestral character of little or no significance (e.g., Gould and Lewontin, 1979; Pagel and Harvey, 1988). The alternative possibility, however, that the functional significance of a feature might be overlooked or underestimated due to retained ancestral characteristics, has received much less attention. Nevertheless, this is a serious problem, especially for extinct species where behavioral data are unavailable. The problem arises because different clades may solve similar functional problems with alternative designs. In many cases, the alternative designs reflect ancestry and the problems of adapting an old structure for a new function. For example, grazing ungulates have more hypsodont (higher-crowned) teeth than browsing ungulates, and this seems readily explained as a consequence of higher rates of dental wear among grazers (Janis, 1986; Stirton, 1947). As Janis (1986) demonstrated, however, grazers of similar diet, such as zebras and wildebeest, may differ significantly in degree of hypsodonty; zebras, like all living equids, are relatively more hypsodont than wildebeest. If these were both extinct species, and hypsodonty were used as an indicator of diet, we might assign the wildebeest to a different diet category than the zebra. These differences in relative hypsodonty may represent retained ancestral characteristics without functional significance or may reflect contrasting requirements of hindgut (equid) as opposed to foregut (bovid) fermentation (Janis, 1986). To minimize the possibility of incorrect behavioral inference, more than one ecomorphological index should be used and careful attention should be paid to trends within clades as well as between them. In the case of hypsodonty, the wildebeest is hypsodont relative to other browsing bovids and the zebra is more hypsodont than extinct browsing equids. Assessment of the relative degree of specialization within their respective clades and the inclusion of additional indicators of diet, such as relative muzzle width, enhance the probability of a correct interpretation of diet.

Goals and Contributions of Paleontological Studies

For the purposes of discussion, ecomorphological studies of extinct species can usually be assigned to one of two overlapping categories, morphological or paleoecological. Morphological studies tend to be focused on questions of adaptation within species or clades, whereas paleoecological studies focus largely on the structure of guilds or communities. The latter, by necessity, incorporate information from the former. Within each category, there are studies that simply use knowledge from present-day organisms to understand the past, while others do this but also use data from the fossil record to better understand the present. The transfer of information is not always, and should not always be, from the present to the past. For example, on the downward side of the figured information flow chart (fig. 7.1), knowledge of the relationship between form and function in living species is used to reconstruct the habits of extinct species. Similarly, the association between particular morphologies and vegetation structures in extant communities (e.g., prehensile-tailed mammals with thick forest) is used to interpret paleoenvironments. On the upward side of the flow chart, fossil studies offer insights into the impact of historical events, such as mass extinction and immigration, on species and communities, as well as documenting long-term trends in community evolution. As will be pointed out below, there are critical issues of community organization and diversity patterns which may be best answered with fossil data (Kitchell, 1985).

Morphological Studies

Within the morphologic category, analyses at the species level are usually directed towards the inference of body size, locomotor and feeding behavior, or other aspects of some species' paleobiology. In such cases, the information transfer is almost entirely downward, from the present to the past, as the habits of extinct species are inferred by analogy with living species. Perhaps the major contribution of such studies is the demonstration that knowledge of soft tissue anatomy is not crucial; data from skeletal and dental anatomy can provide a substantial amount of information concerning body size, dietary, locomotor, and as-

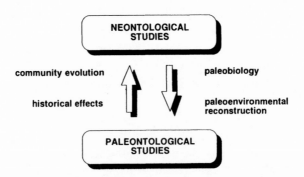

FIGURE 7.1 Reciprocal information transfer between studies of living (neontological) and extinct (paleontological) organisms and communities. Neontological studies provide data that can be used to reconstruct the paleobiology of species and their environments. Paleontological studies provide data on long-term patterns in community evolution and historical effects (extinction, immigration) on diversity levels that can be applied to understanding the present.

sociated behaviors (cf. Norberg, chap. 9, this volume). Indeed, skeletal anatomy can even provide clues about physiology, as exemplified by the previously mentioned studies of bone histology and recent examinations of the role of nasal turbinal structures in mammalian thermoregulation (Bennett and Ruben, 1986; Hillenius, 1992). Some of the most detailed morphological work has been done on primates. For example, body size, tooth facet size and shape, tooth enamel thickness, and dental microwear patterns have been used to discriminate among insectivorous, folivorous, and frugivorous living primates (e.g., Kay and Covert, 1984; Kay and Simons, 1980; Teaford, 1988). Analyses of dental microwear have been applied to the problem of early hominid diets (e.g., Grine and Kay, 1988; Walker, 1981), and recently Covert (1986) used body size and relative molar tooth facet area to examine the diversity of dietary adaptations among early Cenozoic primates. Similarly, studies of limb shape and proportion have enabled investigators to successfully discriminate among several locomotor types, such as leapers, brachiators, and quadrupedal walkers among primates (Aiello, 1981; Fleagle, 1976; Napier and Walker, 1967), and climbers, diggers, and terrestrial trotters among carnivores (Taylor, 1974, 1976, 1989; Van Valkenburgh, 1987). There are numerous ecomorphological studies of the paleobiology of extinct, nonmammalian vertebrates as well, including those on dinosaurs (e.g., Alexander, 1989; Bakker, 1987; Giffin, 1990; Norman and Weishampel, 1985), Mesozoic marine reptiles (e.g., Massare, 1988; Reiss, 1984; Tarsitano and Reiss, 1982), mammal-like reptiles (e.g., Hotton et al., 1986; Jenkins, 1971), and birds (Campbell and Tonni, 1983; Hecht et al., 1985; Ruben, 1991; Witmer and Rose, 1991).

A particularly useful and heuristic approach to studying morphologic evolution within or between clades is the analysis of species' distributions within a morphospace (fig. 7.2). An eco-morphospace is a three-dimensional plot, in which each axis represents a different aspect of the biology of a species, such as body size, diet, and locomotory habits. The units can be multivariate combinations of several measures (derived from discriminant or principle component analyses) or less complex indices of morphology. Thus a morphospace plot of all the species within a particular clade provides a quantitative visual display of the diversity of form (and inferred ecological characteristics) within a clade. If such a plot includes all known living and extinct species within a clade, then it illustrates the morphological diversity expressed within a clade over its evolutionary history. These can then be used to compare the extent of morphological diversification among clades and applied to questions of evolutionary constraint and versatility (cf. Van Valkenburgh, 1991).

Alternatively, if separate morphospaces are plotted for distinct time intervals and these are stacked as in figure 7.2, the pattern of morphological evolution over

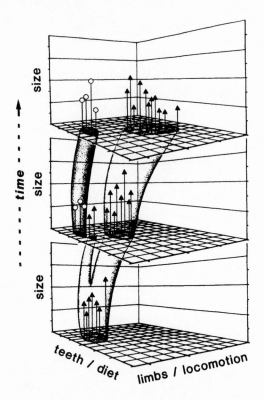

FIGURE 7.2 Hypothetical example of morphospace display of morphological evolution within two coexisting clades over time. Each of the three, stacked three-dimensional plots represents a different time interval, from oldest (bottom) to youngest (top). The axes are the same for all three plots: body size, some dental indicator of diet (teeth/diet), and some skeletal indicator of locomotory habits (limbs/locomotion). Symbols (open circles, closed triangles) represent either individuals or species' means and distinguish members of the two clades. See text for further explanation.

time becomes apparent, and it is possible to track putative adaptive shifts in morphology, or the iterative evolution of similar form. Carleton and Eshelman (1979) used such an approach in their investigation of evolution in grasshopper mice (genus *Onychomys*), demonstrating parallel trends toward insectivorous diets in two lineages. If members of two or more coexisting clades are plotted, as in figure 7.2 (open circles versus solid triangles), the possibility of an evolutionary replacement (or displacement) of one clade by the other can be explored. For example, the clade represented by open circles appears to have replaced a portion of the clade represented by solid triangles in the uppermost and youngest morphospace (fig. 7.2). This application of morphospace analysis to questions of interactions over evolutionary time between clades is described further below.

One of the most important contributions of ecomorphological analyses at the multispecies level is the documentation of the repeated or iterative evolution of a given form under similar environmental conditions. As mentioned above, convergence of form in phylogenetically distinct taxa is very powerful evidence of adaptation. Notably, stochastic simulations of morphological evolution within a clade did not produce parallel evolution, despite the appearance of several other ordered features such as unidirectional trends and specialization of derived forms (Raup and Gould, 1974). An example of a paleontological analysis of iteration of form is Janis's (1982) study of the evolution of horns in ungulates. By examining the ungulate fossil record in the context of paleoclimatic change, she was able to show that horns were acquired independently in ruminant artiodactyls at least three times and that in each case their acquisition is coincident with a habitat shift from closed to open woodland. Moreover, this was predictable (by analogy with living artiodactyls) as a consequence of a change in diet and resource distribution patterns, and probably reflected the evolution of territorial behavior. Other examples of iterative evolution in vertebrates include the evolution of deep-bodied lacustrine semionotid fish in the Jurassic Newark Basin (McCune, 1987), pelvic reduction in lacustrine stickleback fish (Bell, 1987), the repeated tendency of carnivorous mammal species to evolve similarly specialized molars for slicing meat (Butler, 1946; Van Valkenburgh, 1991), and the evolution of a sabertooth-like upper canine tooth first in several lines of mammal-like reptiles (Olson, 1971: 655; Kemp, 1982) and then much later among marsupial and placental mammals (Martin, 1989).

Ecomorphological studies of fossils can be applied to questions of evolutionary tempo and mode and may result in an upward transfer of information, in which insights from studies of fossil taxa can be applied to understanding modern patterns of distribution and abundance (fig. 7.1). For instance, Vrba (1984) explained the difference in diversity between two tribes of African antelopes, Aepycerotini and Alcelaphini, as due to disparities in degree of morphologic and habitat specialization that resulted in differing probabilities of speciation. The tribe Aepycerotini is represented at present by only the impala (*Aepyceros melampus*), a habitat and dietary generalist, and has experienced very low rates of speciation over the last six million years, having produced but a single species. The tribe Alcelaphini, however, has been much more prolific, generating approximately thirty-two species in the same interval, many of which have been or are more specialized than the impala and thus more likely to experience isolation and speciation. With a somewhat different approach, McCune (1981, 1987) analyzed the history of repeated endemic radiations of semionotid fishes preserved within Mesozoic lake deposits of the Newark Basin. Her detailed exploration of four temporally distinct radiations within the genus teased apart aspects of the radiations that could be ascribed to intrinsic variation in fish morphology and those due to extrinsic, environmental effects.

Paleoecological Studies

There are two general sorts of paleoecological studies that rely on ecomorphological analysis: (1) those designed to predict aspects of the paleoenvironment, such as vegetation structure or climate, from the morphologies of members of the paleofauna; and (2) those directed towards the analysis of species interactions and community evolution over time. As in the species-level example above, the first sort relies on documented associations between species' morphologies and their environments in the modern world. For example, Hutchison (1982) used the relationship between extant reptile diversity (both in body size and taxa) and the availability of permanent water to interpret Oligocene environments of North America. Marked declines in the diversity of crocodilians and aquatic turtles in the late Eocene–Oligocene (thirty-six to twenty-four million years ago) suggested increased aridity and seasonality. A number of studies have focused on herbivorous terrestrial mammals, as these are expected to closely track changes in vegetation structure. Habitat preference in bovids is strongly correlated with postcranial limb proportions (Scott, 1985) and femoral shape (Kappelman, 1988). Cursorial bovids inhabit open, savannah environments and have relatively long metapodials and a femoral head shape that restricts hip movement to a fore-and-aft swing (Kappelman, 1988; Scott, 1985). Similarly, feeding behavior in ungulates was predicted based on incisor morphology and relative muzzle width and shape (Janis and Eberhardt, 1988; Solounias et al., 1988) as well as tooth volume, crown height, and aspects of masticatory muscle scars (Janis, 1986; Solounias and Dawson-Saunders, 1988; Webb, 1983). All of these authors have used fossil ungulate morphology to track the worldwide development of extensive grasslands in the Miocene, and some have explored the ecomorphological parallels between the savannah communities of the North American Miocene and East African Recent (e.g., Janis, 1984; Webb, 1983). Other examples of the inference of paleoenvironment from ecomorphological aspects of fossil mammals are to be found in Andrews et al. (1979), Legendre (1986), and Van Couvering (1980). These studies based their inferences on distributions of body size, locomotor types, and feeding types of mammals within distinct extant habitats, such as savannah and rainforest.

Coevolution between herbivores and plants has been examined among dinosaurs and mammal-like reptiles as well as mammals (e.g., Bakker, 1978; Benton, 1979; Krassilov, 1981; Weishampel, 1984a, b; Weishampel and Norman, 1989; Wing and Tiffney, 1987). Based on a review of morphological studies of feeding behavior in Mesozoic vertebrates, Weishampel and Norman (1989) document a significant change in chewing mechanics among Mesozoic herbivores from a predominantly puncture-crushing mode to one utilizing transverse strokes of lower across upper teeth. This occurs alongside a late Triassic–early Jurassic turnover among gymnosperms and the subsequent Cretaceous rise of angiosperms. Although the relationship between food texture and dinosaur chewing behavior is yet to be clarified and is difficult given the absence of close living analogs, the above paleoecological studies highlight critical issues of dinosaur functional morphology to be studied.

One focus of a number of paleoecological studies has been the possible influence of interactions among organisms (competition, predation) on the rise or fall of a clade. For instance, there are a number of studies that consider the question of competitive displacement of one vertebrate group by another over geological time (e.g., Janis, 1989; Krause, 1986; Langer, 1987; Maas et al., 1988; Van Valkenburgh, 1991). All such studies focus on a decline in diversity in one group that overlaps with a rise in diversity of another. Such a pattern could be the result of competition and active displacement of group A by group B, but might instead reflect passive, opportunistic replacement of group A by group B, with the former having declined for reasons unconnected to group B. Alternatively, the shifts in diversity might represent independent responses by each group to some other factor (cf. Benton, 1987). It is very difficult to make a convincing case for competition in the fossil record, but similarity in morphology between the two putative competitors can be used to substantiate the hypothesis of competitive overlap. During the North American early Tertiary, two relatively abundant and diverse clades of small herbivores, the multituberculates and the plesiadapiforms, declined dramatically as the rodents diversified and increased in abundance (Krause, 1986; Maas et al., 1988). This appears to be a classic example of competitive displacement over time but could also reflect independent responses of each clade to a changing environment (Krause, 1986; Maas et al., 1988). To strengthen the case for competition as the cause, morphology was used to establish potential niche overlap in body size, diet, locomotion, and diel activity patterns. Additional support for the significance of competition in controlling diversification rates in one or both groups could come from the discovery of similar patterns of reciprocal diversification among small herbivores at different times and/or geographic settings (e.g., on alternate continents such as South America and Australia). If the pattern is repeated again and again, by phylogenetically distinct taxa in new settings, the hypothesis of competition is strengthened and that of independent responses is weakened.

Paleontological studies of community or guild structure (numbers of species, array of morphologies) over extended periods of time, before and after extinction and immigration events, are likely to produce a great deal of data relevant to theories of community stability, diversity, and evolution. Ecologists are becoming increasingly aware of the potential consequences of historical events on diversity patterns at the local level (Ricklefs, 1987; Roughgarden, 1989). History is truly the domain of the paleontologist and undoubtedly the study of community structure and dynamics is one in which the flow of information can be strongly upward, from the past to the present. Nevertheless, ecomorphological analyses of paleocommunities are relatively rare because they must build on previous species and clade level studies of functional morphology. As mentioned above, there are ecomorphological analyses of paleocommunities designed to predict paleoenvironmental characteristics (e.g., Andrews et al., 1979) but there are also those which emphasize questions of community evolution (e.g., Legendre, 1986; Van Valkenburgh, 1987, 1988).

Both Legendre (1986) and Van Valkenburgh (in press) have used ecomorphological analysis to examine the impact of an extinction event on mammalian community structure. In Legendre (1986), changes in the array of body sizes of species within communities before and after a major extinction event revealed an alteration in community structure that was consistent with environmental cooling and drying. Van Valkenburgh documented both body sizes and feeding morphologies (e.g., meat specialist, omnivore, bone crusher) within the North American predator community before and after a modest late Eocene (circa thirty-four million years ago) extinction event which was associated with a sharp decline in temperature and sea level (Prothero, 1985a, b). It produced an approximately 10% drop in mammalian generic diversity, affecting a range of taxa, including rodents and rhinoceros-like titanotheres (Prothero, 1985a, b). Nevertheless, species richness within the predator community was little affected, dropping from sixteen to fifteen species, but there was considerable taxonomic turnover (ten species extinctions and eight new species appearances). Moreover, the relative representation of different families of carnivores was altered dramatically by the extinction event (fig. 7.3). Despite the upheaval, the ecomorphological configuration of the community underwent relatively little change; as can be seen in figure 7.4, the distribution of species within a morphospace defined by body size and functional measures of the dentition was similar before and after the event (compare figs. 7.4A and B). In fact, more marked changes in predator guild structure appeared about 2.5 million years after the event, as additional species emerged and the morphospace became more densely packed (Figs. 7.4C and 7.4D). Moreover, the dietary emphasis within the guild shifted, away from highly carnivorous forms and toward more omnivorous habits, suggestive of a decline in prey availability. Whether these changes represented a delayed response to the extinction event or a response to then current environmental conditions is as yet unclear.

Apparent delays in adaptive radiation after extinction events are known from

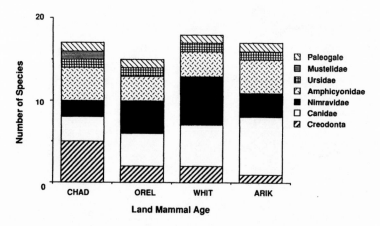

FIGURE 7.3 The impact of the late-Eocene (ca. 35–34 million years ago) extinction event on species diversity and taxonomic composition of the North American predator paleoguild. Stacked-bar plot of species richness (number of species) by family during the mid-Chadronian (36–35 million years ago), subsequent Orellan (34–32 million years ago), Whitneyan (32–29.5 million years ago), and Arikareean (29.5–24 million years ago) land mammal ages. Because the familial affinity of *Paleogale* is not certain, it is presented as a distinct group.

other ecomorphological studies of vertebrates. Archibald (1983) examined body size and trophic diversity within mammalian communities before and after the demise of the dinosaurs and discovered that mammalian communities remained essentially unchanged for 250,000 years after the removal of the dinosaurs. After this delay, mammals began to rapidly invade new ecospace, evolving larger body size and new dental morphologies in many lineages. Similarly, Collinson and Hooker (1987) used both ecomorphological and paleobotanical evidence to demonstrate that browsing mammals (recognized by their dental morphologies) were slow to evolve until the middle Eocene despite the apparent availability of browse ten to fifteen million years before. Wing and Tiffney (1987) also discuss this apparent delayed evolutionary response in their exploration of angiosperm-herbivore interactions from the late Jurassic through the Eocene of North America. This spans the dinosaur extinction, and thus encompasses major shifts in terrestrial herbivore diversity and body size due to the replacement of dinosaurs by mammals. They argue that the evolution of large mammalian herbivores in the Paleocene was retarded by the widespread presence of a new habitat type, dense forest vegetation. Large herbivore species are few in modern forests for a number of reasons relating to resource distribution (Wing and Tiffney, 1987). According to their hypothesis, the diversification of angiosperms typical of dense forests was not possible until after the removal of a primary agent of vegetational disturbance, large herbivorous dinosaurs.

The documentation of delays (on the order of hundreds of thousands or mil-

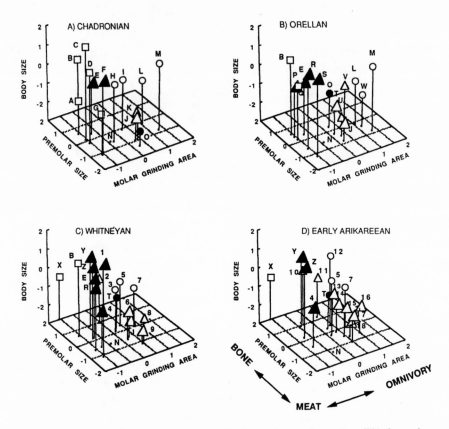

FIGURE 7.4 Body size and dental diversity in the North American predator paleoguild before and after the late Eocene extinction event. (*A*) Pre-extinction (mid-Chadronian); (*B*) approximately one million years post extinction (Orellan); (*C*) approximately 2.5 million years post extinction (Whitneyan); and (*D*) approximately five to ten million years post extinction (early Arikareean). Axes are body size, relative premolar size, and relative molar grinding area, and the two latter suggest dietary habits as indicated below (C) and (D). Each symbol represents a species mean. Open squares, creodonts; solid triangles, nimravids; open triangles, canids; open circles, amphicyonids; solid circles, ursids. Species: A, *Hyaenodon mustelinus;* B, *Hyaenodon horridus;* C, *Hyaenodon megaloides;* D, *Hyaenodon montanus;* E, *Dinictis* sp.; F, *Hoplophoneus mentalis;* G, *Hyaenodon microdon;* H, *Daphoenictis tedfordi;* I, *Daphoenocyon dodgei;* J, *Hesperocyon gregarius;* K, *Hesperocyon paterculus;* L, *Daphoenus hartshornianus;* M, *Daphoenus vetus;* N, *Paleogale* sp.; O, *Camplyocynodon* sp. (Chadronian) or *Parictis* sp. (Orellan–early Arikareean); P, *Hoplophoneus primaevus;* Q, *Hyaenodon crucians;* R, *Hoplophoneus occidentalis;* S, *"Eusmilus" sicarius;* T, *Hesperocyon lippincottianus;* U, *H.* small form; V, *Mesocyon* sp.; W, *Daphoenus minimus;* X, *Hyaenodon brevirostris;* Y, *Nimravus brachyops;* Z, *Pogonodon* sp.; 1, *"Eusmilus" dakotensis;* 2, *Mesocyon sheffleri;* 3, *Brachyrhynchocyon sesnoni;* 4, *Eusmilus cerebralis;* 5, *Pericyon socialis;* 6, *"Nothocyon"* sp. 2; 7, *Paradaphoenus transversus;* 8, *Oxetocyon cuspidatus;* 9, *"Nothocyon"* sp. 1; 10, *Enhydrocyon crassidens;* 11, *Mesocyon geringensis;* 12, *Temnocyon subferox;* 13, *Mesocyon coryphaeus;* 14, *"Brachyrhynchocyon" douglasi;* 15, *"Nothocyon" leptodus;* 16, *Nothocyon geismarianus;* 17, *"Hesperocyon" oreogonensis;* 18, *Cynodesmus cooki.* (For details see Van Valkenburgh, in press.)

lions of years) in the evolution of new morphologies in the presence of apparent opportunity is well known from the marine invertebrate record as well, where delays of up to twelve million years have been observed in the rebuilding of reef communities following mass extinction events (e.g., Copper, 1988; Talent, 1988). Ecosystems recovering from mass extinction, such as those represented by Paleocene reefs or terrestrial vertebrates, appear to have had the wind knocked out of them and only regain their former diversity after a period of relative stasis. Is this because the post–mass extinction environment was still unfavorable (cf. Stanley, 1988), or does it reflect more complex, biotic factors intrinsic to early community succession? Perhaps diversification rate may be partially dependent on the intensity of biotic interactions (competition, predation, commensalism) and these may take a long time to become established. Concerning the evolution of reefs, Talent (1988) argued that the postextinction delay in rebuilding reef systems might reflect the relative difficulty with which new symbiotic relationships were established between frame-builders (e.g., rudists, corals) and photo-synthesizing algae. No matter what the cause or causes of the delays may have been, the consequences are relevant to modern communities. They suggest that modern communities may not represent carefully coevolved systems. Having only recently experienced the extinctions and climatic upheavals of the Pleistocene, many species may have not yet responded to the altered array of opportunities. This is a clear example of paleontological data providing insights into evolutionary patterns that are not deducible from the study of extant communities (upward transfer of information).

In addition to improving our understanding of present-day biodiversity, paleontological data can be applied to the pressing problem of future communities. Ecomorphological analyses of which species appear most susceptible to mass extinction events and which clades recover their diversity most rapidly are likely to yield data of interest to conservation biologists. Of particular importance are studies of the impact of Pleistocene climatic change on both species survival and community stability (e.g., Graham, 1985; Graham and Lundelius, 1984; Owen-Smith, 1987). Because the Pleistocene extinctions affected many more large than small mammals, morphological analyses of the extinctions have focused almost entirely on body size. However, studies are underway to explore the impact of climate change on species within different trophic levels (Graham, pers. comm.), and these will benefit from an ecomorphological approach.

PROMISING DIRECTIONS FOR ECOMORPHOLOGICAL RESEARCH IN THE FOSSIL RECORD

At the species and clade level, there is a great need for ecomorphological analysis of a wider array of vertebrates. As should be apparent from the preceding pages, much work has been done on primates and ungulates, somewhat less on carnivores, and relatively little on other groups. Among mammals, the rodents would be a particularly important group to examine because of their abundance in the fossil record and importance as small herbivores. Of the remaining vertebrates, dinosaurs and the Permo-Triassic communities of reptiles that Olson (1952, 1966) used in his pioneering paleoecological studies deserve to be studied more thoroughly from an ecomorphological and paleocommunity viewpoint.

Ecomorphological studies of fossil vertebrates are likely to have their greatest impact on fields outside of paleobiology when they utilize the record to document broad-scale and long-term trends in morphological and ecological change. The record is rarely good enough to track morphological change over periods of thousands of years or less, but can be superb for periods of millions and tens of millions of years. Paleoecologists would be wise to direct their attention towards questions which can only be answered by the record. For example, is there a predictable impact of mass extinction on ecomorphological diversity within a guild or community? Are there some ecomorphs that are more or less susceptible than others? How rapidly do communities recover from perturbations like mass extinction, not only in species numbers but in morphological and inferred ecological diversity? Invertebrate paleontologists have investigated these kinds of questions with some success, discovering that extinction probabilities within a group such as gastropods or bivalves do appear to have been affected by ecomorphology. Generic longevity was greater among deposit-feeding than suspension-feeding bivalves (Levinton, 1974) and greater among gastropod species with high larval dispersal capability than those with low dispersal capability (Hansen, 1980; Jablonski, 1986c). In a series of papers, Jablonski (1986a, b, 1989) has explored differences in extinction susceptibilities among molluscan taxa during times of normal or background extinction levels and mass extinction events. His results suggest that susceptibilities change and that features that confer extinction resistance during background times (such as larval dispersal mode) are not the same as those that enhance survivorship during a mass extinction episode (such as geographic range). Consequently, mass extinction can result in profound shifts in dominance within communities in which formerly rare taxa rise to prominence. There have been fewer studies of this sort with vertebrates, in part because these analyses are strengthened by the larger sample sizes and better control on species' distribution in time and space that are available for shelled marine organisms. Nevertheless, the Cenozoic record of mammals is impressive and could be more heavily used as a testing ground for macroevolutionary hypotheses.

Most studies of diversity fluctuations over time rely on numbers of taxa (species or genera) but overlook morphological diversity. Although there is obviously some positive relationship between the two, it is probably not direct and may differ greatly in disparate environmental settings. For example, the initial radiation of the angiosperms produced many taxa but a limited range of ecological roles as inferred from seed and wood morphology (Wing and Tiffney, 1987). In other cases, such as the basal radiation of canids in the North American early Miocene, both morphological and taxonomic diversity were high (Van Valkenburgh, 1991). In both of these examples and the one provided by figure 7.2, explanations for diversification were greatly enhanced by the addition of morphological data which suggested ecological causes for the observed pattern. More studies of this type are needed before reaching any general conclusions concerning the controls on rates and extent of morphological diversification.

The intriguing question of energy use within communities over time can be approached with ecomorphological analysis of body size and trophic mode in extinct taxa (cf. Van Valen, 1990). What have been the relative distributions of species in both body size and number within trophic guilds over the last sixty-five million years? Are there secular trends in the body size distribution of species within communities? It is clear that the earliest postdinosaur mammalian communities consisted of relatively small species and that a diversity of large (> 10–20 kg) species did not appear until the Eocene, a time lag of some ten million years (Archibald, 1983; Maas and Krause, in press; Wing and Tiffney, 1987). In a recent analysis of North American mammalian species over the Cenozoic, Stuckey (1990) argues for a diversity maximum (number of genera) in the Eocene, some fifty million years before present. If it is real, was it due to environmental conditions or does the decline reflect differences in energy distribution between early and late Tertiary communities? Perhaps there were simply more species of smaller average size at that time than anytime subsequently.

In sum, the most promising directions in ecomorphological studies of extinct organisms and communities lie in analyzing "the processes of evolution at work in an ecological matrix" (Olson, 1980, p. 10). The processes of evolution are very difficult to observe, except under laboratory conditions, and the larger processes of macroevolution are impossible to observe. The fossil record can provide essential data on the influence of historical events (extinction, immigration) and climate change on patterns and rates of morphological diversification. As the transfer of information upward, from the past to the present, is accelerated and expanded, I look forward to seeing studies of living organisms and communities designed to answer questions generated by fossil data, rather than the other way around. A little bit of "me-too-paleontology" instead of "me-too-ecology" would be a healthy sign for both fields.

ACKNOWLEDGMENTS

Helpful comments on the manuscript were provided by H. Greene, F. Hertel, E. C. Olson, S. Suter, P. Wainwright, R. K. Wayne, and two anonymous reviewers. The work was supported in part by NSF grant BSR 881837.

REFERENCES

Aiello, L. C. 1981. The allometry of primate body proportions. *Symp. Zool. Soc. Lond.* 48:331–358.

Alexander, R. M. 1989. *Dynamics of Dinosaurs and Other Extinct Giants.* New York: Columbia University Press.

Anderson, E. A. 1984. Who's who in the Pleistocene: A mammalian bestiary. In *Quaternary Extinctions,* ed. P. S. Martin and R. G. Klein, 40–89. Tucson: University of Arizona Press.

Andrews, P., J. M. Lord, and E. M. Nesbit Evans. 1979. Patterns of ecological diversity in fossil and modern mammalian faunas. *Biol. J. Linnean Soc.* 11:177–205.

Archibald, J. D. 1983. Structure of the K-T Mammal radiation in North America: Speculations on turnover rates and trophic structure. *Acta Palaeontol. Polonica* 28:7–17.

Bakker, R. T. 1978. Dinosaur feeding behaviour and the origin of flowering plants. *Nature* 274:661–663.

Bakker, R. T. 1987. The return of the dancing dinosaurs. In *Dinosaurs Past and Present,* ed. S. J. Czerkas and E. C. Olson, 38–69. Seattle: University of Washington Press.

Bambach, R. K. 1985. Classes and adaptive variety: The ecology of diversification in marine faunas through the Phanerozoic. In *Phanerozoic Diversity Patterns,* ed. J. W. Valentine, 191–253. Princeton: Princeton University Press.

Behrensmeyer, A. K., and A. P. Hill. 1980. *Fossils in the Making.* Chicago: University of Chicago Press.

Behrensmeyer, A. K., and S. M. Kidwell. 1985. Taphonomy's contributions to paleoecology. *Paleobiology* 11:105–119.

Behrensmeyer, A. K., and R. W. Hook. 1992. Paleoenvironmental contexts and taphonomic modes. In *Evolutionary Paleoecology of Terrestrial Plants and Animals,* ed. A. K. Behrensmeyer, J. D. Damuth, W. A. DiMichele, R. Potts, H. Sues, and S. L. Wing, 15–136. Chicago: University of Chicago Press.

Bell, M. A. 1987. Interacting evolutionary constraints in pelvic reduction of threespine sticklebacks, *Gasterosteus aculeatus* (Pisces, Gasterosteidae). *Biol. J. Linnean Soc.* 31:347–382.

Bell, M. A. 1988. Stickleback fishes: Bridging the gap between population biology and paleobiology. *Trends Ecol. Evol.* 3:320–325.

Bell, M. A., J. V. Baumgartner, and E. C. Olson. 1985. Patterns of temporal change in single morphological characters of a Miocene stickleback fish. *Paleobiology* 11:258–271.

Bell, M. A., M. S. Sadagursky, and J. V. Baumgartner. 1987. Utility of lacustrine deposits for the study of variation within fossil samples. *Palaios* 2:455–466.

Bennett, A. F., and J. A. Ruben. 1986. The metabolic and thermoregulatory status of therapsids. In *The Ecology and Biology of Mammal-like Reptiles,* ed. N. Hotton, P. D. MacLean, J. J. Roth, and E. C. Roth, 207–218. Washington, D.C.: Smithsonian Institution Press.

Benton, M. J. 1979. Ecological succession among late Paleozoic and Mesozoic tetrapods. *Palaeogeo. Palaeoclim. Palaeoecol.* 26:127–150.

Benton, M. J. 1987. Progress and competition in macroevolution. *Biol. Rev.* 62:305–338.

Bik

nevicius, A. R., and C. B. Ruff. 1992a. Use of biplanar radiographs for estimating cross-sectional properties of mandibles. *Anat. Rec.* 232:157–163.

Biknevicius, A. R., and C. B. Ruff. 1992b. The structure of the mandibular corpus and its relationship to feeding behaviours in extant carnivorans. *J. Zool.* 228:479–507.

Brooks, D. R., and D. A. McLennan. 1991. *Phylogeny, Ecology, and Behavior.* Chicago: University of Chicago Press.

Burghardt, G. M., and J. L. Gittleman. 1990. Comparative behavior and phylogenetic analyses: New wine, old bottles. In *Interpretation and Explanation in the Study of Animal Behavior,* ed. M. Bekoff and D. Jamieson, 192–225. Boulder: Westview Press.

Butler, P. M. 1946. The evolution of carnassial dentitions in the Mammalia. *Proc. Zool. Soc. Lond.* 116:198–220.

Campbell, K. E., and E. P. Tonni. 1983. Size and locomotion in teratorns (Aves: Teratornidae). *Auk* 100:390–403.

Carleton, M. D., and R. E. Eshelman. 1979. A synopsis of fossil grasshopper mice, genus *Onychomys*, and their relationships to recent species. Univ. Michigan Papers Paleontol. 21:1–63.

Cheverud, J. M., M. M. Dow, and W. Leutnegger. 1985. The quantitative assessment of phylogenetic constraints in comparative analyses: Sexual dimorphism in body weight among primates. *Evolution* 39:1335–1351.

Collinson, M. E., and J. J. Hooker. 1987. Vegetational and mammalian faunal changes in the early Tertiary of southern England. In *The Origins of Angiosperms and Their Biological Consequences,* ed. E. M. Friis, W. G. Chaloner, and P. R. Crane, 259–304. Cambridge: Cambridge University Press.

Copper, P. 1988. Ecological succession in Phanerozoic reef systems: Is it real? *Palaios* 3:136–152.

Covert, H. H. 1986. Biology of early Cenozoic Primates. In *Comparative Primate Biology,* vol. 1, *Systematics, Evolution, and Anatomy,* ed. D. R. Swindler and J. Erwin, 335–359. New York: Alan R. Liss.

Damuth, J., and B. J. MacFadden. 1990. *Body Size in Mammalian Paleobiology: Estimation and Biological Implications.* Cambridge: Cambridge University Press.

Davis, D. D. 1949. The shoulder architecture of bears and other carnivores. *Fieldiana Zool.* 31:287–305.

Emerson, S., and L. Radinsky. 1980. Functional analysis of sabertooth cranial morphology. *Paleobiology* 6:295–312.

Ewer, R. F. 1973. *The Carnivores.* Ithaca: Cornell University Press.

Felsenstein, J. 1985. Phylogenies and the comparative method. *Amer. Nat.* 125:1–15.

Fisher, D. C. 1985. Evolutionary morphology: Beyond the analogous, the anecdotal, and the ad hoc. *Paleobiology* 11:120–138.

Fleagle, J. G. 1976. Locomotor behavior and skeletal anatomy of sympatric Malaysian leaf-monkeys (*Presbytis obscura* and *Presbytis melalophos*). *Yearbook Phys. Anthro.* 20:440–453.

Fortelius, M. 1985. Ungulate cheek teeth: Developmental, functional, and evolutionary interrelations. *Acta Zool. Fennica* 180:1–76.

Franzen, J. L. 1985. Exceptional preservation of Eocene vertebrates in the lake deposit of Grube Mess (West Germany). Phil. Trans. R. Soc. (London) B311:181–186.

Giffin, E. B. 1990. Gross spinal anatomy and limb use in living and fossil reptiles. *Paleobiology* 4:448–458.

Gittleman, J. L., and M. Kot. 1990. Adaptation: Statistics and a null model for estimating phylogenetic effects. *Syst. Zool.* 39:227–241.

Gould, S. J., and R. C. Lewontin. 1979. The spandrels of San Marco and the Panglossian paradigm: A critique of the adaptationist programme. *Proc. R. Soc.* (London) B205:581–598.

Graham, R. W. 1985. Diversity and community structure of the late Pleistocene mammal fauna of North America. *Acta Zool. Fennica* 170:181–192.

Graham, R. W., and E. L. Lundelius. 1984. Coevolutionary disequilibrium and Pleistocene extinctions. In *Quaternary Extinctions,* ed. P. S. Martin and R. G. Klein, 223–249. Tucson: University of Arizona Press.

Greene, H. W., and G. M. Burghardt. 1978. Behavior and phylogeny: Constriction in ancient and modern snakes. *Science* 200:74–77.

Grine, F. E., and R. F. Kay. 1988. Early hominid diets from quantitative image analysis of dental microwear. *Nature* 333:765–768.

Hansen, T. A. 1980. Influence of larval dispersal and geographic distribution on species longevity in neogastropods. *Paleobiology* 6:193–207.

Harvey, P. H., and A. Purvis. 1991. Comparative methods for explaining adaptations. *Nature* 351:619–623.

Haynes, G. 1982. Utilization and skeletal disturbances of North American prey carcasses. *Arctic* 35:266–281.

Hecht, M., J. H. Ostrom, G. Viohl, and P. Wellnhofer, eds. 1985. *The Beginnings of Birds.* Willibaldsburg: Freunde des Jura-Museums Eichstatt.

Hillenius, W. J. 1992. The evolution of nasal turbinates and mammalian endothermy. *Paleobiology* 18:17–29.

Hotton, N., P. D. MacLean, J. J. Roth, and E. C. Roth, eds. 1986. *The Ecology and Biology of Mammal-like Reptiles.* Washington, D.C.: Smithsonian Institution Press.

Howell, A. B. 1944. *Speed in Animals.* New York: Haffner.

Huey, R. B. 1987. Phylogeny, history, and the comparative method. In *New Directions in Ecological Physiology,* ed. M. E. Feder, A. F. Bennett, W. Burggren, and R. B. Huey, 76–98. Cambridge: Cambridge University Press.

Hunt, R. M., X. Xiang-Xu, and J. Kaufman. 1983. Miocene burrows of extinct beardogs: Indication of early denning behavior of large mammalian carnivores. *Science* 221:364–366.

Hutchison, J. H. 1982. Turtle, crocodilian, and champsosaur diversity changes in the Cenozoic of the north-central region of western United States. *Palaeogeo. Palaeoclim. Palaeocol.* 37:149–164.

Jablonski, D. 1986a. Background and mass extinctions: The alternation of macroevolutionary regimes. *Science* 231:129–133.

Jablonski, D. 1986b. Evolutionary consequences of mass extinction. In *Patterns and Processes in the History of Life,* ed. D. M. Raup and D. Jablonski, 313–329. Berlin, Heidelberg: Springer-Verlag.

Jablonski, D. 1986c. Larval ecology and macroevolution in marine invertebrates. *Bull. Mar. Sci.* 39:565–587.

Jablonski, D. 1989. The biology of mass extinction: A paleontological view. *Phil. Trans. R. Soc.* (London) B325:357–368.

Janis, C. M. 1982. Evolution of horns in ungulates: Ecology and paleoecology. *Biol. Rev.* 57:261–318.

Janis, C. M. 1984. The use of fossil ungulate communities as indicators of climate and environment. In *Fossils and Climate,* ed. P. J. Brenchley, 85–103. London: John Wiley and Sons.

Janis, C. M. 1986. An estimation of tooth volume and hypsodonty indices in ungulate mammals, and the correlation of these factors with dietary preference. In *Teeth Revisited: Proceedings of the VII International Symposium on Dental Morphology, Paris, 1986,* ed. D. E. Russell, J. P. Santoro, and D. Sigogneau-Russell. *Mem. Mus. Natn. Hist. Nat.* (Paris) C53:367–387.

Janis, C. M. 1989. A climatic explanation for patterns of evolutionary diversity in ungulate mammals. *Paleontology* 32:463–481.

Janis, C. M., and D. Eberhardt. 1988. Correlation of relative muzzle width and relative incisor width with dietary preference in ungulates. *Zool. J. Linnean Soc.* 92:267–284.

Jenkins, F. A., Jr. 1971. The postcranial skeleton of African cynodonts. *Bull. Peabody Mus. Nat. Hist.* 36:1–216.

Kappelman, J. 1988. Morphology and locomotor adaptations of the bovid femur in relation to habitat. *J. Morphol.* 198:119–130.

Kay, R. F., and M. Cartmill. 1977. Cranial morphology and adaptation of *Palaecthon nacimienti* and other Paromomyidae (Plesiadapoidea, Primates) with a description of a new genus and species. *J. Hum. Evol.* 6:19–53.

Kay, R. F., and H. H. Covert. 1984. Anatomy and behaviour of extinct primates. In *Food Acquisition and Processing in Primates,* ed. D. J. Chivers, B. A. Wood, A. Bilsborough, 467–508. New York: Plenum Press.

Kay, R. F., and E. L. Simons. 1980. Comments on the adaptive strategy of the first African anthropoids. *Z. Morph. Anthrop.* 71:143–148.

Kemp, T. S. 1982. *Mammal-like Reptiles and the Origin of Mammals.* New York: Academic Press.

Kitchell, J. A. 1985. Evolutionary paleoecology: Recent contributions to evolutionary theory. *Paleobiology* 11:91–104.

Kitchell, J. A., and T. R. Carr. 1985. Nonequilibrium model of diversification: Faunal turnover dynamics. In *Phanerozoic Diversity Patterns,* ed. J. W. Valentine, 277–309. Princeton: Princeton University Press.

Krassilov, V. A. 1981. Changes of Mesozoic vegetation and the extinction of dinosaurs. *Palaeogeo. Palaeoclim. Palaeoecol.* 34:207–224.

Krause, D. W. 1986. Competitive exclusion and taxonomic displacement in the fossil record: The case of rodents and multituberculates in North America. In *Vertebrates, Phylogeny, and Philosophy*, ed. K. M. Flanagan and J. Lillegraven, 95–118. Laramie: Univ. Wyoming Contrib. Geo. Spec. Pap. 3.

Kurtén, B. 1958. Life and death of the Pleistocene cave bear: A study in paleoecology. *Acta Zool. Fennica* 95:1–59.

Kurtén, B., and E. Anderson. 1980. *Pleistocene Mammals of North America*. New York: Columbia University Press.

Langer, P. 1987. Evolutionary patterns of Perrisodactyla and Artiodactyls (Mammalia) with different types of digestion. *Z. Zool. Syst. Evolut.-forsch.* 25:212–236.

Lauder, G. V. 1990. Functional morphology and systematics: Studying functional patterns in an historical context. *Ann. Rev. Ecol. Syst.* 21:317–340.

Legendre, S. 1986. Analysis of mammalian communities from the late Eocene and Oligocene of southern France. *Palaeovertebrata* 16:191–212.

Levinton, J. S. 1974. Trophic group and evolution in bivalve molluscs. *Palaeontology* 17:579–585.

Lockley, M. G. 1986. The paleobiological and paleoenvironmental importance of dinosaur footprints. *Palaios* 1:37–47.

Lockley, M. G. 1987. Dinosaur trackways and their importance in paleontological reconstruction. In *Dinosaurs Past and Present*, ed. S. J. Czerkas and E. C. Olson, 80–95. Seattle: University of Washington Press.

Maas, M. C., D. W. Krause, and S. G. Strait. 1988. The decline and extinction of Plesiadapiformes (Mammalia: ?Primates) in North America: Displacement or replacement? *Paleobiology* 14:410–431.

Maas, M. C., and D. W. Krause. In press. Mammalian community structure in the Paleocene of North America. *Hist. Biol.*

Martin, L. D. 1989. Fossil history of the terrestrial Carnivora. In *Carnivore Behavior, Ecology, and Evolution*, ed. J. L. Gittleman, 536–568. Ithaca: Cornell University Press.

Massare, J. A. 1988. Swimming capabilities of Mesozoic marine reptiles: Implications for method of predation. *Paleobiology* 14:187–205.

McCune, A. R. 1981. Quantitative description of body form in fishes: Implications for species level taxonomy and ecological inference. *Copeia* 1981:897–901.

McCune, A. R. 1987. Lakes as laboratories of evolution: Endemic fishes and environmental cyclicity. *Palaios* 2:446–454.

Napier, J. R., and A. C. Walker. 1967. Vertical clinging and leaping: A newly recognised category of locomotor behaviour of primates. *Folia Primatol.* 6:204–219.

Norman, D. B., and D. B. Weishampel. 1985. Ornithopod feeding mechanisms: Their bearing on the evolution of herbivory. *Amer. Nat.* 126:151–164.

Olson, E. C. 1952. The evolution of a Permian vertebrate chronofauna. *Evolution* 6:181–196.

Olson, E. C. 1966. Community evolution and the origin of mammals. *Ecology* 47:291–302.

Olson, E. C. 1971. *Vertebrate Paleozoology.* New York: Wiley-Interscience.

Olson, E. C. 1980. Taphonomy: Its role in community evolution. In *Fossils in the Making*, ed. A. K. Behrensmeyer and A. P. Hill, 5–19. Chicago: University of Chicago Press.

Ostrom, J. H. 1976. *Archaeopteryx* and the origin of birds. *Biol. J. Linnean Soc.* 8:91–182.

Owen-Smith, N. 1987. Pleistocene extinctions: The pivotal role of megaherbivores. *Paleobiology* 13:351–362.

Oxnard, C. E. 1967. The functional morphology of the primate shoulder as revealed by comparative anatomical, osteometric, and discriminant function techniques. *Amer. J. Phys. Anthrop.* 26:219–240.

Pagel, M. D., and P. H. Harvey. 1988. Recent developments in the analysis of comparative data. *Q. Rev. Biol.* 63:413–440.

Prothero, D. R. 1985a. Mid-Oligocene extinction events in North American land mammals. *Science* 229:550–551.

Prothero, D. R. 1985b. North American mammalian diversity and Eocene-Oligocene extinctions. *Paleobiology* 11:389–405.

Raup, D. M., and S. J. Gould. 1974. Stochastic simulation and evolution of morphology: Towards a nomothetic paleontology. *Syst. Zool.* 23:305–322.

Reid, R. E. M. 1984. The biology of the dinosaurian bone, and its possible bearing on dinosaur physiology. In *The Structure, Development and Evolution of Reptiles*, ed. M. W. J. Fergusson. London: Academic Press.

Reiss, J. 1984. How to reconstruct paleoecology: Outlines of a holistic view and an introduction to icthyosaur locomotion. In *Third Symposium on Mesozoic Terrestrial Ecosystems*, ed. W.-E. Reif and F. Westphal, 201–205. Tubingen: Attempto Verlag.

Ricklefs, R. E. 1987. Community diversity: Relative roles of local and regional processes. *Science* 235:167–171.

Riqlés, A. J. de. 1980. Tissue structure of dinosaur bone: Functional significance and possible relation to dinosaur physiology. In *A Cold Look at the Warm-Blooded Dinosaurs*, ed. R. D. K. Thomas and E. C. Olson, 103–109. Boulder: Westview (A.A.A.S. Symp. 28).

Rollinson, J. M. M., and R. D. Martin. 1981. Comparative aspects of primate locomotion, with special reference to arboreal cercopithecines. *Symp. Zool. Soc. Lond.* 48:377–427.

Rose, K. D., and T. M. Bown. 1986. Gradual evolution and species discrimination in the fossil record. In *Vertebrates, Phylogeny, and Philosophy*, ed. K. M. Flanagan and J. Lillegraven, 119–130. Laramie: Univ. Wyoming Contrib. Geo. Spec. Pap. 3.

Rose, K. D., and A. Walker, 1981. Function of the mandibular tooth comb in living and extinct mammals. *Nature* 289:583–585.

Roughgarden, J. 1989. The structure and assembly of communities. In *Perspectives in Ecological Theory*, ed. J. Roughgarden, R. M. May, and S. A. Levin, 203–227. Princeton: Princeton University of Press.

Ruben, J. A. 1991. Reptilian physiology and the flight capacity of *Archaeopteryx*. *Evolution* 45:1–17.

Rudwick, M. J. S. 1964. The inference of function from structure in fossils. *Brit. J. Phil. Sci.* 15:27–40.

Ruff, C. B. 1989. New approaches to structural evolution of limb bones in primates. *Folia Primatol.* 53:142–159.

Scott, K. M. 1985. Allometric trends and allometry in bovid postcranial proportions. *Bull. Amer. Mus. Nat. Hist.* 179:197–288.

Shipman, P. 1981. Scavenging or hunting in early hominids: Theoretical framework and tests. *Amer. Anthropol.* 88:27–43.

Shipman, P. 1986. *Life History of a Fossil*. Cambridge: Harvard University Press.

Solounias, N., and B. Dawson-Saunders. 1988. Dietary adaptations and paleoecology of the late Miocene ruminants from Pikermi and Samos in Greece. *Palaeogeo. Palaeoclim. Palaeoecol.* 65:149–172.

Solounias, N., M. Teaford, and A. Walker. 1988. Interpreting the diet of extinct ruminants: The case of a non-browsing giraffid. *Paleobiology* 14:287–300.

Stanley, S. M. 1988. Climatic cooling and mass extinction of Paleozoic reef communities. *Palaios* 3:228–232.

Stirton, R. A. 1947. Observations on evolutionary rates in hypsodonty. *Evolution* 1:32–41.

Stucky, R. K. 1990. Evolution of land mammal diversity in North America during the Cenozoic. In *Current Mammalogy*, vol. 2, ed. H. Genoways, 375–432. New York: Plenum Press.

Talent, J. A. 1988. Organic reef-building: Episodes of extinction and symbiosis. *Senckenbergiana Lethaea* 69:315–368.

Tarsitano, S., and J. Reiss. 1982. Plesiosaur locomotion-underwater flight versus rowing. *Neues Jahrb. Geol. Palaont.* 164:188–192.

Taylor, M. E. 1974. The functional anatomy of the forelimb of some African Viverridae (Carnivora). *J. Morphol.* 143:307–336.

Taylor, M. E. 1976. The functional anatomy of the hindlimb of some African Viverridae (Carnivora). *J. Morphol.* 148:227–254.

Taylor, M. E. 1989. Locomotor adaptations of carnivores. In *Carnivore Behavior, Ecology, and Evolution,* ed. J. L. Gittleman, 382–409. Ithaca: Cornell University Press.

Teaford, M. F. 1988. A review of dental microwear and diet in modern mammals. *Scanning Microsc.* 2:1149–1166.

Thomason, J. J. 1985a. Estimation of the locomotory forces and stresses acting on the limb bones of Recent and extinct equids. *Paleobiology* 11:287–298.

Thomason, J. J. 1985b. The relationship of trabecular architecture to inferred loading patterns in the third metacarpals of the extinct equids *Merychippus* and *Mesohippus. Paleobiology* 11:323–335.

Tracy, C. R., J. S. Turner, and R. B. Huey. 1986. A biophysical analysis of possible thermoregulatory adaptations in sailed pelycosaurs. In *The Ecology and Biology of Mammal-like Reptiles,* ed. N. Hotton, P. D. MacLean, J. J. Roth, and E. C. Roth, 207–218. Washington, D.C.: Smithsonian Institution Press.

Van Couvering, J. A. H. 1980. Community evolution in East Africa during the late Cenozoic. In *Fossils in the Making,* ed. A. K. Behrensmeyer and A. P. Hill, 272–298. Chicago: University of Chicago Press.

Van Valen, L. M. 1990. Levels of selection in the early Cenozoic radiation of mammals. *Evol. Theory* 9:171–180.

Van Valkenburgh, B. 1985. Locomotor diversity in past and present guilds of large predatory mammals. *Paleobiology* 11:406–428.

Van Valkenburgh, B. 1987. Skeletal indicators of locomotor behavior in living and extinct carnivores. *J. Vert. Paleont.* 7:162–182.

Van Valkenburgh, B. 1988. Trophic diversity in past and present guilds of large predatory mammals. *Paleobiology* 14:155–173.

Van Valkenburgh, B. 1990. Skeletal and dental predictors of body mass in carnivores. In *Body Size in Mammalian Paleobiology: Estimation and Biological Implications,* ed. J. Damuth and B. J. MacFadden, 181–205. Cambridge: Cambridge University Press.

Van Valkenburgh, B. 1991. Iterative evolution of hypercarnivory in canids (Mammalia: Carnivora): Evolutionary interactions among sympatric predators. *Paleobiology* 17:340–362.

Van Valkenburgh, B. In press. Extinction and replacement among predatory mammals in the North American Late Eocene and Oligocene: Tracking a paleoguild over twelve million years. *Hist. Biol.*

Van Valkenburgh, B., and C. B. Ruff. 1987. Canine tooth strength and killing behaviour in large carnivores. *J. Zool.* 212:1–19.

Van Valkenburgh, B., M. F. Teaford, and A. Walker. 1990. Molar microwear and diet in large carnivores: Inferences concerning diet in the sabretooth cat *Smilodon fatalis. J. Zool.* 222:319–340.

Vrba, E. S. 1984. Evolutionary pattern and process in the sister-group Alcelaphini-Aepycerotini (Mammalia: Bovidae). In *Living Fossil,* ed. N. Eldredge and S. M. Stanley, 62–79. New York: Springer-Verlag.

Walker, A. 1980. Functional anatomy and taphonomy. In *Fossils in the Making,* ed. A. K. Behrensmeyer and A. P. Hill, 182–186. Chicago: University of Chicago Press.

Walker, A. 1981. Diet and teeth: Dietary hypotheses and human evolution. Phil. Trans. R. Soc. (London) B292:57–64.

Walker, A., H. N. Hoeck, and L. Perez. 1978. Microwear of mammalian teeth as an indicator of diet. *Science* 201:908–910.

Wayne, R. K., and S. J. O'Brien. 1987. Allozyme divergence within the Canidae. *Syst. Zool.* 36:339–355.

Webb, S. D. 1983. The rise and fall of the late Miocene ungulate fauna in North America. In *Coevolution,* ed. M. H. Nitecki, 267–306. Chicago: University of Chicago Press.

Weishampel, D. B. 1981. Acoustic analyses of potential vocalization in lambeosaurine dinosaurs (Reptilia: Ornithischia). *Paleobiology* 7:252–261.

Weishampel, D. B. 1983. Hadrosaurid jaw mechanics. *Acta Palaeontol. Polonica* 28:271–280.

Weishampel, D. B. 1984a. Evolution of jaw mechanics in ornithopod dinosaurs. *Adv. Anat. Embryol. Cell. Biol.* 871–110.

Weishampel, D. B. 1984b. Interactions between Mesozoic plants and vertebrates: Fructifications and seed predation. *Neues Jahrb. Geol. Palaont.* 167:224–250.

Weishampel, D. B., and D. B. Norman. 1989. Vertebrate herbivory in the Mesozoic: Jaws, plants, and evolutionary metrics. Geol. Soc. Amer. Spec. Pap. 238:87–100.

Wilson, M. V. H. 1988. Taphonomic processes: Information loss and information gain. *Geosci. Can.* 15:131–148.

Wing, S. L., and B. H. Tiffney. 1987. The reciprocal interaction of angiosperm evolution and tetrapod herbivory. *Rev. Palaeobot. Palynol.* 50:179–210.

Witmer, L. M., and K. D. Rose. 1991. Biomechanics of the jaw apparatus of the gigantic Eocene bird *Diatryma:* Implications for diet and mode of life. *Paleobiology* 17:95–120.

Wolfe, J. A., and H. E. Schorn. 1989. Paleoecologic, paleoclimatic, and evolutionary significance of the Oligocene Creede flora, Colorado. *Paleobiology* 15:180–198.

II. Model Systems

8

Roles of Hydrodynamics in the Study of Life on Wave-Swept Shores

Mark W. Denny

INTRODUCTION

The intertidal zone of wave-swept rocky shores is physically stressful. At high tide, benthic organisms are subjected to the water motions associated with breaking waves (Denny, 1988; Denny et al., 1985). These include velocities that may exceed 14 m s^{-1} (31 mph) and accelerations that may exceed 400 m s^{-2} (about 41 gravities). These water motions are accompanied by large hydrodynamic forces that tend to push organisms downstream (drag and the acceleration reaction) and to pull them away from the substratum (lift). In addition, there are impact forces caused by water-borne missiles such as sand, rocks, and floating logs and the forces accompanying the overturning of boulders upon which the plants and animals live (Dayton, 1971; Shanks and Wright, 1986; Sousa, 1979a, b).

At low tide, intertidal organisms are subjected to the vagaries of the terrestrial environment. The temperature of the substratum can vary from well below freezing to above 50°C, sufficient in either extreme to kill many organisms. Further, the effects of high temperature and wind can rapidly cause death through desiccation, and heavy rains may kill plants and animals through hypoosmotic effects. In the variety and intensity of its physical stresses, the intertidal zone ranks among the most severe on earth.

Despite its physical severity, the intertidal zone of wave-swept rocky shores supports a highly diverse assemblage of organisms. In terms of the number of phyla represented, the intertidal zone is richer than any terrestrial environment, and its species richness exceeds that of adjacent habitats that are less physically stressful (Menge and Sutherland, 1987; Sousa, 1984). Biological diversity is also high in terms of organismal form and function. Intertidal organisms vary in shape from thin encrusting forms to ornate upright structures, in biomaterial properties from rocklike rigidity to extreme flexibility, and in habit from sessile to mobile. The functions associated with this diversity of form are similarly diverse.

The intertidal zone of wave-swept shores can also be very productive. Leigh et al. (1987) have shown that primary production (kg biomass $m^{-2} yr^{-1}$) of intertidal algae on the shores of Washington State exceeds by a considerable margin that of any terrestrial system measured. Indeed, Vermeij (1987) argues that this high productivity is a major contributing factor in the evolution of the morphological and functional diversity characteristic of wave-swept rocky shores.

The unique combination of physical severity, species diversity, and high productivity has attracted the attention of experimental ecologists. Through their efforts over the past thirty years, technology appropriate to the study of this habitat has been devised and innovatively employed, with the result that the intertidal zone has proved to be a useful model system in which new ideas in ecology may be tested experimentally. Working with this system, intertidal ecologists have made important contributions to basic ecological concepts.

The factors that first attracted ecologists to the intertidal environment also appeal to functional morphologists. For instance, the diversity of morphologies and functional strategies among intertidal organisms provides ample raw material for comparative studies, and the physical environment is sufficiently severe to suggest that it may foster clear-cut selection for mechanical design. Here the concept of "design" is used informally to denote the evolved adaptive match between a biological structure and its function (or functions) in a specific environment.

The attractiveness of the intertidal environment is also enhanced by recent advances in technology that make it practical to measure wave heights, water velocities, and hydrodynamic forces in the field (Denny, 1988). In addition, the wealth of information gained by intertidal ecological studies is a potent lure to functional morphologists, tempting them to combine their techniques with those of ecology in the study of intertidal "ecomorphology," to the advantage of both pursuits.

The study of nearshore ecomorphology is still young, however, and it is perhaps timely to cast a critical eye upon the field. In this chapter I explore how the stochastic nature of ocean waves both enhances the potential of rocky shores as a model system for the study of ecomorphology and places limits on the standard biomechanical approach to the study of adaptation. We will see, for instance, why it is difficult to predict wave-imposed fluid mechanical forces operating over short times and small spatial scales, but that the statistics of the random sea can be used as the basis for an accurate prediction of long-term average survivorship. Coupled with the results of ongoing models of global wave patterns, these predictions may prove useful in a mechanistic understanding of community structure on wave-swept rocky shores. The statistics of the random sea also provide an ideal basis for an exploration of the mechanistic connection among behavior, performance, and fitness. And finally, recent experimental results are explored to show

how biomechanics can demonstrate the functional neutrality of some aspects of morphology, thereby defining a limit on the ability of the wave-swept environment to select among structures, and how the stochastic nature of breaking waves limits the utility of "optimality" as an approach to the study of life on rocky shores.

THE ROLES OF HYDRODYNAMICS

This exploration will be limited in two aspects. First, I approach the subject from the perspective of a functional morphologist because that is where my expertise lies. Second, the discussion is confined to the roles played by hydrodynamics. Although water motion is only one of a complicated suite of physical factors that characterize wave-swept shores (temperature, desiccation, and osmotic stress being other critical variables), hydrodynamics appears to play a central role in determining the distribution and abundance of intertidal organisms. Evidence to this effect can be drawn from a variety of sources.

Hydrodynamic Forces Pose a Risk to Survivorship

We first examine evidence that wave-induced hydrodynamic forces (lift, drag, and the acceleration reaction) pose a risk for intertidal organisms. The simple fact that large velocities and accelerations are present on wave-swept shores does not necessarily imply that the resulting hydrodynamic forces pose any danger to the resident plants and animals. If intertidal organisms are sufficiently well designed, the forces they encounter may fall within tolerable limits.

However, one need only walk along the shore after a winter storm to find compelling evidence of large-scale disturbance caused by hydrodynamic forces. Commonly, mounds of algae have been thrown up on the beach by breaking waves, and the remains of limpets, urchins, abalones, and mussels are scattered along the shore. A number of careful studies have documented wave-induced disturbance, notably the work of Paine and Levin (1981), who described the demography of wave-induced bare patches in mussel beds, and that of Sousa (1979a, b), who documented the rate at which intertidal boulders are overturned. Effects of disturbance similar to those described for the intertidal zone have been described for coral reefs (e.g., Connell, 1978).

The Disturbance of Individuals Affects Community Structure

One of the basic premises of nearshore ecology is that physical disturbance is important in structuring wave-swept communities (Paine and Levin, 1981; Quinn, 1979; Sousa, 1984). Consider, for instance, the concept of "wave exposure." The term has never been well defined but is generally accepted to represent an integrated index of both the wave-induced hydrodynamic forces on a site and any ancillary effects on organisms related to water motion. Thus, an "exposed" site is, on average, subjected to large waves and the associated rapid water movement, whereas a "protected" site is, on average, subjected to smaller waves and less rapid water motion. Associated with wave motion are effects on temperature, salinity, and desiccation: at exposed sites, high intertidal organisms benefit from wave spray and can survive higher on the shore without desiccating than can organisms at protected sites. Despite the quantitative fuzziness of wave exposure as a concept, it is nonetheless useful as an ecological index. In much the sense that one may learn to differentiate a good wine from a poor one without being able to quantify the difference or easily explain it to someone else, an association with wave-swept rocky shores provides one with an accurate (though qualitative) "feeling" for the wave exposure of a site.

Correlated with this index is a reproducible trend in species composition and community structure (Ballantine, 1961; Lewis, 1964; Ricketts et al., 1968). If told that you are working at a fully exposed site on the west coast of North America, an intertidal ecologist will be able to describe for you the major species occupying space on the rock, and in many cases will be able to describe the important ecological interactions among these species. If you enquire about a protected site, the answer will entail a different set of species and interactions, but the description may be equally accurate. Certainly there are factors other than wave exposure that have important effects (type of substratum, spatial heterogeneity, history, etc.), but knowledge of wave exposure as an index of flow regime is nonetheless a useful predictor of species composition and community structure.

In some cases, the mechanistic basis of exposure effects is obvious. The presence of severe wave forces may be sufficient to explain why fragile or less tenacious species typical of protected sites are not found in more exposed areas.

In a few cases, a mechanistic link between the flow regime and community structure has been demonstrated experimentally. For instance, the rate at which competitively dominant species are disturbed by physical factors can go far toward explaining the structure of wave-swept intertidal communities. A classic example concerns the mussel *Mytilus californianus*. On the coast of Washington State this mussel is the dominant occupier of space and, in the absence of predation and wave-induced disturbance, forms a monoculture in the mid-intertidal zone (Paine, 1974).

In areas where hydrodynamic forces are relatively low, the sea star *Pisaster ochraceus* is free to prey upon small mussels, and the resulting gaps in the mussel beds form temporary spatial refugia for a wide variety of competitively inferior species. Thus, predation-induced disturbance serves to increase the species richness of the community. For its role in maintaining species richness, *Pisaster* has been dubbed a "keystone" predator (Paine, 1966).

On shores exposed to severe wave action, the sea star may be prohibited from foraging (Quinn, 1979). However, under these conditions wave-induced hydrodynamic forces act to open gaps in the mussel bed and thereby to elevate species richness. In this respect, wave forces play the role of a keystone predator. Similar effects have been noted for the species diversity of benthic algae on boulders subject to wave-induced overturning (Sousa, 1979a, b).

Correlations between Hydrodynamics and Morphology

In addition to correlations between species ranges, community structure, and wave exposure, two ecotypic patterns are apparent in intertidal organisms that suggest the influence of hydrodynamics in the evolution of structure and function.

First, organisms that inhabit wave-swept shores are generally quite small (Denny et al., 1985). For example, the largest animals on the exposed coast of Washington State are sea stars (*P. ochraceus*) that measure at most 20 cm from arm tip to arm tip. The largest algae in the intertidal zone may reach 1–2 m in length. When compared to organisms that inhabit terrestrial environments, or even those that occur subtidally on the same shores, intertidal organisms are small.

Denny et al. (1985) showed that the force imposed by water acceleration in breaking waves (the acceleration reaction described below) scales with size, such that larger individuals of a species are at greater risk of breakage or removal than are small individuals, and proposed that this increase in risk might select for small size in wave-swept organisms. Although the idea is appealing, the absolute size limits predicted (while still small) are larger than those actually observed. The validity of this proposition awaits experimental verification, a process that may be confounded by the simultaneous effects of biological interactions in determining maximal size.

Second, Vermeij (1978) has suggested that the ornamentation of a wide variety of gastropod shells increases along a cline of decreasing average wave action. For example, shells of intertidal gastropods in Alaska, where the average offshore wave height in January is 1.75 m (Gorshkov, 1976), have less ornamentation than do gastropods in Panama, where the average offshore wave height is less than 1 m. As with the trend in size, this is a suggestive correlation. But wave forces have not been demonstrated to be the sole mechanism behind the correlation, and Vermeij (1987) suggests a variety of alternatives ranging from clines in predation to clines in the solubility of calcium carbonate.

In light of this evidence it seems safe to assume that the flow environment of wave-swept rocky shores is important in both the ecology and functional morphology of intertidal organisms. In what fashion can ecomorphological research add to our understanding of this system? In the following pages I explore three roles that seem most promising: (1) to predict on a mechanistic basis the vital rates (birth and survivorship) of individual plants and animals, (2) to quantify the wave exposure of sites, and (3) to assist in defining the limits and functional roles of morphology in ecology.

And finally, I turn around the question posed above and ask instead how our understanding of the wave-swept environment and nearshore ecology can aid in the study of problematic areas in ecomorphology. In particular, I explore the possibility that the statistics of wave forces can serve as a tool for an examination of the mechanistic connection among behavior, performance, and fitness.

PREDICTING VITAL RATES

Flows on wave-swept shores are stochastic and highly complex, and our current understanding is at best a first approximation of the actual flow regime. However, the extent of what we do *not* know should not be allowed to mask that which we *do*. Much of the theory of wave propagation, interaction, and breaking derived in the context of sandy beaches and breakwaters can be applied to the wave climate on rocky shores (Carstens, 1968; Koehl, 1984; Denny, 1988, 1991, 1993; Denny et al., 1985; Denny and Gaines, 1990). In particular, theory regarding the distribution of wave heights allows one to predict the maximal force that will be imposed on an intertidal organism in a given period (Denny, 1991, 1993; Denny and Gaines, 1990).

The Maximal Height of Waves

The argument is as follows. Ocean waves are created by the interaction of wind with the water's surface. The faster the wind, and the longer the distance over which it acts (the fetch), the greater is the wave height (Kinsman, 1965). After waves are formed, they may propagate for great distances. For example, much of the long-period swell that is characteristic of shores on the west coast of North America is due to storms in the northern Pacific Ocean several thousand miles away. Because surface waves can travel long distances, a complex wave pattern is formed as waves from various sources are mixed together near the shore.

The simplest way to describe this "random sea" is in terms of overall "waviness." This is commonly done by measuring the elevation of the ocean surface at one point as a function of time, and expressing the waviness as the standard deviation in surface elevation. For example, if one measures surface elevation, E, every second for N seconds, mean sea level, $\bar{E}$, is,

$$\bar{E} = (1/N) \sum_{i=1}^{N} E_i, \qquad \text{(eq. 8.1)}$$

and the standard deviation in surface elevation (also known as the root mean square wave amplitude) is

$$A_{\text{rms}} = [(1/N) \sum_{i=1}^{N} (E_i - \overline{E})^2]^{1/2}.$$ (eq. 8.2)

Note that this statistic is a description of the entire surface, not just the wave crests and troughs. For a sinusoidal wave (a close approximation to an ocean wave), the peak amplitude (either crest or trough) is $\sqrt{2} \, A_{\text{rms}}$.

Now, each wave has both a crest and a trough, and the overall wave height is the sum of the two amplitudes. Thus, the waviness of the sea can be described in terms of the root-mean-square wave height, H_{rms}:

$$H_{\text{rms}} = 2\sqrt{2} \, A_{\text{rms}}.$$ (eq. 8.3)

At any given time, a substantial fraction of the overall waviness of the sea is due to small waves that are difficult to observe. As a result, H_{rms} is smaller than the wave height one perceives when looking out to sea. This perceived wave height, yet another description of waviness, is better described by the significant wave height, H_s, the average height of the one-third highest waves. Longuet-Higgins (1952) has shown that

$$H_s \approx \sqrt{2} \, H_{\text{rms}} = 4 A_{\text{rms}}.$$ (eq. 8.4)

Please recall that this waviness is due to the interaction of waves from many sources. In a classic analysis of a similar situation involving sound waves, Rayleigh (1894) showed that if the waves that are mixed have similar periods (a condition referred to as being "narrow banded"), the result is a specific distribution of wave heights, the Rayleigh distribution (fig. 8.1).

Unlike the more familiar Gaussian (or normal) distribution, which requires two parameters in its description (the mean and the standard deviation), the Rayleigh distribution can be described completely by the single parameter H_s. For example, the probability is P that a wave chosen at random exceeds a height H when the waviness is H_s:

$$P(H) = \exp -2 (H/H_s)^2.$$ (eq. 8.5)

There is substantial evidence that ocean waves generally conform to a Rayleigh distribution (Sarpkaya and Isaacson, 1981; Thornton and Guza, 1983). When the sea surface is Rayleigh distributed, Longuet-Higgins (1952) has shown that the highest wave expected in a random sample of N waves is

$$H_{\text{max}} = \frac{1}{\sqrt{2}} \, [(\ln N)^{1/2} + 0.28861 \, (\ln N)^{-1/2}] H_s.$$ (eq. 8.6)

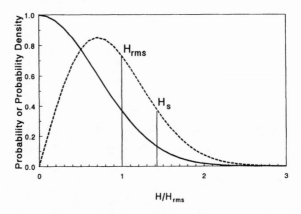

FIGURE 8.1 The distribution of wave heights can often be described accurately by the Rayleigh distribution, characterized by either the root mean square wave height (H_{rms}) or the significant wave height (H_s), both measures of average "waviness."

Note that this equation differs from that used by Denny (1991, 1993), which referred (incorrectly) to the most probable (modal) value of the maximal wave height rather than the expected (mean) value. The difference between the two equations is slight, however.

If waves have an average period of T seconds, $N = \tau/T$ waves pass in τ seconds. Thus,

$$H_{max} = \frac{1}{\sqrt{2}} \left\{ [\ln (\tau/T)]^{1/2} + 0.28861 \, [\ln (\tau/T)]^{-1/2} \right\} H_s. \qquad \text{(eq. 8.7)}$$

The longer the period one considers, the higher the wave one expects to find (fig. 8.2).

If H_s were constant, one could easily predict the maximum height of a wave expected in the course of, say, a year. However, the waviness of the sea varies from day to day. This variation can be taken into account. A comparison of wave data from a variety of sites in Washington and California shows that the distribution of significant wave heights is similar among sites when each site is normalized to the mean H_s for that site (Denny, 1993). The shape of this distribution is such that significant wave heights much larger than the mean (storm waves, for instance) are indeed present, but only for a relatively small fraction of the time. The short time that a large H_s is present means that the largest waves may actually be those associated with less wavy days. These days have smaller waves on average, but are present so often that there is a substantial chance of encountering a very large wave (as predicted by eq. 8.7).

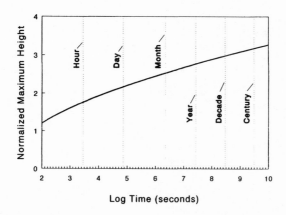

FIGURE 8.2 The maximal wave height as a function of time when H_s is constant. Maximal height is normalized to the prevailing significant wave height and a wave period of 10 s is assumed. (From Denny, 1991.)

Calculations made by Denny (1991, 1993) suggest that the maximum wave height observed in the course of a year is approximately 5.5 times the yearly mean significant wave height at a given site (fig. 8.3). A typical mean significant wave height for an exposed site is 2 m on the Pacific coast of North America (U.S. Army Corps of Engineers, Coastal Data Information Service 1982 to 1987), implying that in the course of a year the site will probably be subjected to a wave 11 m high. The maximum velocity (u_{max}) associated with this wave can be estimated from solitary wave theory (Carstens, 1968; Denny, 1988)

$$u_{max} \approx (2gH_{max})^{1/2}, \qquad \text{(eq. 8.8)}$$

where g is the acceleration due to gravity. For $H_{max} = 11$ m, u_{max} is approximately 15 m s^{-1}. This is in excellent accord with the maximal water velocities estimated from field measurements, which range from 14 m s^{-1} (Denny et al., 1985) to 16 m s^{-1} (Vogel, 1981).

Survivorship

In turn, maximal velocity can be translated into an estimate of the maximal lift and drag imposed on intertidal organisms (Denny, 1988). Lift is a force acting perpendicular to the direction of flow. For an organism with a characteristic planform area, S_p (area projected along the direction in which lift acts), lift can be estimated as

$$\text{Lift} = \tfrac{1}{2}\,\rho\,u^2\,S_p\,C_l \qquad \text{(eq. 8.9)}$$

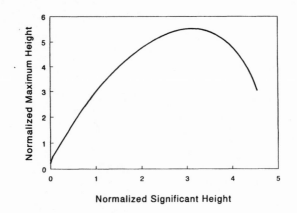

FIGURE 8.3 If H_s varies through time, the expected maximal wave height depends on the yearly mean significant wave height. Both maximal wave height and instantaneous significant wave height are normalized to the yearly mean significant wave height. (From Denny, 1991.)

where u is the water velocity and ρ is the water's density (1025 kg m^{-3} for seawater). C_l is the lift coefficient, primarily a function of the organism's shape and secondarily of its size and the water velocity. The lift coefficient for a given organism can be measured in the laboratory. If one knows the size of a plant or animal (expressed as S_p) and its lift coefficient, one can calculate maximal lift from a knowledge of maximal velocity.

Drag acts in the direction of flow and can be estimated as

$$\text{Drag} = \tfrac{1}{2}\,\rho\,u^2\,S_f\,C_{d'} \qquad\qquad (\text{eq. }8.10)$$

where C_d is the drag coefficient and S_f (the frontal or profile area) is the area projected along the direction of flow. $C_{d'}$ like $C_{l'}$ can be measured in the laboratory. Once C_d is known, drag can be calculated from a knowledge of velocity.

In addition to lift and drag, there is a third hydrodynamic force acting on intertidal organisms, the acceleration reaction mentioned above in connection with the mechanical limits to size. For a sessile organism

$$\text{Acceleration Reaction} = (1 + \alpha)\,\rho\,V\,a, \qquad\qquad (\text{eq. }8.11)$$

where V is the volume of water displaced by the organism, and a is the water's acceleration relative to the organism (Denny et al., 1985). The added mass coefficient, α, is a function of shape (among other things) and can be measured in the laboratory. Given measured values for V and α, one could calculate the acceleration reaction from a knowledge of wave-induced accelerations. Unfortunately, we lack a method to estimate a from wave height, so this force cannot yet be linked directly to wave statistics. However, except for the largest of intertidal organisms, acceleration reaction is likely to be much smaller than lift and drag, and its effects therefore can be neglected (Denny, 1988).

In summary, knowledge of wave statistics, coupled with laboratory measurements of C_d and C_1, allows one to paint an approximate worst-case scenario for the long-term maximal hydrodynamic forces acting on intertidal organisms. Note that this does not imply that the flow regime is predictable on a short-time scale, nor is it meant to imply that the topographic complexity of the substratum does not affect the hydrodynamic forces experienced by benthic organisms. These complications will be discussed in detail in a later section.

The possibility of predicting maximal wave forces suggests a way that hydrodynamics can be integrated with ecological models of disturbance to help explain species interaction and species richness on wave-swept shores. Consider, for instance, the seminal work of Levin and Paine (1974, 1975) and Paine and Levin (1981) in exploring the dynamics of the gaps (or "patches") of free space created in mussel beds by the action of waves. From empirical measurements, they estimated the rate at which patches are "born" and the size distribution of patches at their birth. From further measurements regarding the rate at which mussels invade patches, they were able to estimate the "death" rate of patches. Combining these data, they modeled the demography of patches on Tatoosh Island (Washington State) and tested their predictions against actual patch data. Their predictions closely matched the observed patterns over a period of seven years. Thus, Paine and Levin (1981) have demonstrated that it is possible to predict accurately the dynamics of patches in the space dominated by mussels, given an average rate of patch formation and size-specific reduction rates.

In turn, it may be possible to account for the rate of patch formation on the basis of fluid mechanics. It is possible to estimate from empirical measurements the distribution of tenacity of a population of mussels (Denny et al., 1985; Denny, 1987). From this distribution, the probability of a mussel being removed by a given hydrodynamic force can be calculated. Now, Denny (1987) has shown that water flowing over mussel beds exerts a substantial lift force, and predictions (per eq. 8.9) can be made as to how this force varies with water velocity. Thus, from a knowledge of maximal wave-induced velocity, one can predict the fraction of mussels that will be dislodged (i.e., the birth rate of patches). The maximal velocity can, in turn, be estimated from the mean significant wave height, as outlined above. In sum, it is possible in theory (and perhaps in practice) to predict the long-term rate of hydrodynamically caused patch formation in mussel beds based on a knowledge of the "waviness" of the adjacent ocean.

The accuracy of this kind of mechanistic prediction has yet to be experimentally demonstrated. Preliminary results are encouraging, however. For example, if we assume an annual maximum water velocity of 15 m s^{-1} (as discussed above) information presented by Denny (1987) suggests that the resulting lift would be sufficient to dislodge approximately 30% of the mussels present. Actual removal varies between 2% and 49% (Paine and Levin, 1981), suggesting that predictions made on the basis of hydrodynamics are of the correct order of magnitude. There are a variety of reasons why actual rates of dislodgment are in many cases lower than predicted by the simple model outlined here. In particular, the local topography of a site may be such that individual mussels are protected from the maximal flow.

At least one other study has demonstrated the potential for using a mechanistic approach to predicting survivorship. Carrington (1990) has shown that the high intertidal alga *Mastocarpus papillatus* has blades that grow continuously, but that these are attached to the substratum via a stipe with a fixed cross-sectional area and strength. Because drag acting on a blade is a function of blade area (eq. 8.10), the larger the blade the higher the probability that the stipe will be broken by a given velocity. This situation led Carrington to predict maximum blade size as a function of water velocity. For example, a velocity of 15 m s^{-1} would, on average, break stipes with blades larger than about 7 cm^2 in area. Indeed, this is approximately the size of blades remaining at an exposed site after winter storms.

Although the mechanistic prediction of survivorship appears to hold great promise, there are several practical problems to be solved before this promise may be fully realized. First, accurate measurement of the strength of intertidal organisms is not as straightforward as it may appear. The standard approach is to measure carefully the force required to remove haphazardly chosen individuals at a given site. The resulting distribution of strengths is then used to estimate the probability than an individual has less than a given tenacity, and from this the probability of being dislodged by a given hydrodynamic force can be calculated.

A problem with this approach becomes apparent when one realizes that strength measurements are made in the midst of a dynamic process. In essence, one is measuring the "standing crop" of strengths in a population for which the turnover in individuals may be quite high. For example, individual mussels of a hypothetical species are, in general, strongly attached to the substratum by their byssal threads and can withstand the hydrodynamic forces exerted by all but the largest storm waves. However, the fatigue characteristics of these byssal threads are such that every week 10% of the population suddenly loses all of its attachment strength. Each of these weakened individuals is likely to be carried away by the next large wave, and the size of the adult population thus tends to decrease by 10% each week. The rate of recruitment to the hypothetical population is high enough, however, so that population size is in a steady state.

Given this scenario, what will the standard measurement of strength reveal? The only weak individuals present are those that have lost their byssal strength since the passage of the last large wave, and these form only a minuscule fraction of the population. As a result, the measured strength distribution has the high mean characteristic of unfatigued byssal threads, and the application of wave and hydrodynamic theories will suggest (incorrectly) that very few individuals are ever dislodged by wave-induced forces. In this hypothetical scenario, the weekly disappearance of 10% of the population is invisible to the standard approach to predicting survivorship. It remains to be seen how important this factor is in natural populations.

Measurements of strength can also be affected by behavior. For example, in times of stormy weather mussels lay down more byssal threads, thereby increasing their ability to adhere to the substratum (Price, 1980). Limpets adhere less strongly when foraging, but forage less often when waves are high (Wright, 1978), and sponges can remodel their bodies and the materials from which they are constructed depending on their exposure to waves (Palumbi, 1984a). Because of the ability of such organisms to respond to their environment, extra care must be taken in determining their effective strength.

The mechanistic prediction of survivorship as outlined here also requires knowledge of the long-term wave statistics for a given site. Unless one knows the yearly mean significant wave height, one cannot calculate the maximal expected wave height. Unfortunately, long-term wave data are available only for those few sites where foresightful engineers or biologists have installed recording wave gauges. A potential solution to this problem may be available, however. The National Oceanic and Atmospheric Administration (NOAA) Center for Ocean Analysis and Prediction (COAP) (Monterey, California) supports a continuous effort to predict the state of the ocean's surface based on available data regarding global wind patterns. From 1975 to 1984 this effort took the form of a Spectral Ocean Wave Model (SOWM), capable of predicting various wave parameters (including H_s, period, and average direction) for any point in the Northern Hemisphere. Since 1984, the model has been extended to cover the entire globe (GSOWM). The predictions of these models (archived by COAP) allow one to "hindcast" the wave climate for any point on the coast and thereby to obtain an estimate of the long-term wave climate.

The accuracy of these predictions for this application remains to be demonstrated, and there are some complications (Denny, 1993). For instance, the wave parameters predicted by SOWM are for waves in deep water, and these must be transformed for a given site as waves move onshore (U.S. Army Corps of Engineers, 1984). Nonetheless, the presence of a long-term global data base for wave parameters holds immense potential for intertidal ecomorphology.

Finally, there is a need for further research regarding the link between wave height and maximal water velocity. The prediction of equation 8.8 is at best a first approximation and cannot be expected to apply equally well to all shore topographies (Denny, 1988, 1991; Denny and Gaines, 1990).

Birth Rates

In addition to predicting survivorship, the biomechanical approach may be useful in predicting birth rates. For instance, Pennington (1985) and Levitan (1991) have demonstrated that turbulent mixing in the nearshore can drastically reduce the effectiveness of external fertilization in sea urchins. Calculations by Denny and Shibata (1989) suggest that this problem may be widespread among wave-swept organisms, raising interesting ecological questions regarding reproductive strategy. The application of fluid-mechanical theory to "fertilization ecology" promises to be productive.

QUANTIFYING WAVE EXPOSURE

As noted above, the concept of wave exposure has been used by ecologists as an index of environmental severity. As useful as this index has been, its full potential as a tool for understanding life on wave-swept shores has not been realized because the concept has proven difficult to quantify. For example, species richness on wave-swept shores is often maximal at an intermediate exposure (e.g., Sousa, 1979a, b), and it would be useful to know quantitatively how exposed a site must be before its species richness declines, or how species richness compares for sites of equal exposure in different parts of the world. Similar questions can be asked in regard to the effect of wave exposure on competition and predation (see Menge and Sutherland, 1987). At present, it is not productive to ask these questions because wave exposure has not been quantified. Ecomorphological studies, however, have the potential to quantify wave exposure in some cases.

In part the failure to quantify wave exposure has been due to problems inherent in how the index is defined. For example, when applied at the community level to a given site, it is generally agreed that wave exposure is a complex mixture of environmental factors (e.g., hydrodynamic forces, thermal, osmotic, and desiccation stresses) that correlates both with the species composition and community structure and with the average waviness of the adjacent sea. In this case it is permissible to think of a single exposure value for a given site. However, many different species may inhabit that site, each responding differently to aspects of the environment. A site that is stressful for one species may be benign for another, so there may be many different specific exposures present at a single site. In this respect, "community wave exposure" is fundamentally different from "species wave exposure."

It will be possible to predict community structure on a mechanistic basis only when we understand all of the important physical interactions among species and the interaction of each species with the environment. This is a goal that will not be realized in the near future. As a result, present efforts to quantitatively measure community wave exposure can at best provide data regarding the relative exposure of sites. For example, the rate at which standard objects are bent, broken, or displaced (Dayton, 1971; Harger, 1970) and the rate at which plaster shapes dissolve (Doty, 1971; Muus, 1968) have been used as indicators of wave exposure and do indeed provide numbers that allow the objective ranking of sites on a qualitative exposure gradient. But each of these indicators responds to an unspecified (and often unspecifiable) combination of environmental factors, and this ambiguity as to what is being measured renders these methods useless as means for quantitatively describing the specific mechanisms of exposure.

Of greater potential value are the predictive wave models discussed above. For a "site" large enough to dwarf local variations in microtopography (e.g., a few km of shoreline), long-term hindcasts of mean significant wave height may serve as an appropriate, quantitative index of community wave exposure. The advantages of such an index include the ability to compare readily among many sites around the world and a direct link to the mechanistic predictions of maximal water velocity afforded by the wave theories discussed above. However, large-scale wave height predictions are unlikely to allow one to make meaningful predictions of wave exposure on a scale small enough to be affected by local microtopography. On a small scale, this method (like those described above) is doomed to provide at best a relative measure of exposure.

In contrast, the ecomorphological approach does provide us with tools to quantify species wave exposure and thereby to assist in exploring the interactions which contribute to community structure.

The quantification of species wave exposure is largely an exercise in the appropriate use of technology. The trick is to decide on a single component of the flow environment to measure. Although the results are necessarily limited in scope, they can be precisely interpreted. For example, Jones and Demetropoulos (1968) used a recording spring scale to quantify the maximum water velocity imposed at a variety of sites and correlated these results with the distribution of algal species. Because this technique quantitatively measures a single flow parameter (velocity), it allows for direct, quantitative comparison among sites and for experiments in the laboratory. For instance, it is possible to reproduce field velocities in laboratory flumes, thereby to observe the behavior of organisms subjected to this aspect of the environment. One might notice that a species breaks at a given water velocity in the laboratory. If this velocity corresponds to a critical velocity (noted in the field) above which the species is seldom found, one has strong evidence that water velocity (rather than some other environmental parameter) influences the distribution of this species. Measurement of an ambiguous mixture of environmental parameters can never provide this type of evidence.

The basic approach pioneered by Jones and Demetropoulos (1968) has been refined by several researchers (e.g., Denny, 1983; Palumbi, 1984b). The utility of this approach is exemplified by Denny and Gaines (1990), who provide evidence that the distribution of maximal hydrodynamic forces on wave-swept shores takes a similar form independent of the precise topography of a site and local wave climate. One can infer from these results that a relatively few measurements of maximal force (or, equivalently, velocity) over a short period would allow one to predict the long-term maximal force (or velocity) imposed on a site. Together, these predictions could then serve as an appropriate, quantitative index of species wave exposure.

DEFINING THE LIMITS AND FUNCTIONS OF STRUCTURES

A chronic problem facing ecologists is the necessity to understand what aspects of an organism's structure have ecological importance and, of those that do, how they function. A similar problem is encountered by functional morphologists, who are often faced with the task of understanding whether or not there is a substantive interaction between a structure and a particular environmental factor. For instance, when presented with a streamlined limpet shell, both ecologist and morphologist may ask, Does this shape give the limpet some advantage by reducing drag or avoiding predation by crabs, or is it a result of genetic drift or a ghost of past interactions? Definitive demonstration of the role(s) of a particular structure is difficult in any environment (Endler, 1986; Garland and Losos, chap. 10, this volume), but an ecomorphological approach can be useful in solving these problems.

In some cases, the tools of functional morphology can be employed in straightforward fashion. For instance, in light of the previous discussion of maximal hydrodynamic forces, it is easy to see how knowledge of the shape and strength of an organism can be used in the study of distributional patterns. Consider the case of a hypothetical branching coral. If this species has a sufficiently low strength of skeleton and grows in a high-drag morphology, calculations can show that it is highly unlikely that the species can survive in the wave climate of a windward shore. By providing a means to calculate the physical limits of a species, functional morphology can provide the ecologist with a clue as to whether the distribution of that species is set by hydrodynamic factors, biological interactions, or a combination of the two.

In other cases, functional morphology may show how a previously ignored aspect of the environment can have ecological implications. The effect of turbulence on external fertilization discussed earlier is a prime example, and other cases are reviewed by Denny (1988), Koehl (1984), and Vogel (1981).

LIMITS TO STANDARD METHODS OF FUNCTIONAL MORPHOLOGY

Much of this book is devoted to a demonstration of ways that the standard tools of functional morphology can be used in an ecological context. Rather than review here additional cases that relate specifically to wave-swept shores, I explore instead how this particular environment places limits on the standard approaches used by functional morphology.

There are three traditional methods that functional morphologists have used to demonstrate the adaptive nature of biological structures: (1) morphological convergence among species when faced with similar environments, (2) resemblance of a structure to a theoretical optimum, and (3) the superiority of a structure when compared to an appropriate model. The use of each of these approaches is severely limited in the wave-swept environment.

Convergence Among Species

A general mechanism for demonstrating the functional role of a structural trait is to show that different species have converged on that expression when exposed to the same set of selective factors. Convergence can be implied among homologous structures of related species, or among nonhomologous structures of nonrelated (or very distantly related) organisms. Given the diversity of phyla represented in the intertidal zone, it is particularly tempting to search for evidence that hydrodynamics is a selective factor for morphological convergence among phyla.

However, one striking characteristic of intertidal organisms is the lack of apparent convergence in traits believed to be most important for interaction with hydrodynamic forces. This lack is most apparent in macrophytic algae. For example, on the shores of Washington State it is common to find balloon-shaped *Halosaccion* living next to bushy *Mastocarpus,* both coexisting with encrusting and erect coralline algae.

Similarly, it is common to find mobile, streamlined limpets living next to sessile, relatively bluff acorn barnacles. Both have rigid shells but commonly co-occur with gooseneck barnacles, which have an unfused shell mounted on a flexible stalk. Although we have discussed why it is reasonable to assume that hydrodynamic forces pose a substantial risk to the survivorship of intertidal species, it is at present difficult to imagine how an argument can be made that selection by hydrodynamic forces has resulted in convergence in shape.

This is in contrast to the wings of flying animals, for instance. Aerodynamics clearly plays an important role in determining an animal's ability to fly (and thereby survive), just as hydrodynamics is demonstrably important in the survival of wave-swept benthic organisms. In the case of wings, however, insects, mammals, and birds have converged on a relatively small range of shapes.

The absence of apparent convergence among intertidal species leads one to suppose either that each species has responded to a different set of environmental factors (in which case hydrodynamics is not useful as a means of making general statements about the structure of intertidal organisms), or that (unlike the case of wings) fluid-dynamic forces have not been the predominant selective factor in the evolution of shape in intertidal organisms. In this second case, existing shapes are likely to be a manifestation of important trade-offs among several factors.

Demonstrating Selective Neutrality

Studies of functional morphology provide evidence for this second alternative by demonstrating that a variety of different shapes have the same hydrodynamic characteristics and that as a result, hydrodynamic forces cannot distinguish among them. Two likely examples of such selective neutrality have been documented for intertidal organisms.

Carrington (1990) has shown that the drag coefficients of a wide variety of intertidal macroalgae are indistinguishable at water velocities high enough to pose a risk to the algae (fig. 8.4). This similarity in C_d is in sharp contrast to the dissimilarity in the forms of these algae when they are not subjected to flow (fig. 8.5). In this case, convergence in drag coefficient is due to the flexibility of the organisms involved. At high water velocities, virtually any shape of flexible alga bends into a more-or-less streamlined form. The tendency for flexibility to render different shapes equal in their drag characteristics suggests that once a lineage of algae has become flexible, shape is to a large degree removed from further selection by drag.

If this argument is valid, it can help to explain the vast diversity of form among intertidal algae. In a sense, most of these algae (encrusting forms being the prominent exception) *have* converged in their drag characteristics in that they are flexible. Further convergence cannot easily occur because all the existing forms are selectively neutral to drag.

A similar argument may apply to some rigid structures. Denny (1989) found a limpet shell that, through very minor changes in shape relative to that found in conspecifics, had reduced its drag coefficient by up to 40%. The existence of this avenue to substantial drag reduction raised the question as to why this apparently advantageous trait was not found in all members of the species. The apparent answer is that lift rather than drag is the force most likely to dislodge this species, and the shape that reduces drag does not reduce lift. In other words, even if a

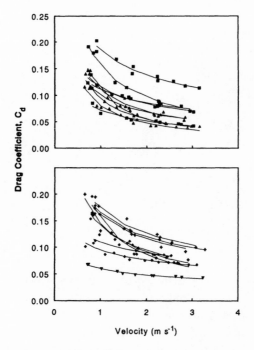

FIGURE 8.4 A wide variety of intertidal macrophytic algae have virtually the same drag coefficients at high water velocities: (■) *Endocladia muricata*, (▲) *Iridaea flaccida*, (♦) *Fucus distichus*, (+) *Gigartina leptorhynchos*, (▼) *Pelvetia fastigiata*. (Data, with permission, from Carrington, 1990.)

limpet could reduce its drag without incurring any concomitant expense, natural selection could not guide it to evolve in this direction. Because lift is the dominant selective force, a wide range of drag coefficients are functionally neutral. In light of this evidence, it is unrealistic to expect convergence among limpets for a shape that reduces drag.

If lift is indeed the dominant hydrodynamic force imposed on limpets, and if hydrodynamic forces are a substantial source of mortality, one might expect convergence among species for shapes that reduce lift. Unfortunately, the relationship between shell shape and lift has not yet been sufficiently studied to allow a search for this convergence.

Problems of functional neutrality extend to issues of materials as well as shape. For example, many sessile plants and animals are glued firmly to the rocks of the shore. The glues of some of these organisms (acorn barnacles are the classic example) approach the strength of the rock. For instance, Denny (1988) reports that of 342 *Balanus glandula* pulled from a rock at Shi Shi Beach, Washington State, 43% came loose because the rock rather than the glue gave

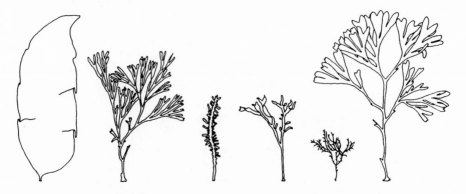

FIGURE 8.5 Despite similarity in drag coefficients, the following algae have very different morphologies: from left to right, *I. flaccida, P. fastigiata, Mastocarpus jardinii, E. muricata,* and *F. distichus.* (Redrawn with permission from Carrington, 1990.)

way. In cases such as this, the strength of the substratum places an upper limit on the ability of hydrodynamic forces to select for a stronger glue. Any glue stronger than the rock is functionally neutral.

Although studies such as these show how difficult it can be to demonstrate the adaptive significance of a structure, they nonetheless serve an important function. By demonstrating selective neutrality, they can help to define specific aspects of shape (for limpets, C_1 but not C_d) where convergence is unlikely to occur and can thereby focus attention on the levels of organization (e.g., flexibility versus rigidity) at which convergence may be examined productively. In this respect, the guidance provided by appropriate study of functional morphology can help to avoid one of the common problems associated with the application of comparative phylogenetic methods: the assumption that there is a one-to-one match between selective regimes and organismal traits (Losos and Miles, chap. 4, this volume).

Optimality

In studies of functional morphology, it is common to calculate the shape of a structure or the property of a material that is in some sense optimal for coping with a given environmental parameter. The criterion used to define optimality depends on the situation (Alexander, 1982), but is typically either the minimization of energy needed to produce the structure or material or the maximization of the return to the organism (either in terms of energy or reproduction) when the structure is used. If the actual structure or material approximates the calculated optimum, it is taken as evidence that one understands the essence of the situation as well as providing provisional evidence that the structure or material has evolved in response to this particular aspect of selection (Oster and Wilson, 1978).

As useful as the optimality approach has been in a variety of biomechanical studies (e.g., Alexander, 1982; Kingsolver and Daniel, 1979, 1983; Denny, 1976; Norberg, chap. 9, this volume), it has its limitations. Several of these (e.g., historical constraints, lack of sufficient variation, selection at one level adversely affecting function at another level) have been discussed in depth by Dawkins (1982) and will not be explored further here. There are three other reasons, however, why optimality is unlikely to be a useful approach to the study of functional morphology in the wave-swept intertidal zone: all relate to the difficulty of determining the optimum.

Temporal Variability. First, the wave-swept environment of rocky shores is extremely variable in both time and space. For example, the heights of individual waves behave in ensemble as if they are statistically independent. In other words, knowledge regarding the height of one wave does not allow one to predict the height of subsequent waves. Because water velocity, and thereby hydrodynamic force, depends on wave height, this means that the magnitude of wave forces is, in the short term, unpredictable. Similarly, the direction from which large wave forces arrive has also been shown to be unpredictable (Denny, 1985).

This unpredictability mediates against any "fine tuning" between hydrodynamic forces and the shape of a rigid intertidal organism, for instance, a limpet. If the direction that water flowed over each individual were constant, it is easy to see how a shape which minimized lift, drag, or acceleration reaction might evolve. For example, limpets that inhabit the stipes of kelps can rely on the flexibility of the kelp to align their shells with the flow. In apparent response, these limpets have a streamlined shape. In contrast, limpets that inhabit rocks are subjected to flows from unpredictable directions. Any shell shape that reduces drag or lift for flow in one direction inevitably increases the lift and drag associated with flow from another direction. In this case, the "optimal" hydrodynamic shape represents a compromise among the risks associated with forces from different directions. As a result, it may not be possible to predict the optimal shape.

Consider a simple example. The overall probability, P_d, that a limpet will be dislodged is

$$P_d = \int_0^{2\pi} P_d'(\phi) P_w(\phi) \, d\phi, \qquad \text{(eq. 8.12)}$$

where ϕ is the angle from which the wave arrives, $P_w(\phi) \, d\phi$ is the probability that the wave arrives from the small range of angles $d\phi$ centered on ϕ, and P_d' is the probability that the animal will be dislodged by a force from this direction. If we assume that the direction of water motion is random, P_w is constant and equal to $1/2\pi$. P_d' is allowed to vary with ϕ, but for every decrease at a given angle, there is an increase at some other angle. If this trade-off is such that $\int_0^{2\pi} P_d'(\phi)d\phi$ is constant among different morphs, the overall probability of dislodgment is constant and an optimal shape cannot be defined.

This notion can be expressed in graphical form (fig. 8.6), where the probability of dislodgment for forces arriving from a given angle is given as the radial distance from the origin. Two hypothetical limpets are shown. One (the solid line) is radially symmetrical, so P_d' is a constant, forming a circle. The other limpet (dashed line) is streamlined such that the directional probability of dislodgment is decreased for waves arriving from the front or rear, but the animal is susceptible to removal by waves arriving from the side. However, because the area within the two curves (i.e., the integral given by eq. 8.12) is the same, the overall probability of dislodgment for the two limpets is the same and they are functionally equivalent to hydrodynamic forces.

At present, we do not know enough about the function $P_d'(\phi)$ to know accurately whether P_d is constant among limpet shapes. Preliminary evidence suggests, however, that it might be. I have measured the lift and drag coefficients for two morphologically dissimilar species, *Lottia gigantea* and *Acmaea mitra* (fig. 8.7). *L. gigantea* appears to be streamlined: its height is only 25% of its length and its apex lies well forward of the center of the shell. In contrast, *A. mitra* is a high-spired cone, with its apex located centrally and a height 63% of its length. The relevant dimensions and force coefficients for the two shells are given in table 8.1.

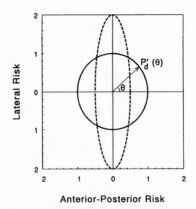

FIGURE 8.6 Risk of removal for hypothetical shapes. The risk of removal may vary with the direction of the imposed force.

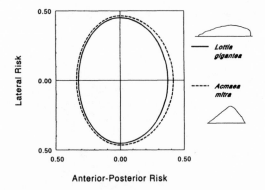

FIGURE 8.7 Risk of removal for two morphologically dissimilar limpets (lateral profiles shown here).

Flow passing over these shells from any direction imposes both a lift and a drag, and the overall force imposed on the limpet's basal adhesive is the vector sum of these two. We may take the ratio between the force per area imposed on the foot and the animal's adhesive tenacity (also measured as force per area) as an index, R, of the risk of dislodgment. This index may be further simplified by dividing by the dynamic pressure ($\frac{1}{2}\rho u^2$):

$$R = \frac{\text{Hydrodynamic Force/Foot Area}}{\text{Tenacity} \times \text{Dynamic Pressure}}$$

$$= \{[C_d(\phi)\, S_f(\phi)]^2 + [C_l(\phi)\, S_p]^2\}^{1/2}. \qquad \text{(eq. 8.13)}$$

where $C_d(\phi)$ and $C_l(\phi)$ are the drag and lift coefficients for angle ϕ, respectively. These values have only been measured for $\phi = 0$, $\pi/2$, π, and $3\pi/2$. However, if we assume that values at other angles can be smoothly interpolated between these known values, we can estimate the directional dependence of R. If we further assume that both species have the same adhesive tenacity (here assigned a value of 1 for simplicity), R is a measure of the relative risk of dislodgment. Thus, a polar plot of R (fig. 8.7) is analogous to the plots of P_d' in figure 8.6.

Despite the gross differences in shell morphology between *L. gigantea* and *A. mitra*, the area within their risk curves is quite similar. In fact, *A. mitra* would need an adhesive tenacity only 7% higher than that of *L. gigantea* to have the same overall risk of dislodgment.

The lack of a clearly optimal shape for limpets can help to explain the diversity of shapes found in these gastropods. If the ability to survive wave-induced forces is not strongly tied to a particular shape, the organism is free (in an evolu-

TABLE 8.1. Drag coefficient (C_d), lift coefficient (C_l), and morphological data for two species of limpets

	Species	
Parameter	Lottia gigantea	Acmaea mitra
S_f ($\phi = 0, \pi$)	6.44 cm²	1.53 cm2
S_f ($\phi = \pi/2, 3\pi/2$)	8.16 cm²	1.52 cm2
S_p	29.20 cm²	3.52 cm2
C_d ($\phi = 0$)	0.24	0.65
C_d ($\phi = \pi/2, 3\pi/2$)	0.59	0.77
C_d ($\phi = \pi$)	0.25	0.60
C_l ($\phi = 0$)	0.37	0.37
C_l ($\phi = \pi/2, 3\pi/2$)	0.42	0.25
C_l ($\phi = \pi$)	0.32	0.22
Height/Length	0.25	0.63
Width/Length	0.78	1.06

Note: S_f is the area projected along the direction of flow, and S_p is the area projected along the direction in which the lift acts. The angle ϕ describes the direction from which flow impinges on the animal; $\phi = 0$ for flow from directly anterior. Shell length is aperture length along the anterior-posterior axis. Width is aperture width measured perpendicular to length. Height is the distance from the plane of the aperture to the top of the shell.

tionary sense) to retain a shape it acquired in another environmental setting or to assume a shape in response to other selective factors. For example, it seems likely that the "streamlined" shape of *L. gigantea* has evolved to provide the animal with an effective anterior plow for evicting interlopers from its territory (Stimson, 1970). In this and other such cases, a knowledge of the organism's history and ecological interactions is as important as a knowledge of its physical environment in interpreting why the organism is shaped as it is (see Losos and Miles, chap. 4, this volume).

The apparent similarity between the limpets discussed here should not be taken too seriously. The simple analysis presented here does not take into account several factors (e.g., the possibility that tenacity also depends on the direction in which force is imposed), and it is based on minimal data for the directional dependence of lift and drag. This example serves, however, to show how variability in the direction of flow can make it very difficult (or even impossible) to calculate an optimal form. In turn, it can be difficult or impossible to estimate how closely an organism approaches a morphological optimum, making it difficult to demonstrate the functional nature of a given morphology.

Spatial Variability. The problem of variability in the environment extends to aspects of spatial as well as temporal variability. To predict accurately the maximal velocity around a given individual, the local topography must be known. For example, a barnacle will be subjected to higher water velocities if it is on a peak or promontory of the substratum than if it is hidden in a valley or bay.

However, peaks and valleys, promontories and bays are terms that can be scale-dependent. A small-scale peak may be contained in a larger-scale valley with a consequent effect on the local flow. At each scale, the local flow depends both on local topography and on larger-scale features. Therein lies a problem.

The nature of topography on rocky shores is such that "fractal" variation within variation occurs at all scales. An example is shown in figure 8.8. Here the intersection of the ocean's surface with the rocky substratum at Pacific Grove, California, defines a shoreline. The topography of the shore is analyzed using a digital filter that acts roughly like a circle of a given diameter rolling along the shoreline. As the circle rolls, it occasionally encounters a valley or bay that is too small to allow entry and the filter records the area of the valley or bay below the circle (fig. 8.8a). The analysis begins with small circles and progresses to larger diameters, keeping track of the cumulative area of valleys and bays.

Figure 8.8b shows the cumulative valley area (per length of shoreline) as a function of circle diameter. As diameter is increased, the area of valleys and bays in which a circle of this size cannot fit increases at a roughly constant rate. For each tenfold increase in diameter, the available embayment area increases by a factor of about 15. Alternatively, we can express this result as the fractal dimension of the shoreline, which is the slope of the line on a log-log plot. For this shoreline, the slope is approximately 1.16, a value typical of shorelines in general (Mandelbrot, 1982). A similar result would be obtained if the filter were inverted so as to quantify peaks.

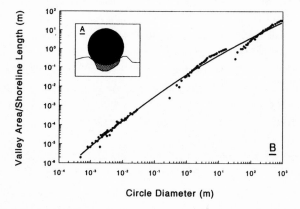

FIGURE 8.8 Variation in shoreline topography occurs at all scales. Here the area of valleys or bays (shown hatched) is measured by a rolling-circle filter as shown in *A*, revealing the fractal nature of the shoreline (*B*). See text for details.

In a practical sense, the fractal nature of the shoreline means that the bigger the organism, the greater the relative variability in local topography. The bigger the plant or animal, the more area of valleys and bays there are in which to hide, but also the more area of peaks and promontories on which the organism might be exposed.

As a result, it is very difficult to describe with any accuracy the topographical nature of a site and its effect on flow. Organisms settling within very small distances of each other may inhabit very different topographic situations and thereby experience different flows. Differences in flow may in turn elicit different functional responses. The net result is that the complex nature of spatial variability may have the same result for the study of function as does temporal variation: difficulty in calculating morphological optima.

Environmental Severity. The second problem with using optimality as an approach in intertidal ecomorphology concerns a question as to whether the physical environment is severe enough to give clear evidence of its selective nature. Is it safe to suppose that the wave-swept environment is *so* severe that nonoptimal forms will long ago have been weeded out?

There are reasons to think that the answer to this question is No. Consider, for instance, the species diversity of intertidal communities. It has been demonstrated both by theory and by measurement that diversity is highest where disturbance is less than maximal (see the review by Menge and Sutherland, 1987). If physical disturbance is too severe, the normal structural variability among individuals may not be sufficient to provide organisms capable of survival. For example, if wave-borne logs batter a shore, it is difficult to imagine how macroscopic representatives of any existing phylum could survive on exposed substrata. The high functional diversity present in most wave-swept intertidal communities is clear evidence, however, that rates of disturbance have been low enough so that a wide variety of organisms have been able to adapt effectively. In fact the diversity is so high that it appears that virtually *any* lineage can evolve a form capable of surviving in the intertidal zone. If the severity of the physical environment is sufficiently low to allow such diversity, it may not be safe to assume that severity is high enough to ensure optimality.

Larval Dispersal. And finally, there is the problem of larval dispersal. The majority of benthic species in the intertidal zone have a planktonic life stage. One effect of this life history strategy is that the offspring from one parent may be widely scattered. In a spatially variable physical environment such as the nearshore, dispersal insures that the young will be subjected to a variety of physical stresses. In this respect, the evolutionary consequences of larval dispersal are much the same as those of a temporally variable environment: the tendency will be for the retention of genetic (and thereby phenotypic) variability rather than fixation of a specific morph optimally adapted to one particular environmental stress.

In light of these factors, it seems unlikely that models for mechanical optima will serve as useful a role in the study of intertidal organisms as they have in others.

Comparison to a Physical Model

When a clear optimum is lacking, it is possible to compare an organism to some other standard. If the organism's structure is better adapted than that of the standard, it may serve as evidence for the structure's function. For example, if one morph of a species of snail survives the imposition of drag forces better than another (all other factors being equal), this serves as evidence that drag can play a selective role in the evolution of shell shape even if the present shape has not reached an optimum. This type of evidence is direct and compelling. In mechanical studies, however, it is tempting to use geometrical shapes or other simple models as a standard rather than another morph of the same or a closely related species. A branching coral is compared to a cylinder or a flat plate, for instance. However, the use of models as standards is fraught with hazards in the wave-swept environment.

Above all, the choice of a standard must be carefully considered. It is always possible to choose an object with particularly bad hydrodynamic performance. In such a case, the fact that a living organism performs better does little to provide evidence of adaptedness. It is also necessary to choose carefully the criteria of performance. Consider, for instance, drag as a criterion of hydrodynamic adaptedness.

At high water velocities characteristic of the intertidal zone, the drag, F, on nonstreamlined, bluff objects (such as a cylinder or flat plate held perpendicular to flow, a cube, or a sphere) can be accurately described by equation 8.10,

$$F = \frac{1}{2} \rho \, u^2 \, S_f C_d \qquad \text{(eq. 8.10)}$$

where the drag coefficient is constant across a wide range of velocities and object sizes. Thus, drag on bluff objects increases with the square of velocity (solid line, fig. 8.9).

Equation 8.14 is often used to define the drag coefficient:

$$C_d \equiv 2 \, F / (\rho \, u^2 \, S_f). \qquad \text{(eq. 8.14)}$$

Objects designed to reduce drag ("streamlined" objects) differ from bluff objects in two respects. First, they have lower drag coefficients; this is the usual

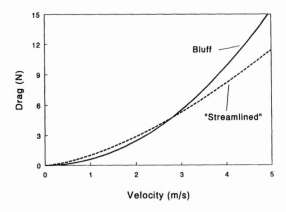

FIGURE 8.9 Drag of a bluff object and a streamlined object as a function of velocity.

criterion for deciding on whether an object is streamlined or not. However, streamlined objects are also characterized by the fact that their drag coefficients decrease with an increase in water velocity.

This variability in C_d is due to the way that the drag coefficient is defined. Consider the example of a hypothetical streamlined organism shown by the dashed line in figure 8.9. In this case, the drag on the object increases as $u^{1.5}$ rather than u^2. If the bluff object and the streamlined organism have the same characteristic area, the drag coefficients for the two are as shown in figure 8.10. At low velocity the drag coefficient of the organism is higher than that of the bluff body; at high velocity it is lower. Thus, the decision as to whether this organism is streamlined or not depends on velocity. In this respect, the drag coefficient as defined by equation 8.14 is less than satisfying. If streamlining is to be used as an index of *shape* it should not be a function of velocity for a rigid organism.

To avoid this problem, we can restate the equation for drag:

$$F = \tfrac{1}{2}\,\rho\,u^{\beta}\,S_f\,C'_d, \qquad\qquad\qquad (eq.\ 8.15)$$

where β is determined by a least-squares regression between drag and velocity, and C'_d (which we will call the shape coefficient) is now a constant. In the case of the archetypical bluff object (a flat plate held perpendicular to flow), $C_d = C'_d$ and $\beta = 2$. The ultimate streamlined object (a flat plate held parallel to flow) has a very low C'_d (about 0.007, in this case calculated using S_p) and $\beta \approx 1.5$. Representative values for β and C'_d are shown in table 8.2 for a variety of standard shapes. These values were measured in a flume with the object secured to the floor and are therefore appropriate for comparison with benthic organisms. Note that while β is generally equal to 2, this is not always the case. A hemisphere

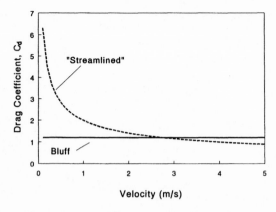

FIGURE 8.10 Drag coefficient for a bluff object and a streamlined object with the same characteristic area as a function of velocity. Because the drag of a streamlined object is lower at high velocities than that of a bluff object, but may be higher at low velocities, the drag coefficient is a problematic index of shape.

(which one would normally consider to be as bluff an object as a sphere) has a β of 1.72.

The biological utility of equation 8.15 lies in its ability to decompose the drag of an organism into its two components, β and C_d'. At high velocities, an organism can experience a reduced drag by having either a small β, a small C_d', or both. By comparing these parameters between organisms and inanimate models, one may hope to find evidence of adaptation to drag.

For example, Vogel (1984) has shown that a variety of plants and animals have the capability to rearrange their structure, with the result that $\beta < 2$. In some cases (a few algae) β is even less than the 1.5 expected for a rigid streamlined shape. The difference between living structure and bluff object is taken by Vogel as evidence of "the pervasive hand of natural selection in tuning the form and flexibility" of the organism.

This conclusion is sensitive, however, to the choice of standard to which these organisms are compared and to the criteria used. In this case, the extremely low values of β are due primarily to the fact that algae have a *high* drag at low velocities. For example, Carrington (1990) has shown that *Mastocarpus papillatus* has a β of 1.62 (nearly as low as that of a rigid flat plate parallel to flow) but has a C_d' of 0.156 (based on S_p), 24 times that of a flat plate measured under similar conditions. In light of this conflicting evidence regarding drag performance, it is difficult to conclude that algae have indeed been designed by natural selection for reduced drag. In fact, it can be argued that they are instead designed to have a high drag shape in slow flows, because this shape (like that of a tree) coinciden-

TABLE 8.2. C'_d and β for standard bluff shapes and two Hawaiian sea urchins

Object	C'_d	β
Flat Plate (⊥)	1.17	2.00
Cube (broadside)	1.28	1.96
Cube (corner-first)	1.22	1.97
Cylinder[a]	0.77	1.99
Cone[b]	0.58	2.07
Sphere	0.55	2.05
Hemisphere	0.49	1.72
Flat Plate (‖)	0.007	1.56
E. mathaei	1.09	1.97
C. atratus	0.42	2.02

Note: C'_d is the shape coefficient, and β is determined by a least-squares regression between drag and velocity. The least-squares fit of the force and velocity data to equation 8.15 was carried out using a simplex algorithm (Caceci and Cacheris, 1984) on untransformed data. All measurements made at Reynolds numbers of 10^4 to 10^5 (based on length in the direction of flow). The area used to calculate C'_d for the flat plate is the planform area of the plate (S_p) rather than its frontal area.
[a]Height/diameter = 2.52
[b]Height/diameter = 0.69

tally increases photosynthetic area or reduces formation of boundary layers (Gerard and Mann, 1979; Wheeler, 1980; Koehl and Alberte, 1988).

Sea urchins provide another example of the difficulty of using a model as a standard for comparison. Two species of urchins occur close to each other on the wave-swept shores of Hawaii. A typically spiny urchin, *Echinometra mathaei,* is found low in the intertidal zone and subtidally where it burrows into the substratum and is protected from mainstream flow. In contrast, *Colobocentrotus atratus* (the shingle urchin) is found in the surf zone on fully exposed substrata. The spines on the aboral surface of *C. atratus* are reduced to form a smooth tiling over the test; around the periphery of the test, flattened spines form a "skirt" that conforms to the substratum. The presence of this "streamlined" species only in exposed areas where other urchin morphs cannot survive strongly suggests that hydrodynamic forces have played a role in selection for test design.

It would be difficult to support this conclusion based on a comparison to a model. For example, of the standard shapes listed in table 8.2, *C. atratus* is closest in shape to the hemisphere, and it would seem reasonable to choose this as a model. The shape coefficient of the urchin is slightly lower than that of a hemisphere (0.423 versus 0.489), but its β is considerably larger (2.016 versus 1.723). Only by comparison to *E. mathaei* is the design of *C. atratus* apparent. In this case, β is virtually the same (2.016 versus 1.974), but the shape coefficient of *C. atratus* is less than half that of its neighbor (0.423 versus 1.089).

Even this result may be misleading, because drag is only one of the forces acting on these urchins. Although the shape of *C. atratus* reduces drag relative to *E. mathaei,* it increases lift (S. Franklin and M. Denny, unpubl.), and the total hydrodynamic force on the two species may be similar. The ability of *C. atratus* to live on exposed substrata is most likely due to its great adhesive tenacity, 11.7 times that of *E. mathaei* (Gallien, 1986).

In summary, the complex and variable nature of the wave-swept environment makes it difficult to use standard biomechanical approaches in establishing the adaptive nature of morphological design. Only when the variability of the environment and the ecological interaction and behavior of the organisms in question is taken into account can appropriate standards be established against which to measure the performance of intertidal organisms.

BEHAVIOR AND THE RANDOM SEA

Let us return briefly to a consideration of the statistics of the random sea. Equation 8.7 and figure 8.2 show that the longer the period under consideration, the higher the wave an organism is expected to encounter. We previously used this statistical relationship to predict the survivorship of wave-swept organisms. A second use is possible, however, one that may enhance the attractiveness of rocky shores for the exploration of the connection between performance and fitness.

Consider, for example, the evolution of locomotory performance. Considerable research attention has been paid to the ecological consequences of locomotion in vertebrates, especially in reptiles (reviewed by Garland and Losos, chap. 10, this volume). These studies have been hampered, however, by the difficulty of determining quantitatively the selective advantage accruing to a given change in locomotory ability. Yes, if a predatory lizard can run a bit faster, it is easy to imagine that it will be more likely to capture motile prey. But the increase in fitness associated with, for instance, a 5% increase in burst speed is difficult to predict quantitatively. Among many other things, it depends on the likelihood of encountering prey of given locomotory capabilities and any disadvantages associated with an increase in burst speed (e.g., a loss of maneuverability), both of which are difficult to measure directly and nearly impossible to predict from first principles.

Complications of this sort can perhaps be avoided in the wave-swept environment. Consider, for instance, a hypothetical subtidal sea star preying upon a sparse population of acorn barnacles. While crawling between barnacles, the adhesive capability of the sea star is reduced, rendering it at risk from dislodgment by waves. Dislodgment will probably not be fatal, but the dislodged animal must right itself and climb back to its feeding grounds. The faster the sea star can crawl during foraging, the less time spent in risky motion while traveling between barnacles, and the larger the probability of settling down on its next meal without being dislodged.

To this point, this example is roughly analogous to that of the lizard cited above: an increase in speed increases the probability of eating. In the case of the wave-swept organism it is possible, however, to predict accurately the consequences of an increase in speed. If a wave 3 m high (for instance) is required to dislodge a moving sea star, equation 8.7 can be used to predict for a given sea state the average period between the occurrence of 3-m-high waves. This time can then be used in conjunction with knowledge of the spatial distribution of barnacles to predict directly the decrease in risk or the increase in effective food availability that accompanies a given increase in speed. In this fashion, simple measurements of crawling speed, adhesive tenacity, caloric content of prey, and nearest-neighbor distances between prey could be used to predict consequences directly in terms of calories consumed, and presumably thereby of reproductive output.

The reproductive behavior of acorn barnacles provides another example of the potential utility of wave statistics in understanding the ecological and evolutionary consequences of behavior and performance. These barnacles utilize internal fertilization, requiring that the penis of one individual be maneuvered into the opercular opening of a neighbor. For subtidal individuals, this process must be accomplished in the presence of wave-induced hydrodynamic forces.

Now the penis of a barnacle is a large, ungainly organ, and one can easily imagine that even relatively benign flows could interfere with its maneuverability. We may further suppose that until the penis is firmly inserted into a neighbor's operculum, the imposition of flows above a critical velocity can prevent copulation. Important questions then arise: For a penis of given capability, what period of time is available in which effective copulation can occur? If it takes a penis four minutes to effect copulation, and supercritical wave velocities are imposed every thirty seconds on average, fertilization may be prevented. What mechanical and behavioral capabilities must a barnacle possess to be able to copulate at all, and how would the likelihood of fertilization vary with a given variation in these capabilities?

Equation 8.7 again provides an ideal tool for the exploration of these questions. For a given sea state (indexed by H_s), it is possible to predict accurately the expected time between the arrival of waves of a given height and, thereby, the time between imposition of a given velocity. If the mechanical capabilities and behavioral proclivities of barnacles can be measured in the laboratory, the properties of the random sea can thus set the stage for a practical, quantitative exploration of the consequences of mechanical performance.

The two examples discussed above are hypothetical, and the specifics of either should not be taken too seriously. The general message they are intended to convey is important, however. By providing an accurate, reliable description of environmental variability, the statistics of the random sea provide a tool of great potential for the study of the connection between performance and fitness, and the wave-swept environment may thus prove to be an ideal system for these studies.

CONCLUSIONS

The intertidal zone of wave-swept rocky shores presents substantial opportunities and problems in ecomorphology. The stochastic nature of ocean waves makes possible the prediction of long-term survivorship with great potential value for our mechanistic understanding of community structure. Similarly, the statistics of the random sea hold potential as a tool for exploring the ecological and evolutionary consequences of behavior. The rapid turnover of individuals in the wave-swept environment allows for an experimental approach to topics of survival and selection that is impractical in other, less stressful habitats, and our present understanding of the ecology of nearshore organisms sets a stage on which these studies may be productively carried out.

Care must be taken, however, when these studies are interpreted in an evolutionary context. Evidence is emerging that many aspects of shape are neutral to selection by hydrodynamic force, limiting our ability to use convergence as a tool for interpreting adaptation. The temporal and spatial variability of the wave-swept environment, in conjunction with the presence of planktonic dispersal stages of many sessile organisms, makes it difficult to choose appropriate optimal shapes against which to compare biological structures, and the choice of alternative standards is similarly fraught with problems. These difficulties are not unique to the wave-swept environment, and their solution is a task of general importance to ecomorphology.

ACKNOWLEDGMENTS

I thank Shawn Franklin for the use of his data regarding the lift and drag characteristics of Hawaiian sea urchins and several anonymous reviewers for their helpful comments. Much of the original data reported here were obtained through the support of NSF grants OCE 87-11688 and OCE 91-15688.

REFERENCES

Alexander, R. McN. 1982. *Optima for Animals*. London: Edward Arnold.
Ballantine, W. J. 1961. A biologically-defined exposure scale for the comparative description of rocky shores. *Field Studies* 1(3):1–19.
Caceci, M. S., and W. P. Cacheris. 1981. Fitting curves to data. *BYTE* 9(5):340–362.

Carrington, E. 1990. Drag and dislodgment of an intertidal macroalga: Consequences of morphological variation in *Mastocarpus papillatus* Kutzing. *J. Exp. Mar. Biol. Ecol.* 139:185–200.

Carstens, T. 1968. Wave forces on boundaries and submerged bodies. *Sarsia* 34:37–60.

Connell, J. H. 1978. Diversity in tropical rain forests and coral reefs. *Science* 199:1303–1310.

Dawkins, R. 1982. *The Extended Phenotype*. Oxford: Oxford University Press.

Dayton, P. K. 1971. Competition, disturbance, and community organization: The provision and subsequent utilization of space in a rocky inertidal community. *Ecol. Mon.* 41:351–389.

Denny, M. W. 1976. The physical properties of spiders' silk and their role in the design of orb-webs. *J. Exp. Biol.* 65:483–506.

Denny, M. W. 1983. A simple device for recording the maximum force exerted on intertidal organisms. *Limnol. Oceanogr.* 28:1269–1274.

Denny, M. W. 1985. Wave forces on intertidal organisms: A case study. *Limnol. Oceanogr.* 30:1171–1187.

Denny, M. W. 1987. Lift as a mechanism of patch initiation in mussel beds. *J. Exp. Mar. Biol. Ecol.* 113:231–245.

Denny, M. W. 1988. *Biology and the Mechanics of the Wave-Swept Environment*. Princeton: Princeton University Press.

Denny, M. W. 1989. A limpet shell shape that reduces drag: Laboratory demonstration of a hydrodynamic mechanism and an exploration of its effectiveness in nature. *Can. J. Zool.* 67:2098–2106.

Denny, M. W. 1991. Biology, natural selection, and the prediction of maximal wave-induced forces. *S. Afr. J. Mar. Sci.* 10:353–363.

Denny, M. W. 1993. Disturbance, natural selection, and the prediction of maximal wave-induced forces. *Contemp. Math.* 141:65–90.

Denny, M. W., T. L. Daniel, and M. A. R. Koehl. 1985. Mechanical limits to size in wave-swept organisms. *Ecol. Mon.* 55:69–102.

Denny, M. W., and S. D. Gaines. 1990. On the prediction of maximal intertidal wave forces. *Limnol. Oceanogr.* 35:1–15.

Denny, M. W., and M. F. Shibata. 1989. Consequences of surf-zone turbulence for settlement and external fertilization. *Amer. Nat.* 134:859–889.

Doty, M. S. 1971. Measurement of water movement in reference to benthic algal growth. *Bot. Mar.* 14:32–35.

Endler, J. A. 1986. *Natural Selection in the Wild*. Princeton: Princeton University Press.

Gallien, W. B. 1986. A comparison of hydrodynamic forces on two sympatric sea urchins: Implications of morphology and habitat. M. S. thesis, University of Hawaii, Honolulu.

Gerard, V. A., and K. H. Mann. 1979. Growth and production of *Laminaria longicruris* (Phaeophyta) populations exposed to different intensities of water movement. *J. Phycol.* 15:33–41.

Gorshkov, S. G. 1976. *World Ocean Atlas*. New York: Pergamon Press.

Harger, J. R. E. 1970. The effects of wave impact on some aspects of the biology of sea mussels. *Veliger* 12:401–414.

Jones, W. E., and A. Demetropoulos. 1968. Exposure to wave action: Measurement of an important ecological parameter on the shores of Anglesey. *J. Exp. Mar. Biol. Ecol.* 2:46–63.

Kingsolver, J. G., and T. L. Daniel. 1979. On the mechanics and energetics of nectar feeding in butterflies. *J. Theor. Biol.* 76:167–179.

Kingsolver, J. G., and T. L. Daniel. 1983. Mechanical determinants of nectar feeding strategy in hummingbirds: Energetics, tongue morphology, and licking behavior. *Oecologia* (Berlin) 60:214–226.

Kinsman, B. 1965. *Wind Waves*. Englewood Cliffs, N.J.: Prentice Hall.

Koehl, M. A. R. 1984. How do benthic organisms withstand moving water? *Amer. Zool.* 24:57–70.

Koehl, M. A. R., and R. Alberte. 1988. Flow, flapping, and photosynthesis of *Nereocystis luetkeana*: A functional comparison of undulate and flat blade morphologies. *Mar. Biol.* 99:435–444.

Leigh, E. G., R. T. Paine, J. F. Quinn, and T. H. Suchanek. 1987. Wave energy and intertidal productivity. *Proc. Natl. Acad. Sci. USA* 84:1314–1318.

Levin, S. A., and R. T. Paine. 1974. Disturbance, patch formation, and community structure. *Proc. Natl. Acad. Sci. USA* 71:2744–2747.

Levin, S. A., and R. T. Paine. 1975. The role of disturbance in models of community structure. In *Ecosystems Analysis and Prediction*, ed. S. A. Levin, 56–57. Philadelphia: Society for Industrial and Applied Mathematics.

Levitan, D. 1991. Influence of body size and population density on fertilization success and reproductive output in a free-spawning invertebrate. *Biol. Bull.* 181:261–268.

Lewis, J. R. 1964. *The Ecology of Rocky Shores*. London: English Universities Press.

Longuet-Higgins, M. 1952. On the statistical distribution of the heights of sea waves. *J. Mar. Res.* 11:245–266.

Mandelbrot, B. B. 1982. *The Fractal Geometry of Nature*. New York: W. H. Freeman and Co.

Menge, B. A., and J. P. Sutherland. 1987. Community regulation: Variation in disturbance, competition, and predation in relation to environmental stress and recruitment. *Amer. Nat.* 130:730–757.

Muus, B. J. 1968. A field method for measuring "exposure" by means of plaster balls. *Sarsia* 34:61–68.

Oster, G. F., and E. G. Wilson. 1978. *Caste and Ecology in the Social Insects*. Princeton: Princeton University Press.

Paine, R. T. 1966. Food web complexity and species diversity. *Amer. Nat.* 100:65–75.

Paine, R. T. 1974. Intertidal community structure: Experimental studies on the relationship between a dominant competitor and its principal predator. *Oecologia* 15:93–120.

Paine, R. T., and S. A. Levin. 1981. Intertidal landscapes: Disturbance and the dynamics of pattern. *Ecol. Mon.* 51:145–178.

Palumbi, S. R. 1984a. Tactics of acclimation: Morphological changes of sponges in an unpredictable environment. *Science* 225:1478–1480.

Palumbi, S. R. 1984b. Measuring intertidal wave forces. *J. Exp. Mar. Biol. Ecol.* 81:171–179.

Pennington, J. T. 1985. The ecology of fertilization of echinoid eggs: The consequences of sperm dilution, adult aggregation, and synchronous spawning. *Biol. Bull.* 169:417–430.

Price, H. A. 1980. Seasonal variation in the strength of byssal attachment of the common mussel, *Mytilus edulis* L. *J. Mar. Biol. Assoc. U. K.* 60:1035–1037.

Quinn, J. F. 1979. Disturbance, predation and diversity in the rocky intertidal zone. Ph. D. diss., University of Washington, Seattle.

Rayleigh, J. W. S. 1894. *The Theory of Sound*. New York: Macmillan.

Ricketts, E. F., J. Calvin, and J. W. Hedgepeth. 1968. *Between Pacific Tides*, 4th ed. Palo Alto: Stanford University Press.

Sarpkaya, T., and M. Isaacson. 1981. *Mechanics of Wave Forces on Offshore Structures*. New York: Van Nostrand-Reinhold.

Shanks, A. L., and W. G. Wright. 1986. Adding teeth to wave action: The destructive effects of wave-borne rocks on intertidal organisms. *Oecologia* 69:420–428.

Sousa, W. P. 1979a. Experimental investigations of disturbance and ecological succession in a rocky intertidal algal community. *Ecol. Mon.* 49:227–254.

Sousa, W. P. 1979b. Disturbance in marine intertidal boulder fields: The nonequilibrium maintenance of species diversity. *Ecology* 60:1225–1239.

Sousa, W. P. 1984. The role of disturbance in natural communities. *Ann. Rev. Ecol. Syst.* 15:353–391.

Stimson, J. 1970. Territorial behavior of the owl limpet *Lottia gigantea*. *Ecology* 51:113–118.

Thornton, E. B., and R. T. Guza. 1983. Transformation of wave height distribution. *J. Geophys. Res.* 88(C10):5925–5938.

U. S. Army Corps of Engineers. 1984. *Shore Protection Manual*. Washington, D.C.: U.S. Government Printing Office.

U.S. Army Corps of Engineers. 1982–1987. *Coastal Data Information Program Yearly Summary.* Washington, D.C.: U.S. Government Printing Office.

Vermeij, G. J. 1978. *Biogeography and Adaptation.* Cambridge: Harvard University Press.

Vermeij, G. J. 1987. *Evolution and Escalation.* Princeton: Princeton University Press.

Vogel, S. 1981. *Life in Moving Fluids.* Boston: Willard Grant Press.

Vogel, S. 1984. Drag and flexibility in sessile organisms. *Amer. Zool.* 24:28–34.

Wheeler, W. N. 1980. Effect of boundary layer transport on the fixation of carbon by the giant kelp *Macrocyctis pyrifera. Mar. Biol.* 56:103–110.

Wright, W. G. 1978. Aspects of the ecology and behavior of the owl limpet, *Lottia gigantea* Sowerby, 1834. *West. Soc. Malac. An. Rep.* 11:7.

9

Wing Design, Flight Performance, and Habitat Use in Bats

Ulla M. Norberg

INTRODUCTION

Flight and echolocation are the characteristic adaptations of bats. Flight has opened up many ecological opportunities for bats as well as birds, because flying permits foraging in otherwise inaccessible places, foraging over large areas, and migration. Most importantly, flight permits specialization upon flying insects as prey. Flight is an expensive mode of locomotion (e.g., Norberg, 1990), so there should be strong selection to minimize its cost. Natural selection may favor a wing design that minimizes the work needed to fly in the manner and at the speeds optimal for the animal. The combination of morphological, ecological, and be-havioral attributes that most benefits a flying animal is related to the type of hab-itat it lives in and its way of exploiting the habitat. Therefore, different flying animals often use different flight strategies, each requiring a particular wing mor-phology. In some bats, variation in echolocation call design are also involved (see Norberg, 1990).

Wings vary in shape and size among flying vertebrates but function aerody-namically in the same way (see Norberg, 1990). The variation in wings can be correlated with different flight modes and speeds: some bats and birds fly contin-uously during foraging, either in open spaces or close to vegetation, whereas others perch between foraging bouts. Some species forage within vegetation, taking insects in air or by gleaning, and some can carry heavy prey. Others, such as nectarivorous bats and birds, can hover, which is one of the most energy-demanding types of locomotion (Pennycuick, 1975). The characteristic flight mode is active flapping, in which the animal flies straight and level. This is a very efficient way to cover distance, although it can require extremely high power out-put (work per unit time). Gliding, which expends minimal energy, is common in large- and medium-sized birds but only rarely used by bats.

The main purpose of this chapter is to relate wing design to flight performance and habitat use in bats. To increase the understandings of the correlations between behavior and morphology in bats I will begin with a brief summary of the probable steps in the evolution of their flight. I will then present the basic bio-mechanical and aerodynamic factors affecting wing design in flying vertebrates and give examples for bats or bat groups with different wing form and habitat use. Finally, I will discuss various adaptive constraints in the ecomorphology of bat flight.

Bat flight morphology as related to flight mode and foraging behavior has been considered by many authors (e.g., Findley et al., 1972; Kopka, 1973; Lawlor, 1973; U. M. Norberg, 1981a, b, 1986a, 1987, 1990; Aldridge, 1985; Aldridge and Rautenbach, 1987; Baagøe, 1987; Norberg and Rayner, 1987; and Norberg and Fenton, 1988). Norberg and Rayner (1987) evaluated the wing morphology of bats in relation to flight performance and flight behavior in order to clarify the functional basis of ecomorphological correlations in bats. Their analysis included 257 species of bats from sixteen of the eighteen extant bat families. The order Chiroptera includes more than 980 species and is divided into two suborders: the Megachiroptera, containing the single family Pteropodidae, and the Microchiroptera, with all the other families.

EVOLUTION OF FLIGHT, WING MORPHOLOGY, AND ECHOLOCATION IN BATS

It has been suggested that the two bat suborders have a diphyletic origin (Pettigrew, 1986; Pettigrew et al., 1989). Neural data (including the visual pathways) and some skeletal and hemoglobin sequence data point to a megachiropteran-primate association, whereas musculoskeletal modifications, cranial morphology, and immunology of serum proteins argue for a monophyletic evolution of bats (Pettigrew et al., 1989; Thewissen and Babcock, 1991).

If bats are monophyletic they may all have developed from nocturnal insectivores that foraged in trees (Bock, 1986) by gleaning insects located by echolocation (Jepsen, 1970; Fenton, 1974). Powered flight probably evolved via gliding from trees, as in birds (U. M. Norberg, 1985, 1986b, 1990). Maximizing net energy gain during foraging in trees might have initiated strong selection for increased gliding performance. Gliding from one tree to another and climbing upwards during foraging is used by modern birds to maximize net energy gain (R. Å. Norberg, 1981a, 1983), and this behavior may have been beneficial for the protobats as well. Once a gliding wing surface had evolved, the bat's energy and time demands for locomotion might have been drastically reduced, thereby increasing foraging efficiency. Initially, small glide surfaces produced low lift-to-drag ratios and made the first glides steep. But even steep parachuting jumps from trees could have reduced the time and energy required for foraging, while increased glide surface area decreased wing loading and gliding speed and allowed safer landings. Parachuting (steep gliding) for escape might also have been important for arboreal animals (Bock, 1965, 1986). Note that gliding in protofliers must have been used for transportation and not for insect catching. Hawking insects in the air requires high maneuverability, which could not have appeared until true flight was established.

Longer wings and more efficient wing profiles resulted in larger lift-to-drag ratios and thrust production. Increased abilities for wing flexion and movement coordination eventually led to horizontal flapping flight. In early gliders, selection pressure for good movement control must have been high, since better control would permit course maneuverability to reach particular destinations. Stability and movement control were probably achieved before flapping flight evolved and may have improved progressively, concurrently with the ability to glide. Stability and control of movements can be achieved by simple wing movements, such as twisting, retraction, and control of the dihedral angle. Eventually, more sophisticated wing characters appeared, and the associated increased maneuverability permitted the early bat to hawk insects in the air.

This new niche favored selection for insect catching in the air, concomitant with the increased ability to maneuver. Radiation of different flight habits led to different wing forms specialized for particular foraging and flight modes.

Wings of fossil bats are short and broad, which is in accordance with the evolution of flight and wing forms; the early bats probably began to glide on enlarged forelimbs and the first "wings" must have been very short and broad. Evidence of their wing design suggests that early bats foraged or lived among vegetation (Habersetzer and Storch, 1987; see "Wing Shape" below), and the ancient microchiropterans may have been perch hunters, whereas the ancient megachiropterans used flight for transportation between trees (Norberg, 1989). No aspect of their wing shapes indicates that these bats were poor fliers. Frugivorous and nectarivorous bats still use flight primarily for transport to food resources and resemble the ancient bats (megachiropteran as well as microchiropteran fossils) in wing design more than other contemporary bats (Norberg, 1989; see fig. 9.3 below). These bats have short, broad wings and have the highest flight power requirements among modern bats.

Not only wing design but also echolocation call structure is adapted to the bat's foraging niche, and there are usually associations between the two structures. Some primitive form of echolocation may have occurred already among ancient bats. Wood and Evans (1980) suggested that broadband clicks (as used by some birds, shrews and tenrecs, odontocete cetaceans, rodents, and the cave-dwelling megachiropteran bats of the genus *Rousettus*) are the "conventional" approach to echolocation. Fenton (1984) further suggested that bat echolocation calls are derived from communication signals, and later, echoes from the calls were used to obtain information about surroundings. Finally, the ability to perceive detailed information about the target was developed, while the communication function was maintained.

The Eocene bats *Icaronycteris* and *Palaechiropteryx* share special basiocranial characters with microchiropteran species, suggesting comparable refinement of ultrasonic echolocation (Novacek, 1985; Habersetzer and Storch, 1989). There is therefore reason to believe that these bats used echolocation. Even if protobats used echolocation for insect detection, earlier forms probably could not make use of such an ability during the gliding stage, because they could not maneuver well enough for aerial hawking before their flight ability was established. During the climbing and gliding stages, the early bats may have used echolocation for communication and to gather information about their surroundings, using it to localize insects only when moving on the ground or among vegetation (Fenton, 1984). When the bats could maneuver well enough for aerial hawking, echolocation may have been specialized for detecting flying targets (Norberg, 1989).

Only a few megachiropteran bats (genus *Rousettus*) echolocate (using broadband clicks) and use echolocation to orient themselves in caves. Other megachiropteran bats rely principally on vision for orientation and to some extent on olfaction to identify food sources. Megachiropterans have not evolved specialized echolocation system, unlike the frugivorous microchiropteran bats (Phyllostomidae).

WING DESIGN AND AERODYNAMIC PERFORMANCE

Some biomechanics and aerodynamics are needed to understand how wing shape is related to flight mode, because optimal wing forms are influenced by mechanical and aerodynamic considerations. Using mathematical and aerodynamic models, several workers have explored the relations between aerodynamics, wing shape, and flight mode (e.g., Pennycuick, 1969, 1975, 1989; U. M. Norberg, 1985, 1987, 1990; Norberg and Rayner, 1987; Rayner, 1988).

Aerodynamics

A flying bat must do work with its flight muscles to move the wings, thereby generating lift and thrust. The rate at which this work is done is the mechanical power, P, required to fly. A large percentage of the metabolic work appears as heat produced in the body. The actual energetic cost of flight depends on the mechanical power requirement and on the mechanical efficiency of the muscles in converting chemical energy into mechanical work. The metabolic power (power input) in bats has been estimated to be from three to eight times the mechanical power (power output), giving a mechanical efficiency of 0.12 to 0.40 (Thomas,

1975) with a mean of about 0.26. However, comparison between estimates of mechanical power derived from aerodynamic theory and empirical data on flight costs indicates that mechanical efficiency scales positively with body mass, the efficiency being about 0.05 for small bats (0.01 kg) and about 0.20 for larger bats (0.8 kg) (U. M. Norberg and T. H. Kunz, unpubl.). Comparison between mechanical power and metabolic power requirements from nectar intake gives a mechanical efficiency of 0.15 for hovering flight and 0.11 for horizontal flight at minimum power speed for a 0.011 kg tropical nectar-feeding bat *Glossophaga soricina* (Norberg et al. 1993).

For sustained forward flight the vertical lift, L, must balance body weight, Mg (mass times acceleration of gravity). The horizontal thrust, T, must balance the overall horizontal drag, D. The mechanical power required to fly is the product of thrust and speed, V,

$$P = TV = DV$$

The cost of transport, C, is the work done by the bat in moving unit weight through unit distance,

$$C = P/MgV,$$

and equals the inverse of the effective lift-to-drag ratio, $1/(L/D)_{eff}$. $(L/D)_{eff}$ is the ratio of the weight to the average horizontal force needed to propel the animal (Pennycuick, 1975).

The mechanical power is the sum of the aerodynamic power (power needed to produce sufficient aerodynamic force to fly) and the inertial power (power needed to oscillate the wings) (fig. 9.1a). We recognize three components of aerodynamic power: the induced power is the rate of work required to generate lift, needed for thrust and weight support, whereas the profile power and parasite power are the powers needed to overcome the form and friction drag of the wings and body, respectively.

The induced power is one of the main power drains in hovering and in slow flight, and to minimize this component the weight should be low and the wing span large. Form and friction drag are high in fast flight, for which minimizing the wing profile power requires a small wing area (at a given speed), while minimizing parasite power requires a slim and streamlined body.

The inertial power is considered to be low in medium and fast flight, since wing inertia is convertible into useful aerodynamic work at the ends of the half-strokes. Inertial power, however, may be important in hovering and in slow flight (Weis-Fogh, 1972, Norberg et al. 1993). To minimize inertial power the wings should be short and light, with a low ratio of arm-wing length to hand-wing length (because most wing mass is concentrated at the arm wing; Norberg, 1979).

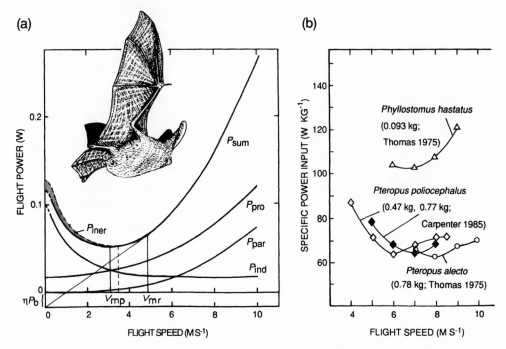

FIGURE 9.1 (a) Flight power versus flight speed for a bat the size of *Plecotus auritus* (body mass 0.009 kg, wing span 0.27 m, wing area 0.0123 m²). The extra drag from the large ears is disregarded. P_b: resting metabolic rate; P_{iner}: inertial power; P_{ind}: induced power; P_{par}: parasite power; P_{pro}: profile power; P_{sum}: sum of mechanical power components (induced, parasite, and profile powers); V_{mp}: minimum power speed; and V_{mr}: maximum range speed. The minimum power speed is the speed at which the animal can fly the longest time on a given amount of fuel. The maximum range speed is the speed where the power/speed ratio (or energy/distance ratio) reaches its minimum; it should be used for maximization of flight distance on a given amount of energy. (Pennycuick, 1975).

(b) Specific power input versus flight speed in some bats, where 1W represents 2.985 ml O_2 min⁻¹. (From Norberg, 1990, with permission.)

The arm wing is the part of the wing proximal to the thumb and fifth digit, whereas the hand wing is the part distal to these digits.

The sum curve of the power components typically has a U-shape (fig. 9.1a). Experimental measurements of oxygen consumption during flight in bats show similar shapes (fig. 9.1b), whereas Torre-Bueno and Larochelle (1978) found the rate of CO_2 production as a function of speed in starlings to be a rather flat curve. The bottom point of the curve defines the minimum power speed, V_{mp}, used to maximize time aloft on a given amount of energy. This is illustrated by nectar- and pollen-feeding bats (*Leptonycteris*) waiting for their turns to get to a flower (Howell, 1979). The tangent to the power curve from the origin defines the maximum range speed, V_{mr}, at which the power/speed ratio (or energy/distance ratio) reaches its minimum. This speed should be chosen whenever the longest distance is to be covered on a given amount of energy, for example, during migration. During nonforaging flights between some central location and foraging areas, the bat should fly faster than the maximum range speed (R. Å. Norberg, 1981b). This should be the case for megachiropteran bats, for example. During foraging flights (as in insectivorous bats), the optimal flight speed is probably higher than the minimum power speed (R. Å. Norberg, pers. comm., cited in U. M. Norberg, 1981b). In the two latter cases, the goal is to maximize the net energy gain per unit time. However, the most economic foraging speed may also depend on the flight characteristics of prey, such as speed, predictability, and habitat selection.

Norberg (1987) compared measured flight speeds of bats with their theoretical optimal speeds (Norberg and Rayner, 1987). Open-field flight speeds of bats are usually higher than the theoretical maximum range speeds, which agrees with the prediction for central-place foraging flights (R. Å. Norberg, 1981b). Speeds of bats flying indoors or of bats released within vegetation accord more with the theoretical minimum power and maximum range speeds.

Wing Shape

The size and shape of the wings can be quantified by three parameters: wing loading, aspect ratio, and tip shape index. Both wing loading and aspect ratio are widely used in aircraft engineering and are valuable in studies of animal flight.

Wing loading relates body weight to wing size and is defined Mg/S, where M is body mass, g is the acceleration of gravity, and S is the wing area. Wing loading varies with size in geometrically similar bats, $Mg/S \propto M^{0.33}$. Because of such allometry, large bats have higher wing loadings than smaller bats. Using Reduced Major Axis (RMA) regression, Norberg and Rayner (1987) found that the exponent varies among the different bat suborders and families; for fifty-one megachiropteran bats the allometric exponent is 0.33, for 208 microchiropteran species the value is 0.54, whereas it is 0.44 for all bats taken together. Separating bats according to feeding categories, wing loadings in frugivores and carnivores increases almost isometrically with body mass, whereas in nectarivores it increases as $M^{0.39}$, and in insectivores as M^{60}. This large discrepancy from geometric similarity in insectivores may be due to different demands on flight speeds and maneuverability in small and large species.

Wing loading ranges from $5.2\ N/m^2$ in the smallest species (butterfly bat, *Craseonycteris thonglongyai*, 2 g) to approximately $60\ N/m^2$ in the largest flying fox (*Pteropus giganteus*, 1.2 kg). Birds span a larger range, from $17\ N/m^2$ in a small hummingbird (2.3 g) to $230\ N/m^2$ in the whooper swan (about 10 kg).

Wing loading is the ratio of a force (weight) to an area and is equivalent to a pressure. The mean pressure and suction over the wings must match the wing loading, and since aerodynamic pressure varies as V^2, then $V \propto (Mg/S)^{1/2}$. Therefore, slow flight is possible with a low wing loading (large wings), while bats with small wings have to fly faster for their body size.

Aspect ratio, AR, a nondimensional number reflecting the shape of the wings, is defined as the ratio of the wing span, b, to the mean chord of the wing, c, and is calculated as $AR = b/c = b^2/S$ (fig. 9.2). The average aspect ratio for bats is approximately 7, ranging from 4.8 in the broad-winged slit-faced bats (Nycteridae) up to 14.3 in the narrow-winged free-tailed bats (Molossidae). Corresponding values for birds range from about 4 in some galliforms and tinamous to 18 in albatrosses. Because wing span does not scale isometrically in bats (U. M. Norberg, 1981a; Norberg and Rayner, 1987), aspect ratio is not quite independent of body mass; for fifty-one megachiropterans aspect ratio scales with body mass raised to the 0.11 power, while the corresponding number for 208 microchiropterans is 0.21 (with large variation among families; Norberg and Rayner, 1987).

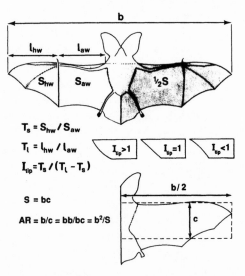

FIGURE 9.2 Definitions of morphological quantities to describe the bat wing, most of which are used also for the bird wing. The wing span b is measured from tip to tip of the extended wings. The wing area S includes the area of the body in between the wings in bird and bats and the tail membrane in bats; S_{hw}: area of the hand wing (distal to the fifth digit in bats and distal to the wrist and including the primary feathers in birds); S_{aw}: area of the arm wing (between the body and the hand wing); and l_{hw} and l_{aw}: corresponding lengths. These quantities are used to define the tip length and tip area ratios, T_l and T_S, and the wingtip shape index, I_{tip}. (Modified from Norberg and Rayner, 1987.)

Aspect ratio can be interpreted as a measure of the aerodynamic efficiency of flight. Low total flight power is obtained with a high aspect ratio wing, particularly in combination with a low wing loading since this permits slow flight with ensuing low wing profile and parasite powers. Slow-flying species should therefore have long high-aspect-ratio wings to reduce induced power and large area to enable the bat to fly slowly. By contrast, fast-flying species benefit from short small wings to reduce profile power. The wings should still be of high aspect ratio for flight economy (Norberg, 1986b). Bats with narrow wings have small wing areas and thus high wing loading, unless their wings are also long. A bat with high wing loading must fly fast (or beat their wings very fast) to obtain sufficient weight support. By flying fast, such a bat reduces induced power but increases parasite and profile powers. Since profile power is much larger than induced power at fast speeds, and increases slightly with wing length (Rayner, 1979), fast-flying species should have a short wing span. Migratory species, whose goal may be minimizing the energy consumed covering a given distance, should have high aspect ratio (narrow wings) for long flight. Their wing loading should be low (long wings) to reduce flight costs, but otherwise high (short wings) for high flight speed if time minimization is important. Therefore, long narrow wings of high aspect ratio are favored for slow forward flight economy, while short narrow wings of high aspect ratio are appropriate for economic fast flight. Hovering and very slow flight with high wingbeat frequency may be an exception, depending on the importance of inertial power. This component increases with increasing wingbeat frequency and wing length; hovering species may therefore benefit by short wings (see "Slow Fliers, Hoverers, Gleaners" below).

Tip shape index is a particularly valuable measure because it is independent of the overall size of the arm wing (chiropatagium) and hand wing (propatagium + plagiopatagium) but is determined by their relative size. This index is defined by Norberg and Rayner (1987) as $I_{tip} = T_s/(T_l - T_s)$, where T_s is the ratio of the hand-wing area to the arm-wing area (tip area ratio) and T_l is the ratio of the length of the hand wing, l_{hw}, to the length of the arm wing, l_{aw} (tip length ratio) (fig. 9.2). High values of I_{tip} indicate rounded or nearly square wing tips, with the hypothetical value infinity corresponding to a completely rectangular wing. The value $I_{tip} = 1$ corresponds to a triangular wing tip, and with lower values the wing tip becomes more pointed.

Maneuverability and Agility

Flying animals perform aerial maneuvers in different ways depending upon their ultimate goal. For example, prey capture, obstacle avoidance, and landing require different types of maneuvers. Maneuverability is usually defined as the minimum radius of turn the animal can attain without loss of speed or momentum, while agility is the maximum roll acceleration during the initiation of a turn and measures the ease or rapidity by which the flight path can be altered (Norberg and Rayner, 1987). It may, however, be convenient to include both types of movement in the term maneuverability (Andersson and Norberg, 1981).

The theory of turning in powered flapping flight has been discussed by various authors (e.g., Pennycuick, 1971; Norberg and Norberg, 1971; Norberg and Rayner, 1987; Norberg, 1990). In a turn the animal loses some of the potential vertical force, because a component of this force is necessary for centripetal acceleration to prevent sideslip. So the animal must bank its wings at some angle to the horizontal, thereby developing more lift than in horizontal flight. At any particular bank angle, $r \propto Mg/S$, where r is the radius of turn. The turn radius is also dependent on the wings' ability to generate lift. For a given value of the lateral acceleration, the radius of turn is least when speed is least because the necessary acceleration is proportional to speed squared, as is lift. Therefore, the ability to make tight turns without stalling is best in bats with low wing loadings and with the ability to control camber of the wings (fore and aft curvature along the wing chord) to enhance lift.

Flying bats may detect insects at ranges of only a few meters (Kick, 1982) so they must make rapid maneuvers to pursue and catch prey. To initiate a turn, a net rolling moment must be produced. This can be done by differential twisting or flexing of the wings, or by unequal flapping of the two wings, giving asymmetrical aerodynamic roll moments (torques). The torque, τ, is proportional to speed squared, wing area, and wing span, $\tau \propto V^2Sb$, so at a given speed, long broad wings with rounded tips will give large aerodynamic torque. But large torque can also be obtained by high flight speeds, which, however, is hardly obtained by species with large wings (low wing loadings). Therefore, large aerodynamic roll moments are obtained by the large size of the wings in slow-flying species (such as species of families Nycteridae, Megadermatidae, Rhinolophidae, Hipposideridae, Natalidae, Thyropteridae, and many vespertilionids), and by the high flight speed in fast-flying species (such as molossids, vespertilionids of genera *Nyctalus* and *Pipistrellus,* and many phyllostomids) (fig. 9.3).

The fastest entry into a turn is achieved at the maximum angular acceleration, α_{roll}, available to the animal. It is the aerodynamic torque divided by the total roll moment of inertia of body and wings J, $\alpha_{roll} \propto \tau/J$. Broad wings and rounded wing tips (large wing-tip shape index I_{tip}) are the important characteristics for maximization of the aerodynamic torque τ in slow-flying species. In fast fliers, low J, and so high α_{roll}, is provided by a thin body, short wings, and pointed wing tips (small I_{tip}). The angular roll acceleration also increases with decreasing body mass (see Andersson and Norberg, 1981; Norberg and Rayner, 1987; Thollesson and Norberg, 1991). Aldridge (1986) recorded roll rates of up to 450 radians/s^2

Low flight speed ← → **High flight speed**

FOR HIGH AGILITY (HIGH ROLL ACCELERATION):

Increased τ:
large l_{tip} (broad, rounded wing tips)
low AR (broad wings)
long wing span

Increased τ:
high v:
high WL (small wing area)

Increased influence on J:
small T_l (short hand wings for high wing flexibility)

Decreased J:
large T_l (long hand wings)
small l_{tip} (pointed wing tips)
high AR (narrow wings)
short wing span, thin body

FOR HIGH MANEUVERABILITY (SMALL RADIUS OF TURN):

Decreased WL:
low body mass
large wing area

Decreased WL:
low body mass

EXAMPLES:

Plecotus auritus
large l_{tip}, small T_l,
low AR, low WL

Pipistrellus pipistrellus
very large l_{tip}, small T_l,
high AR, average WL

Nyctalus noctula
small l_{tip}, large T_l,
high AR, high WL

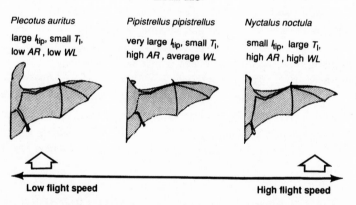

Low flight speed ← → **High flight speed**

FIGURE 9.3 Hypothetical selection pressures for high agility (= high roll acceleration, α_{roll}) and high maneuverability in fast-flying as contrasted with slow-flying insectivorous bats. $\alpha_{roll} = \tau / J \propto v^2 Sb / J$, where τ is the aerodynamic roll moment (torque), v is flight speed, S is wing area, b is wing span, and J is the roll moment of inertia of body and wings combined. I_{tip} is wing-tip shape index, T_l is wing-tip length index, AR is aspect ratio, and WL is wing loading (see text for definitions). Note that high agility is obtained by broad, rounded wings in slow fliers and by small narrow wings in fast fliers. High maneuverability can be obtained by bats with low wing loading (slow fliers) and/or low body mass. The examples represent a slow flier (*Plecotus auritus*), a species flying at moderate speeds (*Pipistrellus pipistrellus*), and a fast flier (*Nyctalus noctula*). (Modified from Thollesson and Norberg, 1991.)

in horseshoe bats that are slow-flying with rounded wing tips and average wing span.

Summarizing, maneuverability is increased by low body mass and low wing loading allowing small turning radii, and high agility (high roll acceleration) can be achieved by low wing moment of inertia and/or large aerodynamic torque (fig. 9.3). For low wing inertia, tip length ratio T_1 should be large (long hand wings), tip shape index I_{tip} should be small (pointed tips), and aspect ratio AR should be high (short narrow wings), characteristics which ensure that wing mass is not concentrated toward the tips. Large aerodynamic torque can be obtained by either high flight speeds or large wings and broad wing tips (large I_{tip} and low AR) but not both, since high speeds are correlated with high wing loadings (small wings). For maximum wing flexure and the greatest control of wing inertia, the arm wing should be proportionally long (low T_1). These different selection pressures for increased agility conflict with each other, and bats with different flight behavior have solved this problem in various ways, as demonstrated in fig. 9.3.

WING DESIGN AND ECOLOGY

Wing shape is a result of varying flight demands and of phylogenetic constraints (Norberg and Rayner, 1987). The observed combination of morphological, behavioral, and ecological traits is usually assumed by ecologists to represent a near-optimal solution that maximizes the fitness of the individual. Optimal foraging models are often built to maximize net energy intake or to minimize foraging time, optimization criteria that are substituted for fitness (e.g., R. Å. Norberg, 1977, 1981b, 1983; Pyke et al., 1977; Krebs, 1978). Since flight is very expensive, one would expect the design of a bat's wing to be under extreme selection pressures. The optimal wing shape (if it exists) is probably dictated by a combination of different factors, such as flight behavior, habitat, and food choice. Selection pressures for these various demands are likely often to be conflicting.

Figure 9.4 is based on a Principal Components Analysis (Norberg and Rayner, 1987) of wing morphology of 215 bat species from sixteen families showing different wing-loading and aspect-ratio categories. This diagram shows the relations between aspect ratio and size-compensated wing loading. It can be used to formulate hypotheses on foraging and flight strategies in relation to flight morphology and for comparisons between species and groups of bats. If only a few key wing characters are known for a bat, its predominant flight performance can usually be predicted.

Ancient bats (megachiropteran as well as microchiropteran fossils; stars and data points enclosed by dashed line labeled "Ancient bats" in fig. 9.4) had short and broad wings and therefore low aspect ratios and high wing loadings. One

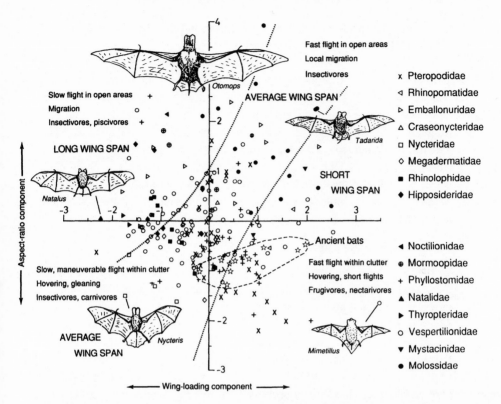

FIGURE 9.4 Scatter plot of the second and third principal components of wing morphology in bats, identified as measures of wing loading (body weight/wing area, Mg/S) and aspect ratio (wing span²/wing area, b^2/S), respectively. Fossil bats are marked with stars and occur within the dashed line (marked "Ancient bats"). Dotted lines separate long wing spans ($+8\%$; upper left) and short wing spans (-8%; lower right) from average wing span. Bats with high wing loadings are fast fliers, and conversely, those with low wing loadings are slow fliers. Bats with high aspect ratios have inexpensive flight (in particular, those with a low wing loading), whereas species with a low aspect ratio have power-demanding flight and usually perch between foraging bouts. The most expensive flight occurs among bats with both high wing loading and low aspect ratio. (From Norberg, 1989, with permission, based on Norberg and Rayner, 1987.)

would expect that the very first fliers among bats had still higher wing loadings and lower aspect ratios. The wing forms of fossil bats would permit foraging or life among vegetation (Habersetzer and Storch, 1987), and perch hunting (Norberg, 1989).

The longest wings in relation to body size are found among sheath-tailed bats (*Emballonura* species, Emballonuridae), naked-backed bats (Mormoopidae) and disk-winged bats (Thyropteridae), while mouse-tailed bats (Rhinopomatidae), the New Zealand short-tailed bat (*Mystacina tuberculata*, Mystacinidae), many free-tailed bats (Molossidae), and most mega- and microchiropteran fruitbats (Pteropodidae and Phyllostomidae) have short wings.

Species with a high wing loading in relation to body size are most New World leaf-nosed bats (Phyllostomidae) and free-tailed bats, some sheath-tailed bats (*Taphozous* species, Emballonuridae), mouse-tailed bats, the New Zealand short-tailed bat, nectarivorous and some frugivorous flying foxes, and particularly the Moloney's flat-headed bat (*Mimetillus moloneyi*, Vespertilionidae). Low wing loadings are found, for example, among false vampire bats (Megadermatidae), slit-faced bats, and Old World leaf-nosed bats (Hipposideridae). Free-tailed, naked-backed, and sheath-tailed bats all have high aspect ratio wings, whereas wings of nycterids, Moloney's flat-headed bat, and many frugivorous bats have low aspect ratios.

Sustained Foraging Flights

Bats with long, high-aspect-ratio wings (narrow wings) have low wing loadings. Their flight is slow and inexpensive because long wings reduce induced power, a prominent power component of slow flight. These bats have slow and enduring foraging flights. They often fly in open spaces, as their long wings would be a hindrance in dense vegetation. This wing design is found in sheath-tailed bats of the genera *Emballonura* and *Rhynchonycteris,* hipposiderids, noctilionids, and some vespertilionids (*Lasionycteris, Nyctophilus, Eptesicus*) (Norberg and Rayner, 1987). The proboscis bat, *Rhynchonycteris naso,* has been observed to forage slowly in straight lines above moving water (Hall and Dalquest, 1963; Bradbury and Emmons, 1974; Bradbury and Vehrencamp, 1976). Some migratory species also belong to this group. Noctilionids (bulldog bats) are partly piscivorous, while the others mentioned above are insectivorous. The low wing loading makes these bats highly maneuverable.

High wing loading and large aspect ratios (short narrow wings) are characteristic for bats with fast sustained flight, such as free-tailed bats, species in the emballonurid genus *Taphozous* (sheath-tailed bats), and the vespertilionids *Tylonycteris* (flat-headed bats) and some *Pipistrellus* species. Most species in this category are insectivorous and hunt flying insects in open spaces. Free-tailed bats have short or average wing span with slightly rounded wing tips, permitting quick agile rolling maneuvers. However, their high wing loadings result in large turning radii.

Locomotion Among Vegetation

Flying within vegetation requires slow flight and short wings. Such wings must be broad (i.e., with a large chord) to compensate for their shortness, with enough wing area to allow slow flight. However, such wings increase power requirements. A large tail membrane and broad wings enable the bat to make rapid changes of direction in slow flight and permit use of the tail membrane to catch insects.

Many bats flying near or within vegetation indeed have short wings, and therefore low aspect ratios. Flight is correspondingly expensive, and these species often are perchers to save energy (see Norberg and Rayner, 1987). As flights are infrequent and brief, with the bat flying only for prey capture, less energy is used than in sustained hawking. Some bats hawk flying insects by making short flights from the perch in a flycatcher style; others use a perch while seeking non-flying prey. Perchers taking nonflying prey include, for example, the false vampire bats *Megaderma lyra, M. spasma, Macroderma gigas,* and *Cardioderma cor* (Megadermatidae) (Vaughan, 1976). "Flycatching" bats have relatively large wing tips and low wing loading, permitting slow flight and high maneuverability, large roll moments of lift, and increased agility. Examples of flycatching bats are the yellow-winged bat *Lavia frons* (Megadermatidae), some horseshoe bats (Rhinolophidae), some of the larger leaf-nosed bats (Hipposideridae), the vespertilionids *Nyctophilus bifax* (northern long-eared bat; Fenton, 1982) and juvenile *Myotis lucifugus* (little brown bat; Buchler, 1980).

Slow Fliers, Hoverers, Gleaners

Bats with low wing loading and low aspect ratio can fly slowly within vegetation and carry heavy prey (Norberg and Rayner, 1987; Norberg and Fenton, 1988). Their broad wings and large tail membrane enable the bats to make tight turns, and several of these species often use the tail membrane for insect catching. Their wing span is average among bats and their wing tips are usually very rounded, enhancing maneuverability.

Many of these bats hover short periods while searching for food or while gleaning from vegetation. Insectivorous and nectarivorous bats known to be gleaners are shown in figure 9.5a. Nycterids, megadermatids, rhinolophids, and many vespertilionids belong to this group; several of these species are carnivores (see below). Norberg and Rayner (1987) distinguish between two forms of gleaning: from vegetation and surfaces such as walls and tree trunks, requiring hovering flight (hover-gleaning: circled dots in fig. 9.5a), and from the ground, requiring the ability to handle prey on the ground and often to take off with it (ground-gleaning: open circles in fig. 9.5a). Hover-gleaning is characteristic of many small vespertilionids of genera *Plecotus, Idionycteris,* and *Myotis* (such as *P. auritus, P. austriacus, P. rafinesquii, P. townsendi, Idionycteris phyllotis, M. auriculus, M. bechsteini, M. evotis, M. keenii, M. myotis, M. nattereri,* and *M. thysanodes;* e.g., Barrett-Hamilton, 1910; Brosset, 1966; Roberts, 1977; Stebbings, 1977; van Zyll de Jong, 1985), some nycterids and hipposiderids, and by

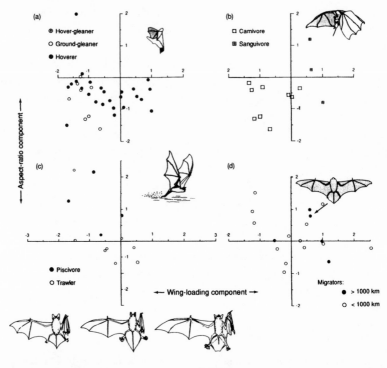

FIGURE 9.5 Scatter plot of the second and third principal components of wing morphology in bats, identified as measures of wing loading (body weight/wing area, Mg/S) and aspect ratio (wing span^2/wing area, b^2/S), respectively. (a) Bats adopting hovering and/or gleaning feeding strategies. (b) Carnivorous (squares) and sanguivorous (plus signs in squares) bats. (c) Piscivores (closed circles) and trawling insectivorous bats (open circles). Wing outlines of *Myotis myotis* (left), a typical slow-hawking insectivorous vespertilionid, of *Pizonyx vivesi* (middle), a piscivorous vespertilionid, and of *Noctilio leporinus* (right), a piscivore of family Noctilionidae. Note the long wing tips and claws on the feet in the two piscivores. (d) Migratory species. Filled circles indicate bats which migrate more than 1000 km, open circles those which migrate between 200 and 1000 km. (From Norberg and Rayner, 1987, with permission.)

many of the smaller insectivorous phyllostomids. Bats known to take insects by ground-gleaning include *Nycteris thebaica* (O'Shea and Vaughan, 1980), *Cardioderma cor* (Vaughan, 1976), *Megaderma lyra* (Fiedler, 1979), *Macroderma gigas, M. californicus* (Bell, 1985), *Antrozous pallidus* (Bell, 1982), and the larger insectivorous phyllostomids (Wilson, 1973; Gardner, 1977; Bonaccorso, 1979).

Hovering is used by many nectarivorous species (filled circles in fig. 9.5a), although many of them often sit on the flower when feeding. Macroglossines (Pteropodidae) and glossophagine and brachyphylline bats (Phyllostomidae) feed primarily on nectar and pollen, but most of them also include fruit and insects in their diet. Glossophagine bats are excellent hoverers (Vogel, 1968; Howell, 1979; Heithaus, 1982; Dobat and Peikert-Holle, 1985; von Helversen and Reyer, 1984), whereas macroglossines perch on flowers when feeding (van der Pijl, 1956).

For energetic reasons hovering bats should benefit from long wings to reduce induced power but short wings to reduce inertial power. Short wings also increase maneuverability amidst foliage. Norberg et al. (1993) estimated the mechanical power required to hover in the nectar-feeding bat *Glossophaga soricina* (Phyllostomidae) and found that the inertial power may represent 55% and the induced power 43% of the total power. It may thus be more important to reduce inertial power in this bat than to reduce induced power, and this can be done by having short wings. Pressure for low hovering costs and maneuverability in dense vegetation seem to have selected for short wings (high wing loadings; fig. 9.6a) in the hovering nectarivorous species. Hover-gleaners and ground-gleaners, on the other hand, have average to long wing spans (low wing loadings).

Detection and localization of prey on surfaces requires not only low wing loading and low aspect ratio for slow and maneuverable flight, but also a very good sense of hearing, which can be increased by having large ears. Only species with a low wing loading, permitting slow and maneuverable flight, can afford to have large, drag-producing ears (U. M. Norberg, 1981a), and species with large ears are usually found among gleaners. Large ears facilitate the acoustically difficult task of detecting and localizing insects on surfaces, such as leaves, and only species with a low wing loading and low aspect ratio (rather short wings) can maneuver to pick insects from vegetation.

Fast Fliers

Species with the combination of a low aspect ratio and high wing loading have very high flight costs. These bats typically make rapid and short flights among vegetation and usually are perchers. The majority of the frugivorous and nectarivorous species (families Pteropodidae and Phyllostomidae) have shorter wings and smaller tail membrane than the slow fliers. Wing loadings are higher, and so are optimal flight speeds. If nectar is sparse or if there are long distances between flowers or fruits, bats would benefit from high wing loadings, furthering fast flight. But the combination of high wing loading and low aspect ratio indicates that the cost of transport is very high. Therefore, most of these bats fly only for short periods. Nevertheless, the small nectarivorous phyllostomids (glossophagines) often hover when feeding from flowers, whereas the larger nectarivorous pteropodids (macroglossines) more frequently perch (see above). Hovering species benefit from short wings (see above) to reduce flight power and to be able to fly close to vegetation. This puts much greater limits on the wing length than for nonhovering insectivorous species that hunt within vegetation but at a greater distance from foliage. Wings are short or of average length in most bats that forage within vegetation. Muscle physiology of some frugivorous bats shows that they use anaerobic metabolism and therefore are well suited to making short bursts of flight (Valdivieso et al., 1968).

Climbing within vegetation for fruits is facilitated by a reduced tail membrane, and this may be a reason why pteropodids and most phyllostomids have small tail membranes. Additionally, such bats may have less need to maneuver to catch insects. Most phyllostomids with an enlarged tail membrane also eat insects (see Norberg and Rayner, 1987). A majority of the phyllostomids have very long handwings with rounded wing tips, which may promote rapid acceleration and hovering, as well as maneuverability.

Carnivory

Only large bats can take large prey, but size alone does not identify carnivorous bats, that is, those which include small terrestrial vertebrates in their diet. Carnivory appears to be restricted to Microchiroptera in the families Nycteridae (slit-faced bats), Megadermatidae (Old World leaf-nosed bats), Phyllostomidae (New World leaf-nosed bats), and Vespertilionidae (vespertilionid bats) (Norberg and Fenton, 1988). Because carnivores need to take off and fly with considerable loads, such bats have a low wing loading and short wings (low aspect ratio) to allow flight within vegetation (Norberg and Rayner, 1987; open squares in fig. 9.5b). Most carnivorous bats feed on a variety of animals, including insects, and use a mixture of ground or foliage gleaning and perch hunting.

Norberg and Fenton (1988) used wing morphology, aerodynamics, foraging, and echolocation behavior to determine general characteristics of carnivorous microchiropteran bats. None of the features alone, namely body mass, relative wing loading ($Mg/SM^{1/3}$), aspect ratio, or orientation behavior, unequivocally identified carnivorous bats. However, a combination of morphological features, including a relatively large body mass (> 0.017 kg), low relative wing loading (< 36), and low aspect ratio (< 6.3), significantly identified carnivorous species from among other animal-eating forms (species to the left of and below the dashed lines in fig. 9.6). Nine of ten investigated carnivorous species use echolocation calls of short duration (< 1 ms), low intensity (to allow the bat to approach the prey without being heard; see page 227), and high frequency (most energy > 50 kHz). The tenth species, the pallid bat (*Antrozous pallidus,* Vespertilionidae), has high-intensity calls. This species stops echolocating after detecting a prey and relies on prey-generated cues to locate its target (Bell, 1982), as do several other carnivorous species.

Two species observed to take large animal prey, *Trachops cirrhosus* and *Phyl-*

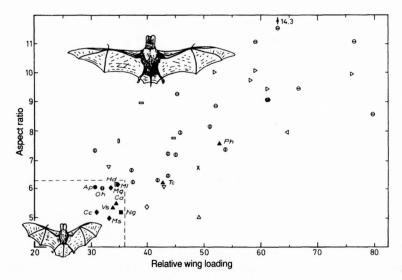

FIGURE 9.6 The nondimensional aspect ratio (wing span^2/wing area, b^2/S) is plotted against relative wing loading ($Mg/SM^{1/3}$) for forty-three species of animal-eating microchiropterans with body masses > 0.017 kg. Relative wing loading is independent of body size, since $Mg/SM^{1/3}$ (mass $\times$ $m\ s^{-2}$)/(mass$^{2/3}$ $\times$ mass$^{1/3}$) $= m\ s^{-2}$ (which is the dimension for gravitational acceleration). The solid symbols represent carnivorous species and the open symbols other animal-eating species. Two species, *Hipposideros diadema* (*Hd*) and *Otonycteris hemprichi* (*Oh*) are not known to be carnivorous, but fall in the aspect ratio and wing loading categories characteristic for carnivorous bats. They were therefore predicted by Norberg and Fenton (1988) to take heavy prey. *Ap* = *Antrozous pallidus*, *Ca* = *Chrotopterus auritus*, *Cc* = *Cardioderma cor*, *Mg* = *Macroderma gigas*, *Ml* = *Megaderma lyra*, *Ms* = *M. spasma*, *Ng* = *Nycteris grandis*, *Ph* = *Phyllostomus hastatus*, *Tc* = *Trachops cirrhosus*. (Based on data from Norberg and Fenton, 1988.)

lostomus hastatus, fall outside the identified aspect ratio and wing loading criteria. *T. cirrhosus* has the low aspect ratio characteristic of carnivores, but a higher wing loading. The frog-eating habits of this species are well documented (Tuttle and Ryan, 1981) and include the use of the calls of male frogs in prey location. *P. hastatus* is mainly frugivorous but occasionally is carnivorous.

Sanguivory

The phyllostomid subfamily Desmodontidae includes the three existing vampire bats, *Desmodus rotundus, Diaemus youngi,* and *Diphylla ecaudata. D. rotundus* feed mainly on blood from large mammals, while *D. youngi* prefers blood of birds and goats, and *D. ecaudata* is presumed to feed only on avian blood (Walker, 1964). Aspect ratio is low in *D. ecaudata,* average in *D. rotundus,* and high in *D. youngi* (plus signs in squares; fig. 9.5b). All three species have high wing loadings, although they, like carnivores, often have to take off with a substantially higher load after feeding. The high wing loading may be related to extensive commuting flights from a communal roost (3.4 km in *D. rotundus;* Wilkinson, 1985); because their feeding is relatively slow, it may be important to decrease travel time by flying faster (i.e., by having a higher loading).

Piscivory and Trawling

Three bats regularly eat fish or crustacea, the bulldog bats *Noctilio albiventris* and *N. leporinus* (Noctilionidae) and the fish-eating bat *Pizonyx vivesi* (Vespertilionidae). All three species have long narrow wings, low wing loading, and high aspect ratio, and are therefore adapted for economic flight away from vegetation, in this case hunting over open water (filled circles in fig. 9.5c). The low wing loading permits slow flight, which is advantageous as the bats have to localize prey in the water, usually by disturbances and ripples on the water surface. Flight over water also allows the bats to benefit from the aerodynamic ground effect. A reduction in induced drag of perhaps 10% can be obtained owing to the proximity of the wing plane to the water surface (Norberg, 1990).

Many insects swarm over water and regularly attract feeding bats. Trawling for aquatic insects on the water surface is more specialized and is characteristic of myotids of the subgenus *Leuconoe* (Findley, 1972). Such behavior has also been reported in several *Myotis* species. Most trawling species have average wing loading and average or slightly below average aspect ratio (open circles in fig. 9.5c). They resemble piscivorous species in their well-developed hind feet, which are often free of the wing membrane (left wing outline in fig. 9.5c), but are more similar to other insectivorous vespertilionids in wing design, indicating that such bats are adapted for flight among foliage as well.

Migration and Commuting

Most temperate bats hibernate during the winter, but some species make regular migrations of at least a few hundred kilometers to warmer locales. Records of migration by bats are reviewed by, among others, Griffin (1970), Yalden and Morris (1975), Fenton and Kunz (1977), Baker (1978), and Aellen (1983). Hibernating bats sometimes also migrate shorter distances to their hibernacula. Many species commute long distances during the night to forage and/or drink.

For economical flight, long-range migrants should have high aspect ratios and pointed wing tips. When time minimization is important, wing loading also should be high, permitting rapid flight at low energy cost. As predicted, migrants tend to have wings of high aspect ratio (fig. 9.5d), and many also have pointed wing tips. But a number of species that migrate shorter distances or are facultative migrators have a lower than average aspect ratio. All of these are relatively small insectivorous bats in which selection pressures on wing design associated with foraging activity may outweigh energy economy during short migrations.

The longest migrations (> 1000 km) are reported in the Mexican free-tailed bat, *Tadarida brasiliensis* (Molossidae); the silver-haired bat, *Lasionycteris noctivagans;* the hoary bat, *Lasiurus cinereus;* the red bat, *Lasiurus borealis;* and the noctule bat, *Nyctalus noctula* (all Vespertilionidae; filled circles in fig. 9.5d). All but the silver-haired bat have a high wing loading, and all but the red bat have a high or average aspect ratio.

Only long-distance fliers can colonize isolated islands by flight, and this is why remote islands have few bat species. For example, the hoary bat is the only species on Hawaii, and this one and a subspecies of the red bat are the only bats on the Galapagos Islands.

Bats migrating fairly long distances (200–1000 km; open circles in fig. 9.5d) include species of genera *Miniopterus* and *Emballonura,* and the phyllostomid *Leptonycteris sanborni* and vespertilionids *Myotis lucifugus, M. grisescens, M. dasycneme, Pipistrellus pipistrellus, P. nathusii,* and *Vespertilio murinus.*

Frugivorous and nectarivorous bats, but also vampires and some insectivores, may need to commute long distances every night to find adequate food resources. These bats would benefit from a high wing loading, which permits fast flights and decreases flight time. Many of these bats have the expected high wing loading.

WING DESIGN, ECHOLOCATION CALL STRUCTURE, AND FORAGING STRATEGY

Wing design and echolocation call structure can usually be correlated with the foraging manner of bats. Associations between the acoustic structure of echolocation calls, foraging zones, and hunting behavior have been treated in detail by several authors and recently summarized by Simmons et al. (1979), Fenton and Fullard (1981), Fenton (1982, 1984, 1986, 1990), and Neuweiler (1984). These associations were extended by Norberg and Rayner (1987) and Norberg (1989), who related them to flight speed and wing design (fig. 9.7).

Among bats adapted for fast aerial-hawking flight, for which long-range prey detection is necessary, there is a close association between wing shape and echolocation call structure. These bats use long, narrow-band echolocation calls of low frequency during searching and cruising flight. These calls appear to be well suited for long-range detection of prey, but give little information about the structure of the target, and are not clutter-resistant (Simmons and Stein, 1980). As the bat approaches the target, it switches to calls of broader bandwidth to gain more information about target detail (e.g., Fenton, 1984, 1990). Such long-range detection calls occur in species with high relative wing loading and high aspect ra-

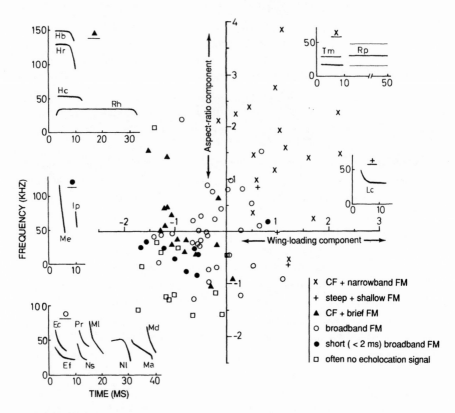

FIGURE 9.7 Scatter plot of second and third principal components, identified as measures of wing loading (body weight/wing area, Mg/S) and aspect ratio (wing span^2/wing area, b^2/S). The inset sonograms are based on Simmons et al. (1979), Fenton and Bell (1981), Thompson and Fenton (1982), Fenton et al. (1983), and Neuweiler (1984). Family symbols as in figure 9.5. (From Norberg, 1989, with permission, based on Norberg and Rayner, 1987.)

tio, such as the free-tailed bats and species in the genera *Taphozous* (sheath-tailed bats) and *Rhinopoma* (mouse-tailed bats, Rhinopomatidae) (Simmons et al., 1978, 1984; Fenton et al., 1980; Habersetzer, 1981; Neuweiler, 1984; Fenton and Rautenbach, 1986; crosses in fig. 9.7). Long-range detection is essential for the following reasons. After detecting a target, the bat must maneuver to catch it, and high wing loading makes the turning radius large. Even though the reaction time is short, a fast-flying bat will travel far before it can initiate a maneuver (Norberg, 1989).

Vespertilionids like *Lasiurus, Nyctalus,* and *Scotophilus* fly fast in relatively open spaces near vegetation. They use echolocation calls which include both steep and shallow frequency modulated (FM) components (Fenton et al., 1983;

Vogler and Neuweiler, 1983; Barclay, 1985; Fenton and Rautenbach, 1986; Aldridge and Rautenbach, 1987). Most of these species have relatively high wing loadings and are fast fliers (plus signs in fig. 9.7).

Slow foraging flight, however, enables the bat to maneuver and catch prey detected at short ranges. Echolocation calls designed for short-range detection among vegetation are usually broadband and ultrasonic. Different combinations of wing design and echolocation call structure represent various solutions to the problems of foraging among vegetation of different density and structure. The echolocation call and wing shape in these bats seem not to be as tightly associated as in fast-flying aerial-hawking species.

Pure, or constant frequency (CF) tones, combined with a brief FM sweep, often in a high frequency range, are used by hipposiderids, rhinolophids, some emballonurids, and the mormoopid *Pteronotus parnellii* to detect fluttering insects by the use of Doppler shifts (e.g., Barclay, 1983; Schnitzler and Ostwald, 1983; Vogler and Neuweiler, 1983; Habersetzer et al., 1984; Fenton and Rautenbach, 1986). These bats have low wing loading and slow and maneuverable flight and forage close to vegetation (triangles in fig. 9.7). Aspect ratio in these bats varies from low (demanding perch hunting) to high (permitting sustained flight).

Broadband (FM) signals, mostly ultrasonic, are used for discriminating target structure, range finding, and short-range detection in clutter (Simmons et al., 1975). These signals include a variety of combinations of short or long, shallow or steep FM calls, and are found in noctilionids and many vespertilionids (e.g., Fenton and Bell, 1979; Thompson and Fenton, 1982; Fenton et al., 1983; Barclay, 1985) with different wing loadings and aspect ratios (open circles in fig 9.7).

Surface gleaners often use very short (< 2 ms) FM calls of low intensity, although some ground gleaners often switch off echolocation and instead listen to and localize the sounds produced by the prey. Habersetzer and Vogler (1983) suggested that these calls might be an adaptation to fine texture discrimination of targets on a surface by spectral differences in the echoes. The very short FM signals allow the bat to receive echoes from close targets without overlapping the emitted sound. Nycterids, megadermatids, phyllostomids, and some plecotines and myotids use these calls (e.g., Simmons et al., 1979; Fenton and Bell, 1979; Fenton and Fullard, 1979; Neuweiler, 1983; solid circles in fig. 9.7). Most of these bats are adapted for slow and maneuverable flight within clutter (low wing loading, low aspect ratio, and rounded wing tips).

In general, wing design and echolocation call structure seem to have responded independently to the selective pressures associated with the different foraging patterns, indicating that correlations between wing shape and the acoustic structure of the calls are secondary effects of adaptation to particular foraging niches. In only one case is there a uniform relationship between echolocation sound and flight morphology, namely among bats adapted for fast-flying aerial hawking (high aspect ratio and high wing loading) for which long-range detection is necessary. But the echolocation call and wing shape of bats foraging in clutter seem not to be related in such a restricted fashion. Nonetheless, different combinations of wing design and call structure may represent various solutions to the exploitation of vegetation of different density and structure.

MORPHOLOGICAL AND BEHAVIORAL CONSTRAINTS

Since flight is energetically very expensive, flying animals have highly advanced morphological adaptations in their wings for efficient flight, implying an ecological limitation of their niches, since highly derived adaptations often involve deteriorated function in other respects.

A bat's choice of food and foraging behavior usually involve constraints in body size and wing shape. For example, a bat hunting moving prey needs high maneuverability and agility, whereas a fruit-eating bat taking food by climbing and hanging in trees does not have the same need for maneuverable and agile flight.

Morphology

Aerial foraging is facilitated by low body mass, large wing area and low wing loading, permitting high maneuverability and slow flight. This helps explain why aerial-foraging insectivorous bats and birds usually are small, since larger flying animals have higher wing loadings (wing loading $\propto$ mass$^{1/3}$ for geometrically similar fliers). Echolocation in bats permits detection of prey only at relatively short range. This puts still higher demands on low mass for slow flight in bats, so that they do not travel too far after prey detection before reacting a turn. Therefore, insectivorous bats in general are smaller than aerial-foraging insectivorous birds (Norberg, 1986b).

Wings must be light to keep inertial forces within reasonable limits, but at the same time must be resistant to bending. The wings of all vertebrate fliers, whether feathered or membraneous, are supported by a framework of skeletal elements controlled by muscles. Several wing arrangements increase the resistance to aerodynamic forces or increase aerodynamic performance (Norberg, 1969, 1972a, 1990).

Bird wings are composed of light feathers. Bending strength resides in the feather shafts themselves, which are robustly supported at their proximal ends, and the feathers are separated from one another but act functionally as a surface during the downstroke (R. Å. Norberg, 1985). The elastic membrane of bat wings is extended by thin arm, leg, and digit skeletons, which resist the large tension transmitted by the membrane when the aerodynamic forces act during flight (Norberg, 1972a, 1990). Because the wing skeleton of bats is relatively thin, it can easily be broken. Damage to the wing is usually more disastrous for a bat than for a bird, and a broken digit or a tear in the bat-wing membrane would adversely affect flight performance much more than would a damaged feather of the bird wing. This may explain why no bats climb in trees to forage for animals, as do many small birds. Protruding foliage could easily damage the thin bat-wing membrane. Most insectivorous bats instead hawk insects in open areas or forest gaps. On the other hand, vegetarian bats take fruits in trees and nectar from flowers, but usually land close to the fruits and flowers, and need not climb about continuously in search for food.

Variation in body size is usually correlated with variation in prey or food-particle size, but also with the type of food taken, as discussed below.

Diet

Differences in diet are influenced by geographical range. Whereas temperate bats are mostly small and insectivorous, tropical communities enjoy a greater variety of food habits (Norberg and Rayner, 1987), reflecting a greater food availability (McNab, 1982). Since fruits and flowers are seasonal in the temperate regions, it is easy to understand why there are no frugivorous or nectarivorous bats. Insectivorous bats are found among Microchiroptera, whereas bats belonging to Megachiroptera are frugivorous and/or nectarivorous.

Insectivory. Most bats (about 70%) and many birds are insectivorous, and the two groups usually utilize the same foraging zones, birds by day and bats at night. Birds evolved more than 100 million years before bats, which may have forced bats to become (or remain) active during nights to avoid competition. In the temperate zone, insects and other food become scarce when the temperature drops in the autumn. Most birds then migrate in autumn to warmer regions with higher food density. But bats hibernate to survive the winter; only a few bat species migrate. Differences in wing shape between birds and bats cannot be correlated with migratory habits, so other factors must be more important.

Bats hunt insects either by hawking or flycatching in the air or by hovering and gleaning among vegetation, which demands high maneuverability and agility. High maneuverability is obtained by low body mass and low wing loading, whereas high agility is obtained by different wing shapes in fast and slow flying bats (see "Maneuverability and Agility" above). Small flying animals also have a larger energy margin (the ratio of the power available from the flight muscles to that required to fly horizontally) because the demands of flight are low compared with other components of metabolism (Pennycuick, 1975, 1986; see also Norberg, 1990). On the other hand, small animals have a narrow margin as regards energy reserves, for example, for surviving the day (bats) or night (humming-birds), and have to be in torpor during the nonforaging periods. Larger bats can-not fly in dense vegetation; their flight is unmaneuverable and they have a smaller energy margin, both factors precluding them from living on insects. Insectivory thus is restricted to relatively small bats; insectivorous bats range from about 0.002 kg (e.g., bats of the genera *Craseonycteris,* butterfly bats, *Thyroptera,* disk-winged bats, *Furiptera,* smoky bats and thumbless bats, *Tylonycteris,* club-footed bats, and *Pipistrellus,* pipistrelles) to about 0.14 kg (*Cheiromeles tor-quatus,* hairless bat). Only seven (4%), however, of the 175 insectivorous bats in the data sample of Norberg and Rayner (1987) are heavier than 0.05 kg.

Large, erect ears would increase the ability of bats to detect and localize in-sects and other prey among vegetation. But large, drag-producing ears can only be evolved in slow-flying species, since drag (and power for flight) increases with speed squared. Therefore, only bats with low wing loadings (enabling slow flight) can afford to have large ears. Gleaning species are usually small and have low wing loadings and slow flight, and most insectivorous gleaners (but not all) have large ears.

Nectarivory. Only small vegetarian bats are able to hover or hang on flowers when taking pollen and nectar. It would be energetically too expensive for a large bat to hover, and impossible to cling on fragile flowers. Therefore, only small bats are nectarivorous. On the other hand, nectarivorous species have less need for maneuverability than do insectivores, and they have higher wing loadings (Norberg and Rayner, 1987). Whereas a low wing loading should be preferable for good hovering performance as it permits greater food intake per flight, a high wing loading permits rapid flight. When nectar supplies are sparse or when there are long distances between suitable flowers, fast flights might be necessary to ensure appropriate rates of food collection (Norberg and Rayner, 1987). There are thus different selection pressures on wing loading in nectarivorous bats de-pending on food availability and food dispersion, but size is the overall limiting factor for them. Microchiropteran phyllostomid bats that are primary nec-tarivores range in size from about 0.0065 kg (*Lichonycteris obscura*) to about 0.034 kg (*Leptonycteris nivalis,* long-nosed bat), whereas nectarivorous mega-chiropterans range from about 0.013 kg to about 0.045 kg (species of subfamily Macroglossinae, long-tongued fruit bats).

Frugivory. In the evolution of flight in bats, those which evolved the ability to hawk insects had to be small, whereas frugivores and bats taking other animal prey than insects had no such size limitation. Primary frugivores are large, al-though their size range overlaps with that of insectivores; they range in size from about 0.01 kg (*Artibeus phaeotis,* a neotropical fruit bat) to about 0.06 kg (*A. lituratus,* big fruit bat) in the microchiropteran phyllostomids, and from about 0.014 kg (*Balionycteris maculata,* spotted-winged fruit bat) to about 1.5 kg (*Pteropus giganteus,* Indian flying fox) in the megachiropteran pteropodids. Whereas the smaller phyllostomids often include some insects in their diet to obtain sufficient protein, pteropodids must take larger quantities of fruit than their energetic requirements to satisfy their protein needs (Thomas, 1984).

Carnivory. As discussed above (see subsection "Carnivory"), carnivorous bats must be large and have a low wing loading to be able to fly slowly and with a heavy load, and their wings must be relatively short for flight within vegetation (Norberg and Rayner, 1987; Norberg and Fenton, 1988). Carnivorous bats range from 0.017 kg (pallid bat, *Antrozous pallidus*) to about 0.160 kg (American false vampire bat, *Vampyrum spectrum*) (Norberg and Fenton, 1988). The pallid bat is mainly insectivorous and has only occasionally been observed flying with small vertebrates (e.g., Bell, 1982). Other carnivorous bats weigh more than 0.025 kg. Some carnivorous species use short, low intensity, high frequency, broadband echolocation calls but rely on prey-generated cues to locate their targets; others are facultative echolocators (Norberg and Fenton, 1988). They also rely on vision for prey localization. In summary, a large body size, low aspect ratio and low wing loading (broad and rather short wings) are necessary adaptations for carnivorous bats.

Mammals

Since bats are mammals, female bats nurse their young and·then have increased nutritional requirements. Lactation and carrying heavy fetuses put bat females under higher stresses than bird females are subjected to during the reproductive period. Birds can build up food reserves for their young (deposited in the eggs) during a shorter period than bats, and bird males often help to rear the young. This may explain why bats give birth to only one or two young at a time, whereas birds can lay numerous eggs. The ability to fly with extra loads is facilitated by a low wing loading. Bats in general have lower wing loadings than do birds (U. M. Norberg, 1981a), and are better adapted to carry heavy loads and to fly more slowly than birds of similar size.

Why Don't Bats Glide?

In birds, gliding is an energy-saving mode of locomotion and is included in soaring and flap-gliding. Many birds (such as gulls, raptors, storks, and cranes) use cross-country soaring during migration; the birds climb in a thermal to a substantial height and then glide in the desired direction, losing height as they go, and then climbing in a new thermal. This is an inexpensive means of moving a unit distance but requires more time to reach a destination than when using horizontal flapping flight. Flap-gliding consists of a flapping phase in which the animal climbs in a straight path, followed by a gliding phase (intermittent gliding). This flight mode is typical of birds larger than large woodpeckers (of mass around 0.2 kg) and of those that are good gliders (Rayner, 1977, 1985).

In general, bats cannot use soaring flight because convective air currents are rare or absent at night. Therefore, bats probably have not been subjected to strong selection forces for soaring (gliding) ability. However, some *Pteropus* species regularly soar on thermal updraughts (see Thomas et al., 1990).

Lavia frons and *Pipistrellus pipistrellus* have been observed to use intermittent gliding with very short gliding phases (U. M. Norberg unpubl.; Thomas et al., 1990). Pennycuick (1971, 1973) studied the gliding flight of the dog-faced bat *Rousettus aegyptiacus* (Megachiroptera) in a wind tunnel and found that its low-speed performance was similar to that of the pigeon. Because the bat could not reduce the area of the wing tip without collapsing the wing, it was less successful at fast gliding. Bat wings are instead adapted for greater control over the profile shape of the manus, so they are more maneuverable than birds in low-speed flight (Pennycuick, 1971, 1973).

Gliding is further confined mainly to medium-sized to large birds, which usually have more than one fiber type in their flight muscles. Slow fibers, which are used to maintain isometric tension, are found in many soaring birds (e.g., George and Berger, 1966; Rosser and George, 1986). Isometric tension is necessary to keep the wings outstretched in the horizontal plane (see McMahon, 1984). Small birds and bats usually have only fast-twitch oxidative fibers, adapted for repetitive movements and sustained flight (e.g., Rosser and George, 1986; Ohtsu and Uchida, 1979a, b). Such fibers are too fast for isometric contraction (e.g., McMahon, 1984). This may explain why only larger bats have been observed to use long gliding sequences.

Why Is the Tail Membrane So Small in Most Vegetarian Bats?

The families Vespertilionidae (insectivores) and Phyllostomidae (frugivores and nectarivores) are the dominant microchiropteran families. In general, the phyllostomids have relatively shorter wing span and higher wing loading than vespertilionids, with relatively longer wing tips that correlate with good hovering performance (Norberg and Rayner, 1987). Hovering is primarily important for the nectarivores. Members of both families have on average low aspect ratio wings. In addition to shorter wings, a further reason for the higher wing loading in phyllostomids is the small size of their tail membranes, which is included methodologically (and functionally) in the total wing area. The higher wing loading and the low aspect ratio result in rapid flight of short duration. Megachiropteran bats also have a reduced tail membrane and thus higher wing loading than vespertilionids.

Insectivorous bats use the tail membrane to catch flying prey, and the large membrane also contributes to maneuverability. Their low wing loadings allow them to carry heavy prey. Because vegetarian bats do not hawk, a large tail is not necessary, and furthermore, a reduced tail is less hindrance in climbing and clinging among vegetation. If nectar is sparse or if there are long distances between fruits or flowers, selection would favor high flight speed (small wing area) that allow foraging over a large area during a single night. This effect evidently outweighs any pressure for low wing loading to fly better with greater nectar loads or with larger single fruits.

FUTURE RESEARCH

Flight mechanics impose significant constraints on behavior, which are responsible for shaping the bat's niche, and the influence of these constraints may be traced in wing adaptation. Echolocation also is a main component of the foraging behavior of the majority of bat species, and may have equal significance to flight in determining the foraging patterns a bat may adopt. Field observations of bat foraging behavior support Norberg's and Rayner's (1987) theoretical predictions on the relationships between wing morphology and flight behavior.

Ecological approaches with more detailed analyses about bat behavior, while foraging and otherwise, would add to our understanding of biotic and abiotic factors important for the evolution of wing design. Comparisons between closely related or sympatric species would be valuable to test if differences in flight morphology can be related to differences in foraging and flight behavior. Deeper knowledge of bat wing morphology, echolocation, and habitat use may provide greater understanding of bat evolution, behavior, and ecomorphology.

Pettigrew's et al. (1989) suggestion that flying foxes evolved from an early branch of the primate lineage, whereas microchiropterans evolved much earlier from small, agile insectivores, has been intensely debated. Megachiropterans share with primates a variety of complex details in the organization of neural (visual) pathways not found in microchiropterans. A few wing characters, such as the difference in the ratio between the lengths of the metacarpal and first phalanx of the digits between the two groups of bats, were also used by the authors as evidence for diphyly in bats. In addition, the flight kinematics differ in Megachiroptera and Microchiroptera; whereas megachiropteran bats (even the small ones) fold their hand wings during the upstroke, microchiropterans keep them extended and straight (Norberg, 1970, 1972b). This may be an additional indication of a diphyletic origin in bats, but the differences in metacarpal lengths and in the wingbeat kinematics would also result from different demands on morphology in large and small bats for optimal wing profile during flight (U. M. Norberg, unpubl.), and are thus adaptive characters. In this monophyletic model, the ancestral bat is assumed to be small- or medium-sized from which mega- and microchiropteran bats evolved, and the small extant megachiropterans (which are nectarivorous) are assumed to have passed a stage as larger frugivorous bats.

An argument against Pettigrew's et al. (1989) results is that visual structures and wing morphology are highly adaptive, and such structures may not be used as evidence for evolutionary relationships. Similarities between Megachiroptera and primates could be due to convergence; both groups have well-developed eyes. In fact, megachiropterans, which are vegetarians, share many flight characters with the microchiropteran, vegetarian bats of the family Phyllostomidae, such as rather short and broad wings and a small tail membrane. Flight muscles and skeletons also are remarkably similar in Mega- and Microchiroptera. Furthermore, distinctive cranial and cervical innervation of wing muscles suggests that bats are monophyletic, but that bats and Dermoptera share a common ancestor that had wings (Thewissen and Babcock, 1991), or rather a gliding surface which in bats evolved into wings. Baker et al. (1991) examined the data and arguments presented by Pettigrew et al. (1989) and concluded that a strong case for diphyly has *not* been established and that chiropteran relationships are complex and require further study of both morphological and molecular data.

ACKNOWLEDGMENTS

I am grateful to Åke Norberg, Robert Dudley, Brock Fenton, and three anonymous reviewers for reading and commenting on the manuscript. This work has been supported by grants from the Swedish National Science Research Council.

REFERENCES

Aellen, V. 1983. Migrations des chauves-souris en Suisse. *Bonn. Zool. Beitr.* 34:3–27.

Aldridge, H. D. J. N. 1985. On the relationships between flight performance, morphology and ecology in British bats. Ph.D. thesis, University of Bristol.

Aldridge, H. D. J. N. 1986. Kinematics and aerodynamics of the greater horseshoe bats, *Rhinolophus ferrumequinum*, in horizontal flight at various flight speeds. *J. Exp. Biol.* 126:479–497.

Aldridge, H. D. J. N., and I. L. Rautenbach. 1987. Morphology, echolocation and resource partitioning in insectivorous bats. *J. Anim. Behav.* 56:763–778.

Andersson, M., and R. Å. Norberg. 1981. Evolution of reversed sexual size dimorphism and role partitioning among predatory birds, with a size scaling of flight performance. *Biol. J. Linnean Soc.* 15:105–130.

Baagøe, H. J. 1987. The Scandinavian bat fauna: Adaptive wing morphology and free flight behaviour in the field. In *Recent Advances in the Study of Bats,* ed. M. B. Fenton, P. A. Racey, and J. M. V. Rayner, 57–74. Cambridge: Cambridge University Press.

Baker, R. J., M. J. Novacek, and N. B. Simmons. 1991. On the monophyly of bats. *Syst. Zool.* 40(2):216–231.

Baker, R. R. 1978. *The Evolutionary Ecology of Animal Migration.* London: Hodder and Stoughton.

Barclay, R. M. R. 1983. Echolocation calls of emballonurid bats from Panama. *J. Comp. Physiol.* A151:515–520.

Barclay, R. M. R. 1985. Long- versus short-range foraging strategies of hoary (*Lasiurus cinereus*) and silver-haired (*Lasiurus borealis*) bats and the consequences for prey selection. *Can. J. Zool.* 63:2507–2515.

Barrett-Hamilton, G. E. H. 1910. *A History of British Mammals*, vol. 1, *Bats*. London: Gurney and Jackson.

Bell, G. P. 1982. Behavioral and ecological aspects of gleaning by a desert insectivorous bat, *Antrozous pallidus* (Chiroptera: Vespertilionidae). *Behav. Ecol. Sociobiol.* 10:217–223.

Bell, G. P. 1985. The sensory basis of prey location by the California leaf-nosed bat *Macrotus californicus* (Chiroptera: Phyllostomidae). *Behav. Ecol. Sociobiol.* 16:343–347.

Bock, W. J. 1965. The role of adaptive mechanisms in the origin of higher levels of organization. *Syst. Zool.* 14:272–287.

Bock, W. J. 1986. The arboreal origin of avian flight. In *The Origin of Birds and the Evolution of Flight,* ed. K. Padian, 57–72. San Francisco: California Academy of Sciences.

Bonaccorso, F. J. 1979. Foraging and reproductive ecology in a Panamanian bat community. *Bull. Florida State Mus. Biol. Sci.* 24:359–408.

Bradbury, J. W., and L. H. Emmons, 1974. Social organization of some Trinidad bats. I. Emballonuridae. *Z. Tierpsychol.* 36:137–183.

Bradbury, J. W., and S. L. Vehrencamp. 1976. Social organization and foraging in emballonurid bats. I. Field studies. *Behav. Ecol. Sociobiol.* 1:337–381.

Brosset, A. 1966. *La Biologie des Chiropteres.* Paris: Masson.

Buchler, E. R. 1980. The development of flight, foraging, and echolocation in the little brown bat (*Myotis lucifugus*). *Behav. Ecol. Sociobiol.* 6:211–218.

Carpenter, R. E. 1985. Flight physiology of flying foxes, *Pteropus poliocephalus.* *J. Exp. Biol.* 114:619–647.

Dobat, K., and T. Peikert-Holle. 1985. *Blüten und Fledermäuse. Bestäubung durch Fledermäuse und Flughunde (Chiropterophilie).* Senchenberg-Buch 78. Frankfurt-am-Main: Waldemar Kramer.

Fenton, M. B. 1974. The role of echolocation in the evolution of bats. *Amer. Nat.* 108:386–388.

Fenton, M. B. 1982. Echolocation calls and patterns of hunting and habitat use of bats (Microchiroptera) from Chillagoe, north Queensland. *Aust. J. Zool.* 30:417–425.

Fenton, M. B. 1984. Echolocation: Implications for ecology and evolution of bats. *Q. Rev. Biol.* 59:33–53.

Fenton, M. B. 1986. *Hipposideros caffer* (Chiroptera: Hipposideridae) in Zimbabwe: Morphology and echolocation calls. *J. Zool.* (London) A210:347–353.

Fenton, M. B. 1990. The foraging behaviour and ecology of animal-eating bats. *Can. J. Zool.* 68:411–422.

Fenton, M. B., and G. P. Bell. 1979. Echolocation and feeding behaviour in four species of *Myotis* (Chiroptera). *Can. J. Zool.* 57:1271–1277.

Fenton, M. B., and G. P. Bell. 1981. Recognition of species of insectivorous bats by their echolocation calls. *J. Mammal.* 62:233–243.

Fenton, M. B., and J. H. Fullard. 1979. The influence of moth hearing on bat echolocation strategies. *J. Comp. Physiol.* A132:77–86.

Fenton, M. B., and J. H. Fullard. 1981. Moth hearing and the feeding strategies of bats. *Amer. Sci.* 69:266–275.

Fenton, M. B., and T. H. Kunz. 1977. Movements and behavior. In *Biology of Bats of the New World Family Phyllostomatidae,* vol. 2, ed. R. J. Baker, J. K. Jones, and D. C. Carter, 351–364. Spec. Publs Mus. Texas Technical University no. 13, Lubbock.

Fenton, M. B., and I. L. Rautenbach. 1986. A comparison of the roosting and foraging behavior of three species of African insectivorous bats (*Rhinolophus hildebrandti*—Rhinolophidae, *Scotophilus borbonicus*—Vespertilionidae, and *Tadarida midas*—Molossidae). *Can. J. Zool.* 64:2860–2867.

236   Ulla Norberg

Fenton, M. B., G. P. Bell, and D. W. Thomas. 1980. Echolocation and feeding behaviour of *Taphozous mauritianus* (Chiroptera: Emballonuridae). *Can. J. Zool.* 58: 1774–1777.

Fenton, M. B., H. G. Merriam, and G. L. Holroyd. 1983. Bats of Kootenay, Glacier, and Mount Revelstoke National Parks in Canada: Identification by echolocation calls, distribution and biology. *Can. J. Zool.* 61:2503–2508.

Fiedler, J. 1979. Prey catching with and without echolocation in the Indian false vampire bat, *Megaderma lyra. Behav. Ecol. Sociobiol.* 6:155–160.

Findley, J. S. 1972. Phenetic relationships among bats of the genus *Myotis. Syst. Zool.* 21:31–52.

Findley, J. S., E. H. Studier, and D. E. Wilson. 1972. Morphologic properties of bat wings. *J. Mammal.* 53:429–444.

Gardner, A. L. 1977. Feeding habits. In *Biology of Bats of the New World Family Phyllostomatidae*, ed. R. J. Baker, J. K. Jones, and D. C. Carter, vol. 2, 293–350. Spec. Publs Mus. Texas Technical University no. 13. Lubbock.

George, J. C., and A. J. Berger. 1966. *Avian Myology.* London: Academic Press.

Griffin, D. R. 1970. Migrations and homing of bats. In *Biology of Bats,* vol. 1, ed. W. A. Wimsatt, 233–264. New York: Academic Press.

Habersetzer, J. 1981. Adaptive echolocation sounds in the bat *Rhinopoma hardwickei. J. Comp. Physiol.* A144:559–566.

Habersetzer, J., and G. Storch. 1987. Klassifikation und funktionelle Flügelmorphologie paläogener Fledermäuse (Mammalia, Chiroptera). *Cour. Forsch. Inst. Senckenberg.* (Frankfurt-a-M.) 91:117–150.

Habersetzer, J., and G. Storch. 1989. Ecology and echolocation of the Eocen Messel bats. In *European Bat Research,* ed. V. Hanák, I. Horácèk, and J. Gaisler, 213–233. Prague: Charles University Press.

Habersetzer, J., and B. Vogler. 1983. Discrimination of surface-structured targets by the echolocating bat, *Myotis myotis,* during flight. *J. Comp. Physiol.* A152:275–282.

Habersetzer, J., G. Schuller, and G. Neuweiler. 1984. Foraging behavior and Doppler shift compensation in echolocating hipposiderid bats, *Hipposideros bicolor* and *Hipposideros speoris. J. Comp. Physiol.* A155:559–567.

Hall, E. R., and W. W. Dalquest. 1963. The mammals of Veracruz. *Univ. Kansas Publs. Mus. Nat. Hist.* 14:165–362.

Heithaus, E. R. 1982. Coevolution between bats and plants. In *Ecology of Bats,* ed. T. H. Kunz, 327–367. New York: Plenum Press.

Helversen, O. von, and H.-U. Reyer. 1984. Nectar intake and energy expenditure in a flower visiting bat. *Oecologia* 63:178–184.

Howell, D. J. 1979. Flock foraging in nectar-feeding bats: Advantages to the bats and to the host plants. *Amer. Nat.* 114:23–49.

Jepsen, G. L. 1970. Bat origins and evolution. In *Biology of Bats,* vol. 1, ed. W. A. Wimsatt, 1–64. New York: Academic Press.

Kick, S. 1982. Target-detection by the echolocating bat, *Eptesicus fuscus. J. Comp. Physiol.* A145:431–435.

Kopka, T. 1973. Beziehungen zwischen Flügelfläche und Körpergrösse bei Chiropteren. *Z. Wiss. Zool.* 185:235–284.

Krebs, J. R. 1978. Optimal foraging: Decision rules for predators. In *Behavioural Ecology: An Evolutionary Approach,* ed. J. R. Krebs and N. B. Davis, 23–63. Oxford: Blackwell.

Lawlor, T. E. 1973. Aerodynamic characteristics of some neotropical bats. *J. Mammal.* 54:71–78.

McNab, B. K. 1982. Evolutionary alternatives in the physiological ecology of bats. In *Ecology of Bats,* ed. T. H. Kunz, 151–200. New York: Plenum Press.

McMahon, T. A. 1984. *Muscles, Reflexes, and Locomotion.* Princeton: Princeton University Press.

Neuweiler, G. 1983. Echolocation and adaptivity to ecological constraints. In *Neuroethology and Behavioural Physiology: Roots and Growing Points,* ed. F. Huber and H. Mark, 280–302. Berlin: Springer.

Neuweiler, G. 1984. Foraging, echolocation and audition in bats. *Naturwiss.* 71:446–455.

Norberg, R. Å. 1977. An ecological theory on foraging time and energetics and choice of optimal food-searching method. *J. Anim. Ecol.* 46:511–529.

Norberg, R. Å. 1981a. Optimal flight speed in birds when feeding young. *J. Anim. Ecol.* 50:473–477.

Norberg, R. Å. 1981b. Why foraging birds in trees should climb and hop upwards rather than downwards. *Ibis* 123:281–288.

Norberg, R. Å. 1983. Optimum locomotion modes for birds foraging in trees. *Ibis* 125:172–180.

Norberg, R. Å. 1985. Function of vane asymmetry and shaft curvature in bird flight feathers: Inferences on flight ability of *Archaeopteryx.* In *The Beginnings of Birds,* ed. M. K. Hecht, J. H. Ostrom, G. Viohl, and P. Wellnhofer, 303–318. Willibaldsburg: Freunde des Jura-Museums Eichstätt.

Norberg, R. Å., and U. M. Norberg. 1971. Take-off, landing, and flight speed during fishing flights of *Gavia stellata* (Pont.). *Ornis Scand.* 2:55–67.

Norberg, U. M. 1969. An arrangement giving a stiff leading edge to the hand wing in bats. *J. Mammal.* 50:766–770.

Norberg, U. M. 1970. Functional osteology and myology of the wing of *Plecotus auritus* Linnaeus (Chiroptera). *Ark. Zool.* 22:483–543.

Norberg, U. M. 1972a. Bat wing structures important for aerodynamics and rigidity (Mammalia, Chiroptera). *Z. Morph. Tiere* 73:45–61.

Norberg, U. M. 1972b. Functional osteology and myology of the wing of the dog-faced bat *Rousettus aegyptiacus* (E. Geoffroy) (Pteropodidae). *Z. Morph. Tiere* 73:1–44.

Norberg, U. M. 1979. Morphology of the wings, legs and tail of three coniferous forest tits, the goldcrest, and the treecreeper in relation to locomotor pattern and feeding station selection. *Phil. Trans. R. Soc.* (London) B287:131–165.

Norberg, U. M. 1981a. Allometry of bat wings and legs and comparison with bird wings. *Phil. Trans. R. Soc.* (London) B292:359–298.

Norberg, U. M. 1981b. Flight, morphology and the ecological niche in some birds and bats. In *Vertebrate Locomotion,* ed. M. H. Day, 173–197. London: Academic Press.

Norberg, U. M. 1985. Evolution of vertebrate flight: An aerodynamic model for the transition from gliding to flapping flight. *Amer. Nat.* 126:303–327.

Norberg, U. M. 1986a. Evolutionary convergence in foraging niche and flight morphology in insectivorous aerial-hawking birds and bats. *Ornis Scand.* 17:253–260.

Norberg, U. M. 1986b. On the evolution of flight and wing forms in bats. In *Bat flight (Fledermausflug) Biona Report 5,* ed. W. Nachtigall, 13–26. New York, Stuttgart: Gustav Fischer.

Norberg, U. M. 1987. Wing form and flight mode in bats. In *Recent Advances in the Study of Bats,* ed. M. B. Fenton, P. A. Racey, and J. M. V. Rayner, 43–56. Cambridge: Cambridge University Press.

Norberg, U. M. 1989. Ecological determinants of bat wing shape and echolocation call structure with implications for some fossil bats. In *European Bat Research,* ed. V. Hanák, I. Horácèk and J. Gaisler, 197–211. Prague: Charles University Press.

Norberg, U. M. 1990. *Vertebrate Flight.* Berlin: Springer.

Norberg, U. M., and M. B. Fenton. 1988. Carnivorous bats? *Biol. J. Linnean Soc.* 33:383–394.

Norberg, U. M., and J. M. V. Rayner. 1987. Ecological morphology and flight in bats (Mammalia; Chiroptera): Wing adaptations, flight performance, foraging strategy and echolocation. *Phil. Trans. R. Soc.* (London) B316:335–427.

Norberg, U. M., T. H. Kunz, J. F. Steffensen, Y. Winter, and O. von Helversen. 1993. The cost of hovering flight in a nectar-feeding bat, *Glossophaga soricina,* estimated from aerodynamic theory. *J. Exp. Biol.* 182:207–227.

Novacek, M. J. 1985. Evidence for echolocation in the oldest known bats. *Nature* 315:140–141.

Ohtsu, R., and T. A. Uchida. 1979a. Correlation among fiber composition and LDH isozyme patterns of the pectoral muscles and flight habits in bats. *J. Fac. Agr. Kyushu Univ.* 24:145–155.

Ohtsu, R., and T. A. Uchida. 1979b. Further studies on histochemical and ultrastructural properties of the pectoral muscles and flight habits in bats. *J. Fac. Agr. Kyushu Univ.* 24:157–163.

O'Shea, T. J., and T. A. Vaughan. 1980. Ecological observations on an East African bat community. *Mammalia* 44:485–496.

Pennycuick, C. J. 1969. The mechanics of bird migration. *Ibis* 111:525–556.

Pennycuick, C. J. 1971. Gliding flight of the dog-faced bat *Rousettus aegyptiacus* observed in a wind tunnel. *J. Exp. Biol.* 55:833–845.

Pennycuick, C. J. 1973. Wing profile shape in a fruit-bat gliding in a wind tunnel, determined by photogrammetry. *Period. Biol.* 75:77–82.

Pennycuick, C. J. 1975. Mechanics of flight. In *Avian Biology,* vol. 5, ed. D. S. Farner and J. R. King, 1–75. London, New York, San Francisco: Academic Press.

Pennycuick, C. J. 1986. Mechanical constraints on the evolution of flight. In *The Origin of Birds and the Evolution of Flight,* ed. K. Padian, 83–98. San Francisco, Calif.: Calif. Acad. Sci.

Pennycuick, C. J. 1989. *Bird Flight Performance: A Practical Calculation Manual.* Oxford: Oxford University Press.

Pettigrew, J. D. 1986. Flying primates? Megabats have the advanced pathway from eye to midbrain. *Science* 231:1304–1306.

Pettigrew, J. D., B. G. M. Jamieson, S. K. Robson, L. S. Hall, K. I. McNally, and H. M. Cooper. 1989. Phylogenetic relations between microbats, megabats and primates (Mammalia: Chiroptera and Primates). *Phil. Trans. R. Soc.* (London) B325:489–559.

Pijl, L. van der. 1956. Remarks on pollination by bats in the genera *Freycinetia, Duabanga* and *Haplophragma,* and on chiropterophyly in general. *Acta Bot. Neerl.* 6:291–315.

Pyke, G. H., H. R. Pulliam, and E. L. Charnov. 1977. Optimal foraging: A selective review of theory and tests. *Q. Rev. Biol.* 52:137–154.

Rayner, J. M. V. 1977. The intermittent flight of birds. In *Scale Effects in Animal Locomotion,* ed. T. J. Pedley, 437–443. London: Academic Press.

Rayner, J. M. V. 1979. A new approach to animal flight mechanics. *J. Exp. Biol.* 80:17–54.

Rayner, J. M. V. 1985. Bounding and undulating flight in birds. *J. Theor. Biol.* 117:47–77.

Rayner, J. M. V. 1988. Form and function in avian flight. In *Current Ornithology,* vol. 5, ed. R. F. Johnston, 1–66. New York: Plenum Press.

Roberts, T. J. 1977. *The Mammals of Pakistan.* London: Ernest Benn.

Rosser, B. W. C., and J. C. George. 1986. Slow muscle fibers in the pectoralis of the turkey vulture (*Cathartes aura*): An adaptation for soaring flight. *Zool. Anz.* 217:252–258.

Schnitzler, H.-U., and J. Ostwald. 1983. Adaptations for the detection of fluttering insects by echolocation in horseshoe bats. In *Advances in Vertebrate Neuroethology,* ed. J.-P. Ewart, R. R. Capranica, and D. J. Ingle, 801–827. New York: Plenum Press.

Simmons, J. A., and R. A. Stein. 1980. Acoustic imaging in bat sonar: Echolocation signals and the evolution of echolocation. *J. Comp. Physiol.* A135:61–84.

Simmons, J. A., M. B. Fenton, and M. J. O'Farrell. 1979. Echolocation and the pursuit of prey by bats. *Science* 203:16–21.

Simmons, J. A., D. J. Howell, and N. Suga. 1975. Information content of bat sonar echoes. *Amer. Sci.* 63:204–215.

Simmons, J. A., S. A. Kick, and M. L. Lawrence. 1984. Echolocation and hearing in the mouse-tailed bats, *Rhinopoma hardwickei:* Acoustic evolution of echolocation in bats. *J. Comp. Physiol.* A154:347–356.

Simmons, J. A., W. A. Lavender, B. A. Lavender, J. E. Childs, K. Hulebak, M. R. Rigden, J. Sherman, B. Woolman, and M. J. O'Farrell. 1978. Echolocation by free-tailed bats (*Tadarida*). *J. Comp. Physiol.* A125:291–299.

Stebbings, R. E. 1977. Order Chiroptera: Bats. In *Handbook of British Mammals,* ed. G. B. Corbet and H. N. Southern, 68–128. Oxford: Blackwell.

Thewissen, J. G. M., and S. K. Babcock. 1991. Distinctive cranial and cervical innervation of wing muscles: New evidence for bat monophyly. *Science* 251:934–936.

Thollesson, M., and U. M. Norberg. 1991. Moments of inertia of bat wings and body. *J. Exp. Biol.* 158:19–35.

Thomas, A. L. R., G. Jones, J. M. V. Rayner, and P. M. Hughes. 1990. Intermittent gliding flight in the pipistrelle bat (*Pipistrellus pipistrellus*) (Chiroptera: Vespertilionidae). *J. Exp. Biol.* 149:407–416.

Thomas, D. W. 1984. Fruit intake and energy budgets of frugivorous bats. *Physiol. Zool.* 57:457–467.

Thomas, S. P. 1975. Metabolism during flight in two species of bats, *Phyllostomus hastatus* and *Pteropus gouldii*. *J. Exp. Biol.* 63:273–293.

Thompson, D., and M. B. Fenton. 1982. Echolocation and feeding behaviour of *Myotis adversus* (Chiroptera: Vespertilionidae). *Aust. J. Zool.* 30:543–546.

Torre-Bueno, J. R., and J. Larochelle. 1978. The metabolic cost of flight in unrestrained birds. *J. Exp. Biol.* 75:223–229.

Tuttle, M. D., and M. J. Ryan. 1981. Bat predation and the evolution of frog vocalizations in the neotropics. *Science* 214:677–678.

Valdivieso, D., E. Conde, and J. R. Tamsitt. 1968. Lactate dehydrogenase studies in Puerto Rican bats. *Comp. Biochem. Physiol.* 27:133–138.

Vaughan, T. A. 1976. Nocturnal behaviour of the African false vampire bat (*Cardioderma cor*). *J. Mammal.* 57:227–248.

Vogel, S. 1968. Chiropterophilie in der neotropischen Flora. *Flora Jena* B157:562–602; 158:185–222, 289–323.

Vogler, B., and G. Neuweiler. 1983. Echolocation in the noctule (*Nyctalus noctula*) and horseshoe bats (*Rhinolophus ferrumequinum*). *J. Comp. Physiol.* A152:421–432.

Walker, E. P. 1964. *Mammals of the World*, vol. 1. Baltimore: Johns Hopkins University Press.

Weis-Fogh, T. 1972. Energetics of hovering flight in hummingbirds and in *Drosophila*. *J. Exp. Biol.* 56:79–104.

Wilkinson, G. S. 1985. The social organization of the common vampire bat. I. Pattern and cause of association. *Behav. Ecol. Sociobiol.* 17:111–121.

Wilson, D. E. 1973. Bat faunas: A trophic comparison. *Syst. Zool.* 22:14–29.

Wood, F. G., and W. E. Evans. 1980. Adaptiveness and ecology of echolocation in toothed whales. In *Animal Sonar Systems*, ed. R.-G. Busnel and J. F. Fish, 381–425. New York: Plenum Press.

Yalden, D. W., and P. Morris. 1975. *The Lives of Bats*. Newton Abbott: David and Charles.

van Zyll de Jong, C. G. 1985. *Handbook of Canadian Mammals*, vol. 2 (*Bats*). Ottawa: National Museum of Natural Sciences.

10

Ecological Morphology of Locomotor Performance in Squamate Reptiles

Theodore Garland, Jr., and Jonathan B. Losos

INTRODUCTION

Relationships between morphology, physiology, or biochemistry, on the one hand, and behavior and ecology, on the other, have been widely documented, as this volume attests. Such relationships provide evidence that most, if not all, organisms are to some extent "adapted" to their current environment. Quantifying how well adapted an organism is, or testing the biological and statistical significance of putative adaptations, may, however, be very difficult (Brooks and McLennan, 1991; Harvey and Pagel, 1991; Losos and Miles, chap. 4, this volume). As well, many studies in ecological morphology, and in the conceptually related fields of physiological ecology (Feder et al., 1987) and comparative biochemistry (Hochachka and Somero, 1984), have ignored the crucial intermediate step of organismal performance (Arnold, 1983; Huey and Stevenson, 1979; Losos, 1990b) when trying to correlate morphology with ecology. In this chapter, we review the literature pertaining to the ecological morphology of locomotor performance in reptiles and relate this knowledge to current paradigms and analytical techniques in organismal and evolutionary biology. We will argue that both maximal whole-animal performance abilities (what an animal can do when pushed to its limits; generally measured in the laboratory, and not to be confused with *efficiency;* see Gans, 1991; Lauder, 1991) and behavior (what an animal actually does when faced with behavioral options; best measured in the field) must be considered when attempting to understand the mechanistic bases of relationships between morphology and ecology.

Locomotion is in many ways ideally suited for studies of ecological morphology. Most behavior involves locomotion, and measures of both locomotor performance (e.g., speed, stamina) and its morphological bases (e.g., limb length, heart size) come easily to mind. Some reptiles are good subjects for measurement of locomotor performance in the laboratory (e.g., with race tracks or treadmills), for quantifying its morphological, physiological, and biochemical bases, and for demographic study and behavioral observation in the field. Measurement of performance is crucial, and reptiles are certainly easier subjects than are some other vertebrate groups, such as birds or bats (Ricklefs and Miles, chap. 2, this volume; Norberg, chap. 9, this volume).

The Morphology → Performance → Fitness Paradigm

"Not infrequently, performance characteristics, measured as maximal speed or endurance, make the difference between eating and being eaten" (Tenney, 1967, p. I–7). The foregoing quotation certainly contains some truth, but actual data indicating the frequency of "close encounters of the worst kind" between predators and prey are few and far between (cf. Castilla and Bauwens, 1991, p. 78; Christian and Tracy, 1981; Hertz et al., 1988; Jayne and Bennett, 1990b).

Studies of ecological morphology implicitly concern fitness and adaptation. Within populations, individual variation in morphology may be related to variation in Darwinian fitness; among populations and higher taxa, morphological variation in Darwinian fitness; among populations and higher taxa, morphological variation may indicate adaptation to different lifestyles. Arnold (1983) proposed a conceptual and statistical—and hence operational—framework for using data on individual variation to study adaptation within populations (fig. 10.1). This paradigm addresses the question of whether natural selection is currently acting on morphology or performance within a single population. Arnold's (1983) discussion considered multiple morphological characters and multiple measures of performance, as well as correlations within these two levels. He pointed out that multiple regression and path analysis could be used to estimate and test the significance of performance gradients (quantifying the effects of morphology on performance), which can be studied in the laboratory (but see Wainwright, chap. 3, this volume), and fitness gradients (quantifying the effects of performance on fitness), which require field studies (see also Emerson and Arnold, 1989; Wainwright, 1991).

This perspective suggests that intrapopulational variation in morphology may have significant influences on fitness only to the extent that it affects performance. Measures of organismal performance capacities thus become central (cf. Bennett, 1989; Bennett and Huey, 1990; Emerson and Arnold, 1989; Pough,

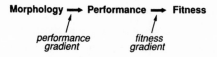

FIGURE 10.1 Simplified version of Arnold's (1983) original paradigm.

1989), and measures of locomotor performance fall easily into the paradigm. The focus on organismal performance as pivotal is in sharp contrast to many previous studies of locomotor ecology, in which the starting point (and sometimes the ending point) has been measurement of limb proportions (see "Case Studies" below). Moreover, the realization that it is easier to study one or the other rather than both gradients simultaneously, and that both parts of the equation are of interest, has stimulated research.

Physiological and biochemical traits may be included within the category of "morphology," and we will subsequently use morphology as shorthand for all three types of traits. This is not to imply that morphology, physiology, and biochemistry are equivalent, nor are we trying to deny the distinction between "form" and "function." The point is simply that all three types of traits are (generally) at a level of biological organization below the whole-organism, and all may influence organismal performance. Calling all three types of traits "morphology" serves to emphasize that similar tools and approaches are useful for studying their effects on organismal performance (cf. Wainwright, 1991).

Arnold's (1983) paradigm was designed specifically to interface with multivariate quantitative genetics theory (Lande and Arnold, 1983). In the quantitative genetic framework (e.g., Boake, 1994; Brodie and Garland, 1993; Falconer, 1989), adaptive phenotypic evolution consists of two parts: natural selection, which is a purely phenotypic phenomenon, and genetic response, which involves inheritance. Some do not like this separation of selection and inheritance (e.g., Endler, 1986), but we agree with Lande and Arnold (1983) that it has considerable operational advantages in allowing the two elements to be studied independently. It also emphasizes that selection may be futile; if a trait is not heritable, then selection cannot lead to or improve adaptation.

How Does Behavior Fit into the Paradigm?

The place of behavior in the paradigm of figure 10.1 is ambiguous. Arnold (1983) did not mention behavior as a distinct level; subsequently, however, Emerson and Arnold (1989; also Schluter, 1989) have included behavior within the category of morphology. We offer an alternative categorization, as depicted in figure 10.2 (modified from Garland, 1994a).

Many biologists imagine that selection acts most directly on what an animal actually does in nature, that is, its behavior. Performance, on the other hand, as defined operationally by laboratory measurements, generally indexes an animal's ability to do something when pushed to its morphological, physiological, or biochemical limits. (Whether animals routinely behave at or near physiological limits under natural conditions is an important empirical issue for which precious few data exist: Daniels and Heatwole, 1990; Dial, 1987; Garland, 1993;

Physiology
Morphology ➡ Performance ➡ Behavior ➡ Fitness
Biochemistry

FIGURE 10.2 Expansion of Arnold's (1983) paradigm to include behavior, as proposed by Garland (1994a).

Garland et al., 1990a; Gatten et al., 1992; Gleason, 1979a; Hertz et al., 1988; MacArthur, 1992; Morgan, 1988; Pough et al., 1992; Seymour, 1982, 1989; van Berkum et al., 1986, Wyneken and Salmon, 1992.) Thus, morphology limits organismal performance, which in turn constrains behavior, and natural and sexual selection act most directly on behavior—what an animal actually does (Garland, 1994a). This modification of Arnold's (1983) original paradigm adds one more level of analysis and places specific emphasis on behavior as the focus of selection. Behavior is seen as a potential "filter" between selection and performance (Garland et al., 1990b).

Further Extensions of the Paradigm

Inserting behavior between performance and fitness seems relatively straightforward (fig. 10.2). But this addition does not necessarily mean the paradigm is complete or general. Many more possible links can be imagined, and a relatively simple chain rapidly becomes a complicated web (e.g., fig. 10.3).

In particular, habitat, broadly defined, is another important factor which may influence behavior, performance capabilities, and even morphology (see also Dunson and Travis, 1991; Huey, 1991). For example, availability of perches or basking sites, their size, and their distribution may affect both what an animal does (e.g., Adolph, 1990b; Grant, 1990; Moermond, 1986; Pounds, 1988; Waldschmidt and Tracy, 1983; see discussion of the "habitat matrix" model below) and what it is capable of doing (e.g., sprint speed in lizards is affected by perch diameter and substrate: Carothers, 1986; Losos and Sinervo, 1989; Losos et al., 1993; Miles and Althoff, 1990; Sinervo and Losos, 1991). Temperature is a habitat characteristic that may affect performance indirectly through its effects on various physiological processes, and by having direct influences on behavior, such as the switches in defensive behavior at low body temperature that occur in some lizards and snakes (Arnold and Bennett, 1984; Crowley and Pietruszka, 1983; Hertz et al., 1982; Mautz et al., 1992, Van Damme, Bauwens et al., 1990; Schiefflen and de Queiroz, 1991). Temperature affects locomotor performance both in absolute terms (Bauwens et al., in press; Bennett, 1990; Garland, 1994b) and, to a lesser extent, relative to other individuals or species. Individual differences in locomotor performance are consistent across temperatures (i.e., fast individuals tend to be fast at all temperatures), but not perfectly so (references in

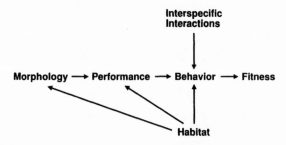

FIGURE 10.3 Inclusion of some other factors that may affect elements of Arnold's (1983) paradigm. Habitat characteristics, such as temperature, may affect basic physiological and biochemical properties as well as behavior (see text). (Of course, behavior and physiology may affect an animal's body temperature; the present diagram is extremely simplified.)

Bennett, 1990; Bennett and Huey, 1990). Thus, the temperature at which an individual happens to be when it encounters a predator may affect its relative fitness (e.g., Christian and Tracy, 1981), and individual differences in thermoregulatory behavior may become crucial (cf. Christian et al., 1985; Waldschmidt and Tracy, 1983).

More subtle habitat effects are also possible. Food in the stomach (Ford and Shuttlesworth, 1986; Garland and Arnold, 1983; Huey et al., 1984), nutritional state (for experiments with mammals, see Brooks and Fahey, 1984, and Astrand and Rodahl, 1986), hydrational state (Moore and Gatten, 1989; Preest and Pough, 1987; Wilson and Havel, 1989, but see Crowley, 1985b; Gatten and Clark, 1989; Stefanski et al., 1989), as well as disease or parasite infection (Schall, 1986, 1990; Schall et al., 1982: but see Daniels, 1985b) all may affect performance ability. Hydrational (Crowley, 1987; Feder and Londos, 1984; Pough et al., 1983; Putnam and Hillman, 1977) or nutritional state may also affect activity levels, that is, behavior. Inter- and intraspecific interactions can also affect behavior in numerous ways (e.g., Fox et al., 1981; Garland et al., 1990a; Henrich and Bartholomew, 1979; Schall and Dearing, 1987; Stamps, 1984). Even hydrational or thermal conditions during incubation or pregnancy can affect locomotor performance of offspring (Miller et al., 1987; Van Damme et al., 1992).

Extending the Paradigm to Population and Species Variation

The paradigm in figure 10.2 can also be applied to understand or predict a relationship between morphology and habitat use among populations or species. The logic of this extension is as follows. First, to the extent that morphological differences among individuals within populations lead to differences in performance abilities that affect fitness, then, assuming the absence of constraints (Maynard Smith et al., 1985), the most "fit" morphology should evolve within any population (Emerson and Arnold, 1989). Second, to the extent that different morphologies function best in different habitats, then natural selection will tend to favor their evolution in the appropriate habitats. If one has an understanding of which morphologies are best suited in given habitats (based on biomechanical or functional analyses, including optimality models, or on empirical studies of natural selection within populations), then one can test the prediction that taxa have adapted to different environments (Baum and Larson, 1991; Bock and von Wahlert, 1965; Losos and Miles, chap. 4, this volume). Caution must be exercised when taking this view, however, as we have little empirical evidence that any given trait(s) in any given population will have reached its selective optimum by the time we study it (Arnold, 1987; Ware, 1982). Moreover, multiple (sub)optimal solutions, which confer equivalent fitness, may exist (Denny, chap. 8, this volume; Feder et al., 1987; Ware, 1982); depending on the shape of the fitness surface, movement from one peak to another may be difficult.

Although Arnold (1983) suggested path analysis for studying the causes (performance gradients, e.g., fig. 10.5 below) and consequences (fitness gradients) of individual variation in performance (and behavior), path analysis might also be employed to study species-level selection processes (cf. Emerson and Arnold, 1989). For example, rather than values for individuals, data points could be population, species, or clade means for morphological, performance, behavioral, or ecological traits. As components of the "fitness" of a population, species or clade (cf. Futuyma, 1986; Vrba, 1989), one might consider geographic range (cf. Jablonski, 1986), evolutionary longevity, and/or number of descendant populations or species (the latter might require paleontological information; but see Nee et al., 1992). Alternatively, some measure of a population's or of a species' "fitness" or "adaptedness" (Michod, 1986) to its current environment might also be possible, such as physiological tolerances, breadth of the Grinnellian niche (James et al., 1984), or demographic traits (e.g., population density, intrinsic rate of natural increase: cf. Baker, 1978; Birch et al., 1963). To quote Stini (1979, p. 388): "A well-adapted population would be . . . one that enjoys a relatively high probability of survival under conditions highly likely to occur." Of course, a path analysis of comparative data would require proper allowance for phylogenetic non-independence (see below). As noted by Emerson and Arnold (1989, p. 302), "there are no strong theoretical grounds for expecting similar performance topographies at the intra- and interspecific levels and there has been virtually no empirical exploration."

Does Morphology Affect Fitness Directly?

Regardless of how complicated a paradigm one wishes to consider, an outstanding conceptual and empirical issue is whether direct paths exist from morphology to fitness (fig. 10.4). Returning to the original formulation, the most

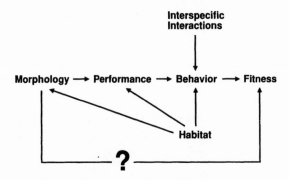

FIGURE 10.4 An outstanding conceptual and empirical issue is whether any direct paths from morphology to fitness are significant (see text).

general path model is one in which all possible effects are depicted, including those directly from morphology to fitness (Arnold [1983, fig. 3] omitted these paths). Consider some hypothetical possibilities. Some individual garter snakes are born with a single or no eyes, an external heart, or a severely kinked tail or spine (Garland, 1988, pers. obs.; Arnold and Bennett, 1988, pers. comm.). These morphological deformities greatly impair locomotor capacities, which in turn limit behavioral options (as compared with "normal" individuals), and would certainly have fitness consequences in nature. In this case, a direct path from morphology to fitness seems unnecessary.

But contemplate two other examples. First, all else being equal (i.e., assuming behavior is unaffected), an albino snake will likely suffer a fitness decrement relative to a normally pigmented individual, because the former will more likely be discovered and eaten by a predator prior to its reproducing. Thus, a direct path appears to exists from morphology (pigmentation) to fitness. Alternatively, if one considers some measure of crypsis as a "performance" variable, then albinism acts through its effects on crypsis (cf. King, 1992), and performance, but not morphology, would seem to have a direct path to fitness, bypassing behavior. But, if an animal could somehow become "aware" that it was differently colored and so alter its behavior to compensate (cf. Morey, 1990), then the effect of albinism might be entirely through the performance → behavior link. Albinism would also affect thermoregulation, making it more difficult for the snake to warm by basking, hence causing it to bask for longer periods of time and increasing its exposure to predators (cf. Andren and Nilson, 1981).

Second, in many species, body size affects the outcome of intraspecific behavioral interactions (Tokarz, 1985; references in Garland et al., 1990a; Faber and Baylis, 1993). This effect may occur simply because size affects strength and stamina, and hence performance at fighting. But in some cases differences in size alone may influence decisions to fight or not, and hence may determine the outcome of an agonistic interaction before any actual fighting occurs. This example suggests a direct effect of a morphological trait (body size) on behavior, and hence on a component of fitness (dominance rank).

Some might consider it logically impossible that morphology can affect fitness other than through its effects on organismal performance (and hence behavior). The idea is that form only matters if it affects function and hence performance; otherwise, morphological variation is selectively neutral. We would prefer to consider the absence of direct morphology → fitness paths as an hypothesis, subject to empirical test. Such tests might involve measurement of vertebral numbers (in snakes) or limb length (in lizards), as well as locomotor performance and survivorship (cf. Arnold and Bennett, 1988; Jayne and Bennett, 1990b; Tsuji et al., 1989). A significant path from morphology to (a component of) fitness (fig. 10.4) would indicate either a direct effect of morphology on fitness or the presence of some unmeasured (latent) performance variable.

LEVELS AND METHODS OF ANALYSIS

Interpopulation

Although analyses relating morphology, performance, behavior, and fitness (broadly defined) most commonly involve interspecific comparisons (to be discussed below) or, more recently, individual variation, studies of interpopulation differences are essential to evolutionary analyses (Garland and Adolph, 1991; James, 1991). Most previous studies of population (geographic) variation focus on morphometric or allozymic characters (e.g., Zink, 1986), although studies of variation in mitochondrial DNA are now common (e.g., Avise et al., 1987; Lamb et al., 1989). Consequently, phylogenetic analyses of population differentiation cannot be far off (cf. Schluter, 1989; Snell et al., 1984), and we encourage such studies of population differences in locomotor performance and its correlates. If possible, such studies should include a "common garden" approach, in which animals are raised in the laboratory for at least one generation to maximize the probability that observed phenotypic differences are actually genetically based (Garland and Adolph, 1991). Common garden controls are important for studies of different species as well, although most biologists seem less concerned at this level. Population differences in locomotor performance may be consistent across years (Huey and Dunham, 1987), but year-to-year variation in performance exists (Huey et al., 1990) and may confound attempts to correlate morphology with ecology (cf. Wiens and Rotenberry, 1980).

Intrapopulation: Individual, Ontogenetic, and Sexual Variation

Arnold (1983, fig. 10.1) considered studies of individual variation within populations, including effects of morphology on performance (e.g., mechanistic physiology) and the effects of performance on fitness (e.g., direct studies of natu-

ral selection in the wild). Quite a few such studies of reptilian locomotion have
been completed since 1983. A major conclusion of these studies is that measures
of locomotor performance show substantial and repeatable individual variation
within single populations (Bennett, 1987; Bennett and Huey, 1990; Huey et al.,
1990; Jayne and Bennett, 1990a; see also Djawdan, 1993; Friedman et al., 1992,
on mammals). This variation and repeatability is, of course, a prerequisite for
attempts to quantify relationships between morphology and performance (performance gradients) or between performance and behavior, fitness, or ecology
(e.g., fitness gradients).

One advantage of studying individual variation is that phylogenetic effects are
not a concern. So, for example, the effects of body size can be studied by examining an ontogenetic series (e.g., Garland, 1984, 1985; Jayne and Bennett, 1990a;
Pough, 1977, 1978) rather than multiple species. Similarly, the mechanistic
correlates of performance variation can be studied (e.g., Garland, 1984, 1985;
Garland and Else, 1987; Losos et al., 1989; Tsuji et al., 1989) without concern
that phylogenetic effects may confound the results (cf. Losos, 1990a, b, c).

Quantitative Genetic Analysis

Individuals within a population may not provide statistically independent data
points, because they are related to varying extents. Quantitative genetics uses this
fact to partition observed phenotypic variances and covariances into genetic (due
to inheritance, which is analogous to phylogenetic descent; cf. Lynch, 1991) and
environmental sources, each of which can be more finely partitioned (Boake,
1994; Brodie and Garland, 1993; Falconer, 1989; Garland, 1994a).

Quantitative genetic analyses are not a traditional part of ecological morphology. They must become an integral part, however, if we are to move towards an
understanding of the mechanisms of microevolution. We will not consider quantitative genetic analyses of reptilian locomotor performance in detail. Only a few
studies have been completed, all on garter snakes (*Thamnophis*) or lizards
(*Sceloporus, Lacerta;* reviews in Bennett and Huey, 1990;, Brodie and Garland,
1993; Garland, 1994a). All studies to date have relied on analyses of presumed
full-sibling families to estimate heritabilities. For many reptiles, gravid females
can be captured in relatively large numbers in the field. After offspring are born
or hatched in the laboratory, measurements of locomotor performance are made
on each. Unfortunately, heritability estimates from sets of full-siblings represent
neither a "narrow-sense" nor a "broad-sense" heritability; in addition, multiple
paternity will lead to an underestimation of additive genetic effects in full-sibling
data sets (Brodie and Garland, 1993; Falconer, 1989; Garland, 1994a; Schwartz
et al., 1989). Thus, significant among-family variance in studies of full-siblings
suggests heritability, but does not prove it. With one exception (Bauwens et al.,

1987), all studies to date have found significant among-family variance for measures of locomotor performance in reptiles.

Experimental Approaches

Experimental approaches can be used in several ways, for example: (1) to examine the mechanistic bases of performance variation; (2) to mimic the effects of short- or long-term changes that may occur naturally within individuals; (3) to examine the effect of conditions during development on morphology and performance abilities; and (4) to increase the range of variation in organismal performance and so increase statistical power to detect its ecological and selective importance. Experimental approaches have the advantage that they can isolate and study the effects of variation in one variable independent of correlations with other variables (cf. Lande and Arnold, 1983; Mitchell-Olds and Shaw, 1987; Slinker and Glantz, 1985; Wade and Kalisz, 1990). Experimental approaches have been underutilized for analyzing links in the morphology → locomotor performance → behavior → fitness chain (or web) and in ecomorphology in general (but see Benkman and Lindholm, 1991; Carothers, 1986; Hanken and Wake, 1991; Hillman and Withers, 1979; Huey et al., 1991; James, 1991; Jayne and Bennett, 1989; Lauder and Reilly, 1988; Ruben et al., 1987; Webster and Webster, 1988).

Causal mechanistic relationships suggested by correlative studies of individual variation in locomotor abilities (fig. 10.5: e.g., Garland and Else, 1987; Gleeson and Harrison, 1988; John-Alder, 1984a, b, 1990) can be tested with such physiological techniques as blood doping (cf. Withers and Hillman, 1988), but this has scarcely been attempted in reptiles (Gleeson, 1991). Hormonal (John-Alder, 1990; Joos and John-Alder, 1990; Moore and Marler, 1987; Moore and Thompson, 1990) or pharmacological (e.g., John-Alder et al., 1986b) manipulation to change metabolism and performance is also possible. (Levels of some hormones fluctuate rapidly in reptiles, whereas some measures of locomotor performance are quite repeatable, which suggests that the former may have little effect on the latter.) With respect to morphology, the importance of tail length and loss, toe loss, toe fringes, and skin flaps for sprinting and gliding performance in lizards and snakes has also been assessed experimentally (e.g., Arnold, 1984a; Carothers, 1986; Daniels et al., 1986; Formanicwiz et al., 1990; Huey et al., 1990; Jayne and Bennett, 1989; Losos et al., 1989; Marcellini and Keefer, 1976; Pond, 1981).

Within-individual variation in reptilian locomotor performance has been examined as a consequence of several factors, such as physical conditioning (training), feeding, reproductive state, and hormonal state. Physical conditioning studies of the type so common in mammalian exercise physiology (e.g., Brooks

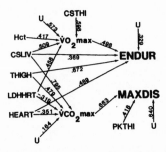

FIGURE 10.5 Path analysis of performance gradients for treadmill endurance at 1.0 km/h (ENDUR) and maximal distance running capacity (MAXDIS) around a circular track in the lizard *Ctenosaura similis* (data from Garland, 1984). For this analysis, only those variables that entered into multiple regression equations as significant predictor (independent) variables and/or that could be explained to a significant extent as dependent variables were considered (see Table 4 of Garland, 1984: SMR was also excluded). Path coefficients were estimated in two ways: first, from standardized partial regression coefficients as described in Nie et al. (1975); second, from the standardized solution output by LISREL Version 4 (Joreskog and Sorbom, 1978), an iterative, maximum likelihood fitting procedure. These two approaches yielded virtually identical results; the figure shows the LISREL results. A variety of path analytic models were fitted with LISREL in order to obtain the model which was judged to best fit the data based on an approximate chi-square-goodness-of-fit statistic and contained no nonsignificant (i.e., $P > .05$) path coefficients (approximate 2-tailed t-tests with 14 degrees of freedom). The path analytic model shown here had a chi-square of 22.4 (df = 20, $P = .3199$), indicating an acceptable fit to the data. Individual path coefficients had t-values of between 2.57 and 14.8, which, by comparison with $t_{(14,.05)} = 2.145$, suggests that all paths are significant. CSTHI = thigh citrate synthase activity (per gram of tissue), Hct = hematocrit, CSLIV = liver citrate synthase activity, THIGH = total mass of right thigh muscles, LDHHRT = lactate dehydrogenase activity in the heart, HEART = heart mass (including atria), PKTHI = thigh pyruvate kinase activity, U = unexplained variation. (See Arnold, 1983; also Sokal and Rohlf, 1981; Bulova, in press; and assumptions in Emerson and Arnold, 1989, p. 299.)

CSLIV is significantly related to ENDUR, $\dot{V}O_2$max, and $\dot{V}CO_2$max (see also multiple regressions in Table 4 of Garland, 1984). Deleting CSLIV from these predictive models resulted in lower coefficients of determination for the multiple regression equations or a higher chi square for the LISREL-fitted path analytic model. Garland (1984) interpreted these results (and data for mammals) as suggesting that liver oxidative capacity plays a significant role in the activity metabolism of ctenosaurs, perhaps via conversion of metabolites during or after activity. Recent studies, however, suggest that the liver is not an important site of lactate metabolism during recovery in amphibians or reptiles (Gleeson and Dalessio, 1989; Gleeson 1991).

and Fahey, 1984; Astrand and Rodahl, 1986) have been attempted only twice with reptiles. These two studies employed very different training regimens and species from different families, yet both failed to improve organismal performance (speed, stamina, maximal oxygen consumption: Garland et al., 1987; Gleeson, 1979b; but see Gleeson, 1991, p. 189). On the other hand, captivity and the accompanying relative inactivity may decrease maximal oxygen consumption ($\dot{V}O_2$max) (Bennett and John-Alder, 1984; Garland et al., 1987; but see John-Alder, 1984b). Training studies definitely deserve further attention; unfortunately, they can be quite labor-intensive because training regimens cannot be automated as easily as they can with mammals. An outstanding issue is the extent to which "natural training" occurs in the wild (Burghardt, 1984; Garland et al., 1987). Acclimation and acclimatization of reptilian locomotor performance has been studied only rarely (Gatten et al., 1988; Hailey and Davies, 1986; Kaufmann and Bennett, 1989; Payne and Gatten, 1988), as has seasonal variation, which is in some cases significant (Garland, 1985; Garland and Else, 1987; Gleeson, 1979b; Huey et al., 1990; John-Alder, 1984b). Infection with pathogens or parasites could also be used to lower performance (Schall, 1990; Schall and Dearing, 1987; but see Daniels, 1985b; Schall, 1986).

Body size, which often correlates with locomotor performance (see below), can be manipulated in a variety of ways. For example, variation in diet or in thermal regimen (Sinervo and Adolph, 1989) may affect growth rate and hence age-specific body size; such experimentally induced variation may be useful in studies of static allometry (i.e., within an age class). Sinervo and Huey (1990; Sinervo, 1990; Sinervo et al., 1992; see also Bernardo, 1991; Hahn and Tinkle, 1965; Janzen, 1993; Sinervo and Licht, 1991) have used experimental manipulation of egg size in an attempt to separate the effects of body size per se from other factors that may affect speed or stamina. Embryo manipulation studies are common in mammals (e.g., Atchley et al., 1993; Cowley et al., 1989; Hill and Mackay, 1989; Kirkpatrick and Rutledge, 1988) but apparently have not been attempted in reptiles.

Hydric and thermal conditions during incubation can affect locomotor performance of reptiles (Miller et al., 1987; Van Damme et al., 1992), and such effects may not be uncommon (references in Garland and Adolph, 1991). For example, thermal conditions during pregnancy can affect the number of body and tail vertebrae developed by garter snakes (Fox, 1948; Fox et al., 1961; Osgood, 1978; C. R. Peterson and S. J. Arnold, pers. comm.), which in turn may affect locomotor performance (Arnold and Bennett, 1988; Jayne and Bennett, 1989; M. R. Dohm and T. Garland, in preparation). Many other factors may affect maternal size and/or condition and in turn affect offspring size and/or performance; some of these effects can be controlled for statistically via regression analysis and computation of residuals (Brodie, 1989b; Brodie and Garland, 1993; Garland, 1988; Garland and Bennett, 1990; Tsuji et al., 1989).

Truly evolutionary experiments, involving organismal performance or components thereof, are possible using artificial selection (e.g., Bennett et al., 1990; Garland and Carter, 1994; Hill and Caballero, 1992; Huey et al., 1991; Rose et al., 1987; Schlager and Weibust, 1976), but such experiments have not yet been reported for locomotor performance in any organism. Relatively long generation times may preclude such possibilities for reptiles, although experiments with mice are now being conducted (T. Garland, unpubl.). Direct manipulation of the germ line (e.g., genetic engineering to produce transgenic mice) is now routine in many animals (see Hill and Mackay, 1989) but has not been attempted with reptiles.

CONFOUNDING ISSUES IN THE STUDY OF PERFORMANCE AND ECOLOGICAL MORPHOLOGY

Measuring "Performance" as Opposed to "Behavior"

Arnold (1983, p. 352) defined performance as "the score in some ecologically relevant activity, such as running speed. . . ." Most estimates of maximal locomotor performance in reptiles are made in the laboratory, although some field estimates are available (e.g., Belkin, 1961; on mammals see Djawdan and Garland, 1988; Garland et al., 1988). In either laboratory or field, however, definition and measurement of "performance" as opposed to "behavior" is not always simple (cf. Friedman et al., 1992; Garland, 1994a, b). For example, if maximal sprint speed is measured by chasing an animal along a race track, how can one be sure that each individual actually runs at its morphological, physiological, or biochemical limits? Animals may vary in their response to stimuli (their "motivation"), such that some run at their physiological limits and others do not. Thus, behavioral variation, just like morphological or physiological variation, can affect laboratory measurements of performance (see also Wainwright, chap. 3, this volume).

In the laboratory, repeated testing of individuals and use of the fastest trial(s) as an index of maximal speed (e.g., Bennett, 1980; Formanowicz et al., 1990; Garland, 1984, 1985, 1988; Gleeson and Harrison, 1988; Huey 1982a; Losos et al., 1989; Marsh, 1988; Marsh and Bennett, 1985, 1986; Sinervo et al., 1991; but see Jayne and Bennett, 1990a, b) may help circumvent motivational problems. (It is well known in human, horse, and dog racing that performances of individuals vary significantly with the competition and setting.) For some performance measures, it may be possible to verify by supplementary tests that physiological limits have been reached. Thus, physiological exhaustion in endurance trials can be supported by testing for loss of righting response (e.g., Huey et al., 1984, 1990), or by measuring whole-body (Arnold and Bennett, 1984) or blood (Djawdan, 1993) lactic acid concentrations. Alternatively, measures of "race quality" can be used in statistical analyses (Tsuji et al., 1989). In any case, what some workers term "performance" others term "behavior" (e.g., Bennett, 1980).

Another possibility is to test for correlations between individual (or interspecific) differences in performance and traits thought to affect performance. If underlying morphological, physiological, or biochemical traits explain (statistically) a large fraction (e.g., 47–89%; Garland, 1984, fig. 10.5; Garland and Else, 1987) of the variance in locomotor performance, then it is unlikely that variation in performance is due solely to differences in motivation or willingness to run. To date, published studies of individual variation have been somewhat more successful in identifying physiological correlates of endurance than of sprint speed (see "Case Studies" below), which suggests that it may be easier to obtain measures of physiologically limited performance capacities in stamina- than in sprint-type activities.

Some studies of individual variation indicate that measures of "behavior" may show correlations with measures of "performance." For example, antipredator display (Arnold and Bennett, 1984), scored at the end of treadmill endurance trials, showed significantly positive correlations with both treadmill endurance and sprint speed in the garter snake *Thamnophis sirtalis* (Garland, 1988; see also Arnold and Bennett, 1988, on *T. radix* and Brodie, 1992, on *T. ordinoides* concerning the correlation between speed and distance crawled prior to assuming an antipredator display). Arnold and Bennett (1984) previously showed that whole-body lactic acid concentrations of *T. radix* exhibiting antipredator displays (at the end of stamina trials) were similar to those of snakes forced to exercise for thirty minutes. Thus, one might expect the antipredator display to be partly dependent on, and hence limited by, physiological capacities. However, Garland, et al. (1990b) found that, whereas speed and endurance showed significant (although weak) correlations with lower-level morphological, physiological, or biochemical traits, antipredator display did not (see also Arnold and Bennett, 1988). Thus, an alternative interpretation is that underlying variation in some axis of "motivation" (Bolles, 1975) has effects on measures of speed, endurance, and antipredator display (higher scores are more offensive and seem to require more physical exertion), leading to some positive correlation.

The foregoing examples emphasize that caution must be exercised when designing or interpreting measures of locomotor "performance." Our discussions of Arnold's (1983) paradigm and extensions thereof assume that true measures of morphologically or physiologically limited performance can be obtained.

Allometry and its Importance

Body size affects many traits, including locomotor performance (e.g., Dunham et al., 1988; Garland, 1984, 1985; Garland and Huey, 1987; Losos, 1990a, b, c). Variation in body size may therefore obscure or enhance relationships between other traits (Emerson et al., chap. 6, this volume). Unfortunately, much of the older ecomorphological literature has attempted to remove the effect of size by using ratios, which is generally ineffective and potentially misleading (cf. Packard and Boardman, 1988).

The importance of considering allometry can be illustrated with a hypothetical example. Suppose that sprint abilities determine habitat use in lizards. Many studies of lizards have noted correlations between relative limb length (expressed as a proportion of snout-vent length) and various habitat variables (see "Limb Length and Habitat Use" below), and have implicated differences in locomotor ability as the underlying cause of the relationships. Both within and between species, limb length rarely scales isometrically with snout-vent length (i.e., as individuals or species increase in size, limb length either becomes relatively longer or shorter; see below, figs. 10.7c, 10.8c). Further, within and between species, sprint ability usually increases with body size. Consequently, a relationship may exist between habitat use and body size due to the effect of size on sprint speed. Because relative limb length is partly a function of body size (except when limb length scales isometrically with size), a spurious relationship would exist between relative leg length and habitat use. Our reanalysis of Pianka's (1969, 1986) data illustrates this problem (see discussion below and figs. 10.7, 10.8).

Confounding effects of body size can be controlled in a variety of ways. Perhaps the most common way is to regress each variable of interest (e.g., sprint speed, limb length) on some measure of body size (e.g., body mass, snout-vent length) and then compute residuals. These residuals can then be used in correlation or regression analyses or various multivariate techniques, such as principal components analysis (e.g., Garland, 1984, 1985, 1988; Jayne and Bennett, 1990a, b; Losos, 1990a, b, c). If the effects of additional covariates (e.g., temperature) or categorical variables (e.g., sex, season, population) need to be removed as well, then residuals can be computed from multiple regressions including dummy independent variables (e.g., Garland and Else, 1987; Gatten et al., 1991; Jackson, 1973b; Packard and Boardman, 1988; Sokal and Rohlf, 1981). Regressions to compute residuals need not be restricted to linear models (cf. table 3 of Garland and Else, 1987; Jackson, 1973b; Jayne and Bennett, 1990a; see also Chappell, 1989). With interspecific data, methods that allow for statistical complications due to phylogeny must be used when computing residuals (Garland et al., 1992; Harvey and Pagel, 1991; Losos and Miles, chap. 4, this volume; Martins and Garland, 1991).

The foregoing approach is not without problems, however. We will mention three here (cf. Huey and Bennett, 1987; Tracy and Sugar, 1989). First, least-squares regression analysis assumes that the independent variable contains no measurement error. This assumption is not true of measures of body size, resulting in underestimates of true structural relationships (Harvey and Pagel, 1991; LaBarbera, 1989; Pagel and Harvey, 1988; Riska, 1991; Sokal and Rohlf, 1981). Unfortunately, alternatives to least-squares regression slopes (e.g., reduced major axis, major axis) are not easily employed where multiple independent variables need to be considered. Second, using the same individual measurement of body size (e.g., each animal weighed or measured a single time) as the independent variable for a series of dependent variables may result in correlated errors being introduced into all residuals. Such correlated errors can be avoided by taking several measurements of body size (e.g., when speed is measured, when stamina is measured, when limb length is measured; Garland, 1984; Garland and Else, 1987). Third, when correlating the residuals, one degree of freedom should perhaps be lost for each dependent variable for which residuals are computed.

One alternative to the residual approach outlined above is to simply use multiple regression of the dependent variable (e.g., sprint speed) on both a measure of body size and, say, limb length (e.g., Snell et al., 1988). The problem here is that body size and limb length will generally be highly correlated, and the results of multiple regression analyses are unreliable in the face of such multicollinearity (Slinker and Glantz, 1985). We believe that regression of both speed and limb length on body size, then testing for correlation between their residuals, is a more reliable procedure. Alternatively, experimental manipulations that change mass or limb length—but not both—could be helpful in reducing the correlations between independent variables (cf. Lande and Arnold, 1983; Mitchell-Olds and Shaw, 1987; Slinker and Glantz, 1985).

In some cases, the actual value of the allometric exponent is of interest, perhaps in relation to theoretical models of scaling (e.g., Emerson, 1985; Garland, 1985; Harvey and Pagel, 1991; LaBarbera, 1989; Marsh, 1988). Unfortunately, how best to estimate allometric relationships is unclear. As noted above, the independent variable in allometric studies always incorporates some "error variance," which means that slopes will tend to be underestimated. Moreover, for comparisons of population and/or species means, allometric slopes should be estimated phylogenetically, not merely by a regression involving values for tips of a phylogeny (Garland et al., 1992, 1993; Garland and Janis, 1993; Harvey and Pagel, 1991; Losos, 1990c; Lynch, 1991; Martins and Garland, 1991; Purvis and Garland, 1993).

Phylogeny and its Importance

Inheritance of a phenotypic trait cannot be studied without knowledge of the relatedness of individuals. Analogously, the evolution of a phenotypic trait cannot properly be studied without knowledge of phylogenetic relationships. That all organisms are descended in a hierarchical fashion from common ancestors means that no set of taxa can be assumed to be biologically or statistically independent. Phylogenetic non-independence has implications for all aspects of statistical analyses, including hypothesis testing, power to detect significant relationships between traits, and estimation of the magnitude of such relationships (Felsenstein, 1985; Harvey and Pagel, 1991; Losos and Miles, chap. 4, this volume; Lynch, 1991; Martins and Garland, 1991; Pagel, 1993). Several methods now exist for incorporating phylogenetic information into comparative analyses, and various examples exist in which phylogenetic analyses lead to qualitatively different conclusions (Garland et al., 1991, 1993; Harvey and Pagel, 1991; Nee et al., 1992). It should also be noted that phylogenetic methods for estimating and testing, for example, character correlations, can sometimes *increase*—not just decrease—statistical significance as compared with an inappropriate nonphylogenetic analysis.

As the vast majority of previous comparative studies have been analyzed with inadequate allowance for phylogenetic non-independence, conclusions drawn from them must be viewed with caution. For example, many of the allometric studies we discuss were done nonphylogenetically; practical constraints (e.g., lack of suitable phylogenies: see figs. 10.7, 10.8 below) and time limitations have precluded our trying to redo all of them! Nevertheless, future population- or species-level examination of the morphology → performance → behavior → fitness paradigm should be done with appropriate allowance for phylogenetic non-independence (e.g., Losos, 1990b). An interspecific path analysis, comparable to Arnold's (1983) paradigm for microevolutionary studies (cf. Emerson and Arnold, 1989), would be particularly desirable.

CASE STUDIES

Morphology → Performance

Interspecific differences in locomotor performance are well established in reptiles (e.g., Bennett, 1980; Garland, 1994b; Huey and Bennett, 1987; Losos, 1990b,; van Berkum, 1988; references therein). Population differences have been shown a number of times as well (e.g., Garland and Adolph, 1991; Huey et al., 1990; Sinervo et al., 1991; Snell et al. 1988; but see Bennett and Ruben, 1975). A somewhat surprising finding has been the substantial variation in performance among individuals within single populations (Bennett, 1987; Bennett and Huey, 1990; Huey et al., 1990; Pough, 1989). Sex differences in performance exist and are in some cases due to sex differences in body size; unfortunately, few studies have actually tested for sex differences with adequate sample sizes (e.g., Garland, 1985; Huey et al., 1990; Jayne and Bennett, 1989; Tsuji et al., 1989). Some individual variation in performance ability is due to differences in age and/or size (Garland and Else, 1987; Hailey and Davies, 1986; Marsh, 1988; Pough, 1977, 1978); their effects have been thoroughly separated in only two studies of reptiles (Huey et al., 1990; Sinervo and Adolph, 1989).

Variation in performance calls for both proximate and ultimate explanations. In this section, we consider the former—studies examining the mechanistic bases of performance variation. Note that studies of individual variation in performance and morphology constitute attempts to quantify performance gradients

(Arnold, 1983), although special assumptions are required when individuals of multiple ages are studied (Emerson and Arnold, 1989).

Endurance. Variation in endurance has been less studied than has variation in sprinting ability (see "Sprint Speed" below). Most commonly, endurance is measured as running time to exhaustion on a motorized treadmill. Interspecific comparisons of lizards indicate that treadmill endurance capacity has evolved in concert with both body mass and body temperature (Autumn et al., 1994; Garland, 1994b). Interspecific correlates of endurance have not been studied in detail, but appear to include the energetic cost of locomotion (lower cost leads to higher stamina at a given speed), $\dot{V}O_2$max, and indices of blood oxygen carrying capacity (Autumn et al., 1994; Bennett et al., 1984; Garland, 1993; Gleeson, 1991; Gleeson and Bennett, 1985; Gleeson and Dalessio, 1989; John-Alder et al., 1983; John-Alder et al., 1986a; Secor et al., 1992). Two populations of *Sceloporus merriami,* which differ in maximal sprint speed, apparently do not differ in treadmill endurance (Huey et al., 1990).

Within populations, treadmill endurance increases ontogenetically in most species of lizards (Garland, 1984, 1994b, unpubl.; Huey et al., 1990; see also Daniels and Heatwole, 1990) and in the two species of snakes that have been studied (Jayne and Bennett, 1990a; Secor et al., 1992), although not necessarily in a linearly allometric fashion (Garland and Else, 1987; Jayne and Bennett, 1990a). Positive static allometry occurs in garter snakes (Garland, 1988; Jayne and Bennett, 1990b).

Morphological, physiological, and biochemical correlates of individual differences in treadmill endurance have been studied in the lizards *Ctenosaura similis* (fig. 10.5) and *Ctenophorus nuchalis* (Garland, 1984; Garland and Else, 1987). Correlations of each variable with body mass were removed by computing residuals from regression equations. After this procedure, several underlying variables were shown to correlate significantly with endurance (e.g., $\dot{V}O_2$max, thigh muscle mass, enzyme activities). Correlations with $\dot{V}O_2$max and with thigh muscle mass occur in three of five species studied to date (Garland, 1984; unpublished data on *Callisaurus draconoides* and *Cnemidophorus tigris;* Garland and Else, 1987; John-Alder, 1984b). These studies were the first to document performance gradients for reptilian locomotion.

Treadmill endurance does not correlate with residual hindlimb length in *Sceloporus merriami* (Huey et al., 1990) or in hatchling *S. occidentalis* (Tsuji et al., 1989). However, a small (r = .218) but significant correlation exists between treadmill endurance and residual tail length in hatchling *Sceloporus occidentalis* (Tsuji et al., 1989). Treadmill endurance correlates positively with $\dot{V}O_2$max in *Thamnophis sirtalis* (Garland and Bennett, 1990; Garland et al., 1990b).

Stamina can also be measured by chasing animals around a circular track at top speed until exhaustion and recording total distance and/or time run. Several species of lizards have been so tested, but generalities are not yet apparent (Bennett, 1980, 1989; Garland, 1993; Mautz et al., 1992). Studies of individual variation have also documented morphological and physiological correlates of maximal distance running (or crawling) capacity in both lizards (Garland, 1984, unpub.) and garter snakes (Arnold and Bennett, 1988; Dohm and Garland, unpub.) (see also Gatten et al., 1991, on alligators). Less useful measures of stamina, in terms of comparability and ecological relevance, can be obtained by holding animals in any type of container, prodding them to struggle, and recording the time until cessation of activity, loss of righting response, etc. (Daniels and Heatwole, 1990; Snyder and Weathers, 1977).

Sprint Speed. Sprint speed is the most commonly studied aspect of reptilian locomotor abilities, and a considerable body of research addresses the mechanistic basis of variation in sprinting. At this point, differences in body size and in relative limb length seem to be the most important causal factors.

Within populations of lizards and snakes, sprint speed generally increases with body size (mixed samples: Daniels and Heatwole, 1990; Garland, 1985; Huey, 1982a; Huey and Hertz, 1982; Huey et al., 1990; Losos, 1990c; Losos et al., 1989; Marsh, 1988; Secor et al., 1992; Sinervo, 1990; Snell et al., 1988; static allometry: Arnold and Bennett, 1988; Garland, 1988; Garland and Arnold, 1983; Jayne and Bennett, 1990b; Sinervo, 1990; Sinervo and Adolph, 1989; Sinervo and Huey, 1990; Tsuji et al., 1989). Several exceptions exist (Garland, 1984, unpub.; Brodie, 1989a), however, and the snake *Thamnophis sirtalis* shows a curvilinear allometry (Jayne and Bennett, 1990a).

Evidence for a relationship between speed and size is more equivocal in interspecific comparisons. In a phylogenetic analysis of fifteen Caribbean *Anolis* species, Losos (1990b) has shown that sprint speed and snout-vent length evolved together. By contrast, van Berkum (1986) found no relationship between sprint speed and size among seven species of Costa Rican *Anolis*. At higher taxonomic levels, no simple linear relationship between speed and size appears to exist (Garland, 1982, unpub.; but see Marsh, 1988, p. 131).

Biomechanical models predict a positive relationship between limb length and sprint speed (discussions in Garland, 1985; Losos, 1990b; Marsh, 1988). Indeed, most "cursorial" mammals have elongated legs resulting primarily from increased length of the distal elements (Garland and Janis, 1993; Hildebrand et al., 1985; Janis, in press; references therein). Cursorial lizards also exhibit elongated limbs, but, by contrast, all limb elements seem to increase in length, with no apparent regularity as to which element increases the most (Rieser, 1977).

Many studies have investigated whether the predicted positive relationship between limb length and sprint speed exists in lizards. Given the oft-observed correlation between size and speed, most studies remove the effect of size (see above) on both limb length and speed to examine whether relative limb length correlates with relative sprint speed. Intrapopulational studies have been evenly split: a correlation between relative limb length and sprint speed exists in *Tropidurus albemarlensis* (Snell et al., 1988; but see discussion below), *Sceloporus occidentalis* (Sinervo, 1990; Sinervo and Losos, 1991), *Sceloporus merriami* (Huey et al., 1990), and *Urosaurus ornatus* (D. B. Miles, pers. comm.), but not in *Ctenosaura similis* (Garland, 1984), *Ctenophorus* (*Amphibolurus*) *nuchalis* (Garland, 1985), *Leiolepis belliani* (Losos et al., 1989), or hatchling *Sceloporus occidentalis* (Tsuji et al., 1989). Positive interpopulational or interspecific correlations have been reported several times (Bauwens et al., in press; J. Herron and B. S. Wilson, pers. comm.; Losos, 1990c; Miles, 1987, pers. comm.; Sinervo and Losos, 1991; Sinervo et al., 1991; Snell et al., 1988).

Although several theories (e.g., geometric similarity, elastic similarity, dynamic similarity) have been proposed that predict the relationship between size and sprint speed, none adequately explains the available data for mammals or lizards (Chappell, 1989; Garland, 1985). The relationship between limb length, stride length, and sprint speed is perhaps simpler and more intuitive than these theories imply. Snell et al. (1988) argue that the relationship between body size and sprint speed in *Tropidurus albemarlensis* (pooling individuals of both sexes and from two populations) results from the correlation of both variables with hindlimb length, rather than there existing a direct relationship between body size and sprint speed. Stepwise multiple regression analysis indicated that, after allowing for the positive correlation between sprint speed and snout-vent length, individual variation in hindlimb length predicts a significant amount of variation in sprint speed. However, in a second analysis, after hindlimb length is removed, no relationship exists between size and sprint speed. The strong correlation ($r = .93$) between hindlimb length and mass in *Tropidurus* suggests extreme caution in interpreting these results (cf. Slinker and Glantz, 1985). Nonetheless, a reanalysis of data for fourteen species of *Anolis* (from Losos, 1990b, c) reveals a similar pattern: when residuals are taken from regressions on snout-vent length, hindlimb length and sprint speed are still significantly related, but when residuals are taken from regressions on hindlimb length, snout-vent length and speed are not significantly correlated (see also Bauwens et al., in press).

The relationship between other morphological variables and sprint speed has been less studied. Lizards are renowned for their ability to drop their tails to thwart predation; a number of studies have experimentally assessed the effect of tail loss on sprint speed (reviewed in E. N. Arnold, 1988; see also Russell and

Bauer, 1992). Many lizards with experimentally reduced tails run more slowly (Arnold, 1984a; Ballinger et al., 1979; Formanowicz et al., 1990; Pond, 1981; Punzo, 1982), but not *Phyllodactylus marmoratus* (which does not use its tail as a counterbalance: Daniels, 1983), *Sphenomorphus quoyii* (Daniels, 1985a), *Sceloporus merriami* (Huey et al., 1990), or the snake *Thamnophis sirtalis* (Jayne and Bennett, 1989). Although a negative effect on sprint performance may have long-term repercussions, the importance of tail loss in a given predator-prey encounter is probably more a function of predator distraction than of altered lizard escape speed (Dial and Fitzpatrick, 1984; Cooper and Vitt, 1991).

The biomechanics of legless locomotion are poorly understood (Gans, 1975; Jayne and Davis, 1991; Secor et al., 1992). Vertebral numbers may relate to inter-specific differences in locomotor performance in snakes (Jayne, 1985, 1986, 1988a, b). Within a population of *Thamnophis radix,* numbers of body and tail vertebrae correlate in an interactive fashion with speed in juveniles (Arnold and Bennett, 1988). Relative tail length also may affect snake sprint speed (Jayne and Bennett, 1989).

The effect of muscle size and composition on sprint performance would seem to be an obvious area for study, but little work has been done to date. Gleeson and Harrison (1988) report significant inverse correlations between sprint speed and muscle fiber areas in desert iguanas (*Dipsosaurus dorsalis*). The low maximal speeds of chameleons appear at least partly related to their having slow-contracting muscles (Abu-Ghalyun et al., 1988; see also Peterson, 1984).

Sprint speed is also affected by temporary changes in body condition. Both gravidity and recent ingestion of food result in increased mass and decreased flex-ibility and so may have similar effects on sprint performance. Gravidity generally lowers sprint performance in both lizards and snakes (e.g., Brodie, 1989a; Coo-per et al., 1990; Garland, 1985; Huey et al., 1990; Van Damme, Bauwens, and Verheyen, 1989). Population differences in the effect of gravidity on maximal sprint speed exist in *Sceloporus occidentalis* (Sinervo et al., 1991). Similarly, a full stomach has, in some cases, been shown to lower speed and/or endurance in lizards and snakes (Ford and Shuttlesworth, 1986; Garland and Arnold, 1983; Huey et al., 1984). Apparently, burst speed may be less sensitive to such effects than is endurance (cf. Cooper et al., 1990; Garland, 1985; Garland and Else, 1987). Further studies of the effects on locomotor capacities of recent feeding, especially in relation to models of optimal foraging, allocation of time and en-ergy to foraging versus reproduction, and associated costs and trade-offs, should prove interesting. Brodie and Brodie (1990) used sprint speed as a bioassay for the effects of tetrodotoxin on garter snakes.

Two caveats must be kept in mind when considering investigations of sprint speed. First, some animals appear not to sprint at top speed in a race track. For example, maximal reported speeds of desert iguanas (*Dipsosaurus dorsalis*) in race tracks (single fastest individuals) are 10.1 km/h (Bennett, 1980; higher speeds were reported for some other species), 18 km/h (Marsh, 1988; Marsh and Bennett, 1985), and 15.0 km/h (Gleeson and Harrison, 1988), whereas Belkin (1961) reports a maximal field speed of almost 30 km/h and J. A. Peterson (pers. comm.) reports observing similar speeds on a high-speed treadmill (compare also speeds recorded for *Uma notata* in the laboratory [Carothers, 1986] and the field [Norris, 1951]). Why some species do not perform well in a race track is unclear. Alternative techniques for measuring sprint speed, such as high-speed treadmills (J. A. Peterson, pers. comm.; Garland, unpubl.) may circumvent such problems.

Second, measuring acceleration in racetracks is much more difficult than obtaining sprint speed alone, because lizards must sit motionless just in front of the first photocell, then burst along the track when startled. Many species are not so cooperative (but see Carothers, 1986; Huey and Hertz, 1984a), instead struggling and running as soon as being placed in the track. It is unfortunate that measures of acceleration are difficult to obtain, because acceleration, as opposed to just maximal steady-state sprint speed, may be of prime importance in predator-prey interactions (Elliott et al., 1977; Huey and Hertz, 1984a; Webb, 1976).

Jumping. Biomechanical models suggest that body and muscle mass, limb length, location of center of mass, muscle composition, and behavior all can affect jumping performance (Alexander, 1968; Emerson, 1985; Losos, 1990b; Pounds, 1988). The biomechanics of jumping have been investigated only in the legless pygopodid lizard *Delma tincta* (Bauer, 1986) and in *Anolis carolinensis* (Bels and Theys, 1989; Bels et al., 1992).

Considerable variation in jumping ability exists both within and among lizard species (Losos, 1990c; Losos et al., 1989; Losos et al., 1991). Among fifteen species of *Anolis,* body size and jumping ability are positively related (Losos et al., 1991; Losos, unpubl.). Within species, jumping ability increases with size in *Leiolepis belliani* (Losos et al., 1989) and in seven species of *Anolis* (Losos et al., 1991, unpubl.). In addition, positive but nonsignificant relationships also exist in nine of eleven other species of *Anolis* (Losos, unpubl.).

With the effect of size removed, relatively long-legged species of *Anolis* jump relatively farther than do shorter-legged species. Lesser and negative effects of tail and forelimb length on jumping ability also exist in anoles (Losos, 1990c). By contrast, no relationship between relative limb length and jumping ability is evident in *Leiolepis belliani* (Losos et al., 1989). No differences have been detected between males and females in *Leiolepis belliani* (Losos et al., 1989) or in *Anolis frenatus* (Losos et al., 1991), although sample sizes were small in the latter study.

Gliding and Parachuting. Gliding reptiles, including the enchantingly named *Icarosaurus* and *Daedalosaurus,* date to the Permian (Ricqles, 1980). Gliding abilities have been noted in a considerable number of snakes (e.g., *Chrysopelea:* Shelford, 1906; Heyer and Pongsapipatana, 1970) and lizards (*Ptychozoon:* Mertens, 1960; Heyer and Pongsapipatana, 1970; Marcellini and Keefer, 1976; *Holapsis:* Schiotze and Volsoe, 1959). The premier reptilian gliders are the members of the Southeast Asian genus *Draco,* which have evolved flight membranes formed from a patagium stretched over elongated and movable ribs. These lizards can glide for considerable distances; "flights" of over 20 m have been observed in nature, and 60 m in experimental trials (see Colbert, 1967). In other reptiles, gliding ability appears to be enhanced by the presence of flaps of skin along the sides of the body, neck, and tail and between the toes; the ability to increase ventral surface area; and the tendency to adopt an outstretched posture (Mertens, 1960; Oliver, 1951; Russell, 1979). Because the distinction is often not clear, we will use "gliding" throughout the paper to refer to both gliding and parachuting from one arboreal position to another or to the ground (see Rayner, 1981, 1987).

Several studies have experimentally investigated the role of morphology and behavior on gliding ability. Larger individuals of *Leiolepis belliani* have greater wing loading (i.e., mass/surface area) and fall more rapidly (Losos et al., 1989). Wing loading and glide performance were also inversely correlated in *Ptychozoon lionatum* (Marcellini and Keefer, 1976). When surface area is experimentally decreased (by preventing dorsoventral flattening in *Leiolepis belliani* [Losos et al., 1989] and tying lateral cutaneous folds to the body in *Ptychozoon lionatum* [Marcellini and Keefer, 1976]), gliding performance was diminished. The importance of body posture on rate of descent was suggested in *Leiolepis belliani*. Dead lizards, which tumbled rather than falling in a horizontal, outstretched position, fell faster than did live lizards. Further, for a given wing loading, live lizards fell slower, which suggests that behavioral adjustments, such as creating a concave (rather than flat) ventral surface while falling, enhance parachuting abilities (the snake *Chrysopelea ornata* also adopts a similar concave posture when falling [Shelford, 1906]). The importance of gliding as a means of moving through the environment is obvious, particularly in forests in which movement from tree to tree would otherwise require a lizard to climb into the crown or to the ground. Further, Oliver (1951) observed *Anolis carolinensis* avoiding predators by jumping from trees, gliding to the ground, and running away. Similar behavior is displayed by a number of geckos (Russell, 1979), and perhaps, by spiny-tailed iguanas (*Ctenosaura similis*) over 1 kg in body mass (Garland, pers. obs.).

Conclusions. A variety of ecologically relevant measures of performance have been studied in lizards and snakes. Most attention has been focused on sprint speed, perhaps because it is relatively easy to measure. Although the effects of morphology and physiology on sprinting (and on gliding) capability are theoretically the simplest, we seem to have more empirical information on the mechanistic correlates of variation in endurance. For all measures of performance, the effect of many variables remains to be assessed. Other important aspects of locomotor performance, such as burrowing (Gans et al., 1978), climbing (Losos and Sinervo, 1989; Sinervo and Losos, 1991; see also Thompson, 1990, on rodents), and swimming ability (Bartholomew et al., 1976; Daniels and Heatwole, 1990; Gans, 1977; Gatten et al., 1991; Schoener and Schoener, 1984; Tracy and Christian, 1985; Turner et al., 1985; Vleck et al., 1981) have received little attention.

Performance → Behavior or Fitness/Ecology

Studies using quantitative data to correlate interspecific variation in locomotor abilities with ecology or behavior are relatively rare. In this section, we briefly summarize the data relating endurance and sprint capabilities to behavior, fitness, and ecology, and then address two topics in which the relevance of performance has been more extensively examined: the context-specificity of performance and the relationship of variation in performance capabilities to foraging mode.

Endurance, $\dot{V}O_2max$, and Anaerobic Metabolism. The relevance of endurance capacity for squamate natural history is just becoming apparent (Garland, 1993, 1994). Treadmill endurance at 1.0 km/h appears to be related to average daily movement distance (cf. Garland, 1983) among nine species of lizards (table 2 of Hertz et al., 1988). Similarly, Loumbourdis and Hailey (1985) suggested a correlation between both active (not verified $\dot{V}O_2max$) and resting metabolic rates and "lifestyle" of lizards, though further corroboration is necessary (see also Kamel and Gatten, 1983). Various adaptive explanations for the high aerobic capacities ($\dot{V}O_2max$) of *Cnemidophorus tigris, Heloderma suspectum,* and *Varanus* species have been offered (Bennett, 1983; Bickler and Anderson, 1986; Garland, 1993; John-Alder et al., 1983). The uses of anaerobic metabolism in nature, and possible correlations between anaerobic capacities and lifestyle in reptiles, have been reviewed elsewhere (e.g., Bennett et al., 1985; Gatten, 1985; Pough and Andrews, 1985a, b; Seymour, 1982, 1989; Ultsch et al., 1985).

Sprint Speed. Population differences in wariness have been documented in lizards (Bulova, in press; Shallenberger, 1970), but possible performance correlates have been studied only once. In more open areas on the eastern end of Isla Plaza Sur, Galapagos lava lizards (*Tropidurus albermarlensis*) are warier (i.e., they flee further and earlier when approached by humans) than are lizards from the more vegetated western end of the island (Snell et al., 1988). Lizards living in the more open areas are presumed to be more vulnerable to predation, although data on predation rates are lacking. Males from the eastern population are, indeed, faster than western males, but females do not differ; the morphological basis for the difference in speed among males is unclear.

Four studies have attempted to measure natural selection acting on individual variation in locomotor performance in the field (see Bennett and Huey, 1990). These constitute direct attempts to quantify fitness gradients (fig. 10.1) for reptilian locomotion. Jayne and Bennett (1990b) found that survivorship of garter snakes (*Thamnophis sirtalis*) was positively related to laboratory measures of both speed and distance crawling capacity, although not during the first year of life. R. B. Huey and colleagues (see also Tsuji et al., 1989; van Berkum et al., 1989) conducted a similar study of hatchling fence lizards (*Sceloporus occidentalis*), but the analyses are not yet complete. In both cases, comparisons of families of presumed full-siblings suggest that speed and stamina may be heritable (Garland, 1988; Jayne and Bennett, 1990a; Tsuji et al., 1989; van Berkum and Tsuji, 1987). Miles (1989, pers. comm.) reports significant directional selection on speed in *Urosaurus* lizards. Finally, A. E. Dunham, R. B. Huey, K. L. Overall, and colleagues are continuing a study of selection on locomotor performance in *Sceloporus merriami,* with preliminary results suggesting no significant selection on locomotor performance (pers. comm.). As with almost all studies of selection (Endler, 1986; Wade and Kalisz, 1990), interpretation of these studies is problematical because the causal basis for selection on sprint speed is not understood. In the absence of data on whether lizards and snakes actually use maximum abilities in nature and, if they do, whether variation in maximum abilities is biologically significant, a story can be devised for any result. This is not to suggest that selection studies are unimportant, but, rather, to urge that they be coupled with detailed analyses of natural history and possibly experimental manipulations (Greene, 1986; Hews, 1990; Pough, 1989; Sinervo et al., 1992).

Some information on fitness gradients can be obtained through laboratory studies or controlled trials in seminatural field enclosures (e.g., Schwartzkopf and Shine, 1992). For example, in laboratory enclosures, Garland et al. (1990a) studied whether social dominance might be related to capacities for speed or stamina in *Sceloporus occidentalis*. In paired encounters between males competing for access to a basking site, winners were significantly faster than were losers, but did not have higher treadmill endurance (see also Hews, 1990; Wilson and Gatten, 1989; Wilson et al., 1989).

Context-Specificity of Performance. Evolutionary specialization to a particular "niche" may come at the expense of lowered fitness (including lower "effectiveness" [Gans, 1991] or energetic efficiency [e.g., Andrews et al., 1987]) in other "niches" ("the jack-of-all trades is master of none" idea [Huey and Hertz, 1984b; Jackson and Hallas, 1986]). If so, then one might predict that performance capability would be context-specific; species, for example, might perform a task best under conditions most similar to those they experience most often in nature (see also Bauwens et al., in press).

To test this idea, Losos and Sinervo (1989) measured sprint speed of four species of *Anolis* on different diameter rods. In nature, the long-legged *A. gundlachi* uses wide structures, whereas the short-legged *A. valencienni* often moves on narrow twigs; the other two species are intermediate in both respects. On wide perches, sprint speed and leg length were directly related among the species; however, on narrow rods, all four species had similar sprint abilities (fig. 10.6). It is understandable why the long-legged species do not use narrow structures in nature: their locomotor capabilities would be impaired. But why should short

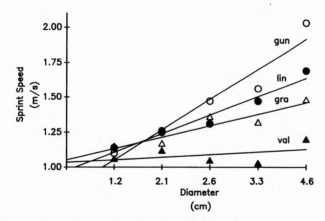

FIGURE 10.6 Interaction between locomotor performance ability and species identity for four species of *Anolis* lizards running on rods of varying diameter (modified from Losos and Sinervo, 1989). Species differences in maximal sprint speed are apparent only on larger rods. gun = *Anolis gundlachi*, lin = *A. lineatopus*, gra = *A. grahami*, val = *A. valencienni*.

legs evolve to utilize narrow structures? *Anolis gundlachi* can run as fast as *A. valencienni* on narrow structures without incurring the latter's 50% sacrifice in speed on broad structures. The answer probably lies in the species' ability to move without difficulty on narrow surfaces ("surefootedness" *sensu* Sinervo and Losos, 1991). In over 75% of the trials on the narrowest rod, *A. gundlachi* stumbled or fell off, whereas *A. valencienni* was much less affected. Consequently, *A. valencienni* seems to have traded the ability to move rapidly on broad surfaces for the ability to move without trouble on narrow ones.

A similar trade-off is apparent among four populations of *Sceloporus occidentalis* (Sinervo and Losos, 1991) and between *S. occidentalis* and *S. graciosus* (Adolph and Sinervo, in preparation). By contrast, relative sprint performance in two chameleon species is consistent across a range of surface dimensions (Losos et al., 1993). In both the *Anolis* and *Sceloporus* cases, field studies are needed to evaluate the relative ecological importance of maximum speed versus surefootedness. Biomechanical studies should enlighten relationships between limb length, perch diameter, and sprinting (Alexander, 1968; Emerson and Koehl, 1990; Hildebrand et al., 1985; Pounds, 1988).

The foregoing studies underline the importance of measuring performance over a variety of appropriate conditions, of measuring several different aspects of performance, and of considering a variety of behaviors that may relate to performance abilities (cf. Bennett, 1989; Garland, 1993, 1994a, b; Huey and Hertz, 1984a; Sinervo and Losos, 1991).

Foraging Mode and Relative Clutch Mass. Lizards have traditionally been classified either as "sit-and-wait" or as "active" foragers (Pianka, 1966; Regal, 1983; Schoener, 1971), although the distinction represents extremes of a continuum rather than an actual dichotomy (appendix I of Garland, 1993; McLaughlin, 1989; O'Brien et al., 1990; Pietruszka, 1986). A variety of attributes distinguish the two foraging modes among terrestrial lizards. Sit-and-wait (or ambush) foragers tend to be stocky, have short tails, and carry relatively large clutches, whereas active (or "widely foraging" or "cruising") foragers often are slender, with long tails and relatively small clutches (Huey and Pianka, 1981; Vitt and Congdon, 1978; Vitt and Price, 1982; Perry et al., 1990). Relatively little intrafamilial variation in foraging mode exists among lizards; hence, interspecific comparisons of foraging types often are confounded by phylogeny (Dunham et al., 1988). However, intrageneric (Huey and Pianka, 1981; Huey et al., 1984) and even intraspecific (Robinson and Cunningham, 1978) variation in foraging mode exists in lacertids (Perry et al., 1990).

The two foraging modes tend to differ in sprint speed and endurance in an expected manner; active foragers have greater endurance, but sit-and-wait foragers, which often capture prey by a quick lunge from an ambush site, often show greater sprinting ability (Huey et al., 1984). Further, in interspecific comparisons, foraging mode appears to be associated with differences in resource acquisition and reproductive rates (Anderson and Karasov, 1981, 1988; Karasov and Anderson, 1984; Nagy et al., 1984). Three studies have revealed some morphological and physiological correlates of differences in foraging mode and locomotor performance (Bennett et al., 1984; Garland, 1993, 1994b).

Nonetheless, with the following exceptions, few examples indicate tight relationship between morphology, performance, and foraging mode. Active foraging lizards appear to experience higher predation, and their longer tails may represent an adaptation by increasing the likelihood that a predator will grab the detachable tail, leaving intact the rest of the lizard (Huey and Pianka, 1981; Vitt, 1983). Among snakes, active foragers tend to have longitudinal stripes and flee when approached; longitudinal stripes can give the impression that a moving snake is stationary, making it easy for a predator to lose sight of a fleeing snake. By contrast, more sedentary species often have broken patterns and rely on crypsis or active defense to thwart predators (Jackson et al., 1976; Pough, 1976). Further, antipredator behavior (fleeing versus crypsis) and color pattern (blotched versus striped) are genetically correlated in *Thamnophis ordinoides* (Brodie, 1989b, 1993).

The relationships of clutch size, performance, and foraging mode are better established. Gravidity leads to decreased sprint speed and/or endurance in lizards (e.g., Cooper et al., 1990; Garland, 1985; Garland and Else, 1987; Sinervo et al., 1991; Van Damme, Bauwens, and Verheyen, 1989; but see Huey et al., 1990) and snakes (Brodie, 1989a; Jayne and Bennett, 1990a; Seigel et al., 1987). Thus, active foragers, which rely greatly on sustained locomotion, generally have evolved smaller relative clutch masses (RCM), which minimize these effects (Ananjeva and Shammakov, 1985; Dunham et al., 1988; Huey and Pianka, 1981; Magnusson et al., 1985; Vitt and Congdon, 1978; Vitt and Price, 1982; but see Henle, 1990; again, phylogeny is a confounding factor [Dunham et al., 1988]). Shine (1980), Seigel et al. (1987), and Van Damme, Bauwens, and Verheyen (1989) suggested that locomotor impairment correlated with RCM, but Brodie (1989a) argued that gravidity per se, rather than RCM, is the major cause of decreased locomotor capacities (see also Sinervo et al., 1991). In a similar vein, Shine (1988) argued, based on biomechanical considerations, that a given RCM would hinder snake locomotion more in aquatic than in terrestrial species. He thus interpreted the lesser RCM of aquatic species as adaptive.

A linear relationship between gravidity, decreased sprinting abilities, and increased mortality cannot be assumed. For some species, gravid females may indeed be more vulnerable to predation (Shine, 1980; references in Cooper et al.,

1990), but in other species behavioral shifts may compensate for decreased loco-motor performance. For example, the lizards *Lacerta vivipara* (Bauwens and Thoen, 1981) and *Eumeces laticeps* (Cooper et al., 1990), and the snake *Thamnophis ordinoides* (Brodie, 1989a), alter their behavior when gravid, becoming less active and more reliant on crypsis and aggression for defense. Indeed, no increase in mortality was observed for gravid *Lacerta vivipara* (Bauwens and Thoen, 1981; see also Schwartzkopf and Shine, 1992).

Conclusions. Relationships between performance ability and ecology or be-havior have been little explored, either intra- or interspecifically. When studies have been conducted, they often are correlational and do not directly examine the causal basis of any correlation. They are thus not able to distinguish selection and sorting (*sensu* Vrba and Gould, 1986). As well, phylogeny confounds most an-alyses to date. Relationships of foraging behavior, reproductive state, and loco-motor performance, however, do seem reasonably well established.

Morphology → Behavior or Fitness/Ecology

Biologists comparing species have long noted correlations between form and lifestyle or habitat (e.g., Bock and von Wahlert, 1965; Gans, 1988; Luke, 1986; Rayner, 1981, 1987; Ricklefs and Miles, chap. 2, this volume; Van Valkenburgh, chap. 7, this volume; references in Wainwright, 1991) and have taken such rela-tionships as evidence that particular morphologies are adaptations (*sensu* Gould and Vrba, 1982) for particular lifestyles or habitats (Harvey and Pagel, 1991). Indeed, such comparisons and interpretations are at the heart of traditional eco-logical morphology (e.g., Norberg, chap. 9, this volume). Rarely, however, does evidence exist demonstrating that morphological differences actually lead to dif-ferences in performance abilities that are appropriate for different habitats. Here, we review studies that have related reptilian locomotor morphology to behavior or ecology, and evaluate the extent to which locomotor performance represents the mechanistic basis of such relationships.

Limb Length and Habitat Use. Several studies have correlated interspecific or interpopulation variation in limb length with differences in microhabitat use or behavior. Usually, however, the effect of variation in limb length on performance and the relevance of differences in performance to variation in habitat or behavior are not investigated. In addition, differences in limb proportions (often expressed as ratios) can be confounded by differences in size (see "Allometry and its impor-tance" above). We first offer a brief discussion of locomotor allometry in lizards.

Within most lizard species, hindlimb length shows negative allometry (e.g., Garland, 1985; Kramer, 1951; Marsh, 1988; Pounds et al., 1983), but positive allometry occurs in a few species (Dodson, 1975; Pounds et al., 1983). Interspecifically, negative allometry of hindlimb length is common among lizards (e.g., figs. 10.7c, 10.8c below; Losos, 1990b). A number of adaptive explanations have been offered to explain these patterns. For example, Dodson (1975) proposed that the positive ontogenetic allometry in two species of *Sceloporus* represented an adaptation for increased home range size and movement in adults, although the underlying mechanism (e.g., increased mobility) for such a relationship was neither specified nor measured. Pounds et al. (1983) hypothesized that positive ontogenetic allometry in *Sceloporus woodi* might relate either to a shift to more arboreal habitats or to reliance on sprinting to avoid predation. Because sprint speed usually increases ontogenetically (see above), Pounds et al. (1983) interpreted the negative allometry observed in four other iguanid species as a means of compensating for the lesser sprinting ability of smaller individuals by increasing the length of their hindlimbs (cf. Grand, 1991, on antelope). Sprint speeds, however, were not measured, and Garland (1985) has argued that data on scaling of limb dimensions alone are inadequate to infer scaling of sprint speed.

The importance of explicitly considering allometry is apparent in a study of fourteen species of diurnal *Ctenotus* skinks in Australia. Pianka (1969) found that relative hindleg length (length of one hindleg/snout-vent length) correlated positively with use of open space (fig. 10.7a). Pianka (1986) inferred that long legs must enhance sprint speed and thus are beneficial in open spaces, whereas, in dense vegetation, long legs actually impede efficient locomotion (see also Jaksic' and Nuñez, 1979). (In fact, some *Ctenotus* do fold their legs against the body and use serpentine locomotion to move through clumps of spinifex grass [James, 1989].) Unfortunately, the relationship between relative limb length and habitat is confounded by body size. Larger species use less open habitats (fig. 10.7b) and have relatively shorter legs (fig. 10.7c). When these correlations with body size are removed by regression analysis, a relationship between residual use of open space and residual hindleg length is suggested, but nonsignificant ($P < .12$; fig. 10.7d).

Among species of nocturnal Australian geckos, use of open spaces and relative hindleg length are uncorrelated (data from Pianka, 1986; contrary to analyses of a smaller data set in Pianka and Pianka, 1976; fig. 10.8a). In this case, however, use of open space is not significantly related to svl ($r^2 = .025, P = .6652$; fig. 10.8b). Analysis of residuals from regressions on svl (a conservative procedure) also indicates that open-space use is not significantly predicted by relative leg length (figs. 10.8c, d). Both of these examples (*Ctenotus,* nocturnal geckos) deserve phylogenetic reanalyses when suitable phylogenies become available.

A number of other studies have investigated whether a relationship exists between limb proportions and habitat. Longer-legged taxa use more open (includ-

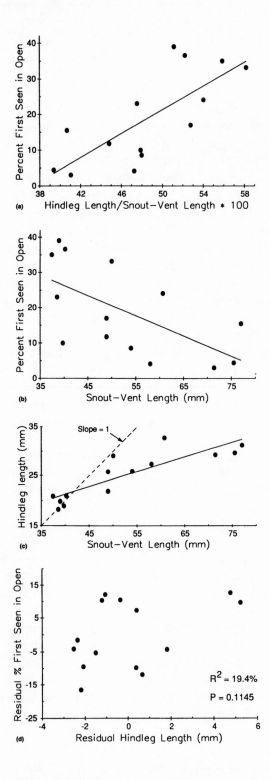

ing large boulders) and/or terrestrial habitats among populations of *Sceloporus woodi* (Jackson, 1973b), among populations in a cline connecting *Liolaemus platei* and *L. lemniscatus* (Fuentes and Jaksic', 1980), and between species of *Sceloporus* (Jackson, 1973a, b) and *Lacerta* (Darevskii, 1967). By contrast, in a study of North American temperate lizard communities, Scheibe (1987) found that large, bulky, and long-legged sit-and-wait foragers occupied extensively vegetated habitats, whereas small, slender, and short-legged active foragers used more open habitats. However, this study is a prime example of the importance of conducting comparative analyses in a phylogenetic context: of the three families represented in the study, only the Iguanidae (twenty-one of twenty-nine species in the study) exhibit substantial variation, most of which is distributed among genera. This phylogenetic non-independence leads, in effect, to a substantial inflation of the inferred degrees of freedom, possibly leading to spurious significant results (cf. Garland et al., 1991). No relationship between limb length and habitat use is evident among *Liolaemus* (Jaksic' et al., 1980) or Brazilian cerrado lizards (Vitt, 1991).

Several other studies have looked for correlates of limb length in lizards. Ananjeva (1977) found that among five species of Russian *Eremias*, limb proportions were related to locomotor behavior (e.g., climbing, burrowing). In addition, insular populations of lacertid lizards in the Adriatic Sea generally have shorter legs than do mainland populations, except on islands with steep cliffs. Kramer (1951) suggested that the lack of predators on islands and the need for clinging and jumping ability on cliffs is responsible for these patterns. Carlquist (1974) discussed other examples of variation in limb proportions.

Interpreting the foregoing results is difficult, because the functional consequences of morphological variation are unknown. Further, the ecological relevance of functional differences (assuming they exist) is usually speculative. Nonetheless, several studies have related limb morphology, functional capabilities, and habitat use. Laerm (1974) compared two species of basilisks which differed in their use of aquatic habitats. The more aquatic *Basiliscus plumifrons,* which has longer legs and wider toe fringes (see below), can run more quickly on water than can *B. vittatus,* which is of similar size.

FIGURE 10.7 (a) Positive relationship between use of open microhabitats and relative hindleg length for 14 species of diurnal *Ctenotus* skinks in Australia (Pianka, 1969; revised data from Pianka, 1986, appendices C.3 and G.3). (b) Negative relationship between percentage first seen in open and snout-vent length, a measure of body size. (c) Deviation from geometric similarity (dashed line) for relationship between hindleg length and snout-vent length. Solid line is least squares linear regression. (d) Nonsignificant relationship between residual percentage first seen in open and residual hindleg length.

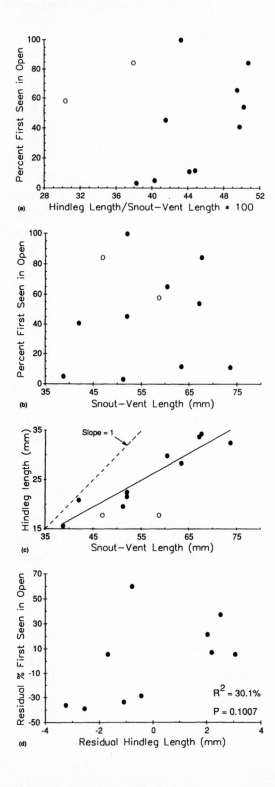

(a) Hindleg Length/Snout−Vent Length * 100

(b) Snout−Vent Length (mm)

(c) Slope = 1
Snout−Vent Length (mm)

(d) $R^2 = 30.1\%$
P = 0.1007
Residual Hindleg Length (mm)

Williams (1972) coined the term "ecomorph" to refer to the radiation of *Anolis* lizards in the Greater Antilles. A set of morphologically distinctive species (termed ecomorphs and named for the microhabitat they most commonly use; e.g., "trunk-crown") occurs on each of the Greater Antilles (i.e., Cuba, Hispaniola, Jamaica, and Puerto Rico [Mayer, 1989; Williams, 1972, 1983]). Comparison of island faunas indicates that not only are the same morphological types present on each island, but that morphologically similar species also are similar in ecology and behavior (Losos, 1990a, b; Moermond, 1979a, b; Rand and Williams, 1969; Williams, 1972, 1983). Although *Anolis* phylogeny is controversial (Burnell and Hedges, 1990; Cannatella and de Queiroz, 1989; Guyer and Savage, 1986, 1992; Williams, 1989), at least three independent radiations have occurred on Jamaica, Puerto Rico, and Cuba-Hispaniola (Williams, 1983).

The ecomorph types differ morphologically in a number of characters, and figure 10.9 confirms that members of each ecomorph type are more similar morphologically to other members of that ecomorph on other islands than they are to more closely related species on the same island (Loso, 1992). Thus, parallel or convergent evolution has occurred. The ecomorphs also differ ecologically (perch height, perch diameter, and distance to the nearest available perch) and behaviorally (frequency of movements, display rate, distance jumped, and relative proportion of runs, jumps, and walks; Losos, 1990a, b; Moermond, 1979a, b; Rand, 1964, 1967; Schoener and Schoener, 1971a, b).

The "habitat matrix" model (Moermond, 1979a, b, 1986; Pounds, 1988) suggests that ecomorphological radiation is driven by adaptation in locomotor morphology and behavior: *Anolis* alter their behavior in different microhabitats; as species evolve habitat specializations they also evolve morphologies appropriate for the behaviors they use. More specifically, the habitat matrix model predicts that as the distance to the nearest perch increases, lizards should jump less frequently but over longer distances; as the surface becomes narrower, lizards should be forced to move more slowly and carefully. Consequently, limb length should correlate with perch diameter and with distance to the nearest perch. Data from twenty-eight species of Costa Rican, Hispaniolan, Jamaican, and Puerto

----

FIGURE 10.8 (a) Nonsignificant relationship between percentage first seen in open and relative hindleg length for 12 species of nocturnal geckos in Australia (Pianka and Pianka, 1976; revised data from Pianka, 1986, appendices C.3 and G.3). Following Pianka and Pianka (1976, p. 131), two species (open circles) are excluded from subsequent regression analyses. (b) Nonsignificant relationship between percentage first seen in open and snout-vent length. (c) Deviation from geometric similarity (dashed line) for relationship between hindleg length and snout-vent length. Solid line is least squares linear regression. (d) Nonsignificant of relationship between residual percentages first seen in open and residual hindleg length.

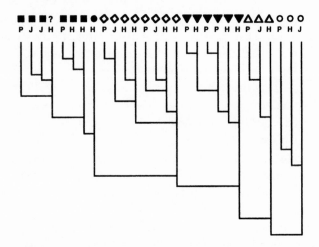

FIGURE 10.9 UPGMA phenogram of the position of *Anolis* lizard species in a multidimensional morphospace (from Losos, 1992). Variables analyzed were forelimb length, hindlimb length, tail length, mass, and lamellae number (all with the effect of snout-vent length removed via computation of residuals from least squares linear regressions), and snout-vent length itself. Members of each ecomorph type (indicated by the different symbols) are more similar morphologically to other members of that ecomorph on other islands than they are to more closely-related species on the same island. Thus, parallel or convergent evolution has occurred.

Rican *Anolis* support these predictions (Losos, 1990b, c, unpubl.; Moermond, 1979a, b; Pounds, 1988).

The predictions of the habitat matrix are predicated on the assumption that variation in morphology affects performance, which is related to differences in ecology and behavior. Studies summarized above (see "Morphology → Performance") support this assumption. Relative sprinting and jumping abilities (with the effect of size removed) are related to relative limb length. In turn, variation in performance capability correlates with differences in locomotor behavior and microhabitat use (Losos, 1990a, b).

Nonetheless, it is surprising that only a weak relationship was found between absolute performance ability and ecology or behavior. One might expect that an organism's absolute running and jumping abilities should be important in determining its ecology and behavior, rather than whether it can run or jump well for its size (cf. Huey and Bennett, 1987, p. 1106; "Discussion" in Garland, 1994a, b). Perhaps relative limb length affects ecology and behavior via a relationship with some other aspect of performance, such as endurance or energetic costs of locomotion (Autumn et al., 1994; Full et al., 1990; Full, 1991; Garland and Janis, 1993; John-Alder et al., 1986a; Strang and Steudel, 1990; references therein). In this scenario, relative jumping and sprinting abilities would also be correlated with limb length but not necessarily causally involved in the morphology → ecology relationship. A phylogenetically based path analysis (cf. Arnold, 1983) might help to tease apart these relationships.

More generally, the relationships between morphology, performance, and ecology may not be constant across microhabitats (see "Context-Specificity of Performance" above). Rather, species might be adapted to perform best in their own microhabitat, and different microhabitats might select for the optimization of different performance capabilities. Performance tests made under homogeneous conditions might thus obscure why particular morphologies are appropriate for particular "niches."

A final caveat concerning the functional and adaptive significance of limb length is in order. Muth (1977, p. 718), in a study of *Callisaurus draconoides*, which occupies open areas in often extremely hot North American deserts, concluded that: "Long legs also may function to increase the height of the body in the elevated posture, and to exploit the convective heat loss regime" (see also Arnold, 1984b; Greer, 1989, p. 33). Thus, long limbs may be advantageous both for reducing the threat from predators and for reducing heat load, both of which are likely to be higher in open as compared with closed microhabitats. Given that limb length may affect several abilities (e.g., sprinting, climbing, jumping, moving through grass, thermoregulation, push-up displays), that it varies among individuals and among populations (e.g., Bulova, in press), and that it correlates strongly with both body size and phylogeny, integrated analyses of the evolution of limb length are warranted.

Limblessness. Elongation of the body and loss of the fore- and/or hindlimbs has evolved at least twelve times among squamate reptiles (Edwards, 1985). With the exception of snakes, most legless squamates are either fossorial, use burrows, or shelter under objects on the ground (Gans, 1975); Walls (1942) has argued that snakes evolved from burrowing forms. Several forms of limbless locomotion exist; lateral undulation, however, is common to most limbless vertebrates (Gans, 1975). Undulatory locomotion is also used by some limbed squamates, which fold their legs against the body while moving through fluid (e.g., water, sand) or tangled environments. The evolutionary origin of limblessness appears usually to be associated with the utilization of narrow openings and crevices (Gans, 1975; Gans et al., 1978; Shine, 1986), although in some circumstances nonfossorial locomotor performance may be enhanced by using undulatory locomotion (as in some large Australian skinks; John-Alder et al., 1986a; Garland, pers. obs.). An analysis of the performance capabilities of a series of related species varying in limb dimensions might shed valuable light on the evolution of limblessness.

In snakes, the energetic cost of locomotion during lateral undulation is similar to that predicted for a quadrupedal lizard (or mammal) of similar size (Autumn et al., 1994; John-Alder et al., 1986a), whereas the cost of concertina locomotion is significantly higher (Walton et al., 1990; but see Dial et al., 1987) and the cost of sidewinding significantly lower in at least one species (Secor et al., 1992). As determined on a motorized treadmill, maximal aerobic speeds, and hence endurance capacities at certain speeds, may be rather low in snakes (Garland, 1988; Jayne and Bennett, 1990a, b; Walton et al., 1990; but see Secor et al., 1992) as compared with some lizards of similar size (e.g., Garland, 1984, 1994b; Garland and Else, 1987; John-Alder and Bennett, 1981; John-Alder et al., 1986a). Part of this difference may be related to temperature (cf. Garland, 1994b): existing data for snakes have been taken at temperatures (30°C) somewhat lower than those at which lizards have been measured (35–40°C); however, species from both groups have been measured at or near their mean field-active body temperature.

Toe Fringes. Toe fringes, composed of laterally projecting elongated scales, have evolved at least twenty-six times within seven lizard families (Luke, 1986). The evolution of fringes has been associated primarily with the occupation of windblown sand or water habitats (Luke, 1986). Presumably, fringes work by providing increased surface area and hence less slippage on fluid substrates (Laerm, 1973), and may be more important during acceleration than at maximal speed, when slippage may be minimal (Carothers, 1986, p. 872). Carothers (1986) demonstrated that when the fringes of *Uma scoparia* were removed, the lizards ran more slowly and with lower acceleration on sandy, but not on rubber, surfaces (also see discussion of Laerm's [1974] study of basilisks under "Limb Length and Habitat Use" above).

Conclusions. Although a wealth of data indicate that locomotor morphology and ecology are related among reptiles, we have little hard evidence about the causal basis of this relationship. Information on the ecological consequences of variation in performance abilities and whether the morphology-performance-ecology relationship is constant among habitats is necessary to help elucidate these correlations. Such integrative studies conducted within populations are particularly notable in their absence.

DISCUSSION

Biologists have for many years gathered data documenting relationships between morphology and ecology, as several chapters in this volume attest. Some studies merely describe correlations without considering directly the mechanism by which morphology, physiology, and biochemistry are transduced into behavior and ecology. Other studies have argued, sometimes through the use of biomechanical or physiological models (e.g., Emerson et al., chap. 6, this volume; Norberg, chap. 9, this volume; Wainwright, chap. 3, this volume), that morphology must relate to ecology through its effect on organismal performance abilities, but have not actually measured performance. The idea that morphology affects fitness and ecology only through its effects on performance has been formalized in the paradigms discussed above (Arnold, 1983; Emerson and Arnold, 1989; "Introduction" above; fig. 10.1).

The Importance of Behavior

Studies integrating measurements of morphology, performance, behavior, and ecology are rare but increasing in frequency. Those now available indicate that the morphology → performance → fitness paradigm is indeed a useful one for examining natural selection and adaptation in the ecological morphology of reptilian locomotor performance. The simplicity of the paradigm makes it experimentally tractable, but also incomplete. In particular, the role of behavior and habitat must be considered (see "Introduction" above; figs. 10.2, 10.3).

By way of example, consider the role of behavior relative to morphology in gliding reptiles. Numerous arboreal species have evolved structures that increase surface area and hence increase gliding capabilities. However, the most widespread adaptation for increasing gliding ability is a behavioral one: when falling, arboreal species of squamate reptiles, amphibians, and mammals adopt an outstretched posture that maximizes surface area and stability (Oliver, 1951; Russell, 1979). In an early study, Oliver (1951) compared the falling behavior of the arboreal *Anolis carolinensis* to that of the more terrestrial *Sceloporus undulatus*. The anole immediately adopted the outstretched position and landed relatively unharmed; the fence lizard, by contrast, cycled its limbs in an attempt to run and landed forcefully. In a considerably more elegant study, Emerson and Koehl (1990) compared the role of posture and morphological modifications, such as flaps of skin between the legs and toes, in model treefrogs in a wind tunnel. They found that the effect of behavioral and morphological features exhibited by flying frog species, when tested independently, often had quite different effects on the performance of the models. Further, the effect of posture was dependent on morphology.

We have argued that natural selection does not act directly on performance ability, but rather on what an organism does (see "Introduction" above). Selection can act on performance only when individuals are behaving in the same way (e.g., trying to escape from a predator by running away as quickly as possible) and some individuals are better than others, due to their higher performance abilities; otherwise, selection will act on the differences in behavior. Thus, behavior is a filter through which performance is related to fitness (Garland et al., 1990b), and performance is itself a filter through which morphology is transduced (cf. Ricklefs and Miles, chap. 2, this volume) to determine behavioral options (Garland, 1994a). Behavior can affect the performance → fitness link in two general ways: by negating advantages in performance ability, and by compensating for decrements in ability.

An example of behavior making performance differences irrelevant is presented in a study of crypsis in the Pacific treefrog, *Pseudacris regilla,* which occurs in two color morphs, green and brown. Morey (1990) investigated whether frogs would choose to sit on substrates matching their color, and whether substrate matching affects susceptibility to predation by garter snakes, *Thamnophis elegans,* in laboratory trials. Individual frogs tended to select substrates matching their own color. Laboratory trials were conducted in which one frog matched the substrate and a second did not. Indeed, in the ten trials in which frogs did not move, the snake chose the nonmatching frog nine times. However, in eighteen trials, one of the frogs moved and was immediately attacked regardless of whether its color matched the background. In this example, color gives a performance advantage, crypsis, which leads to increased prey survival only when coupled with the appropriate behavior (nonmovement).

That alterations in behavior can compensate for decreased performance ability has been demonstrated in several contexts. Lizards compensate for diminished locomotor abilities at lower body temperatures either by increasing their approach distance (i.e., the distance at which they will flee from a potential predator) (Rand, 1964; Shallenberger, 1970; but see Bulova, in press) or by switching from flight to aggressive defense (Crowley and Pietruszka, 1983; Hertz et al., 1982: see also Arnold and Bennett, 1984; Schieffelin and de Queiroz, 1991). Lizards and snakes with decreased locomotor abilities due to tail loss (Ballinger, 1973; Dial and Fitzpatrick, 1984; Formanowicz et al., 1990), gravidity (Bauwens and Thoen, 1981; Formanowicz et al., 1990), a full stomach (Herzog and Bailey, 1987), or exhaustion (Arnold and Bennett, 1984; Garland, 1988) similarly alter their behavior. In some cases, these behavioral shifts may suffice to eliminate completely the effect of decreased performance on survival (references in Brodie, 1989a; but see references in Cooper et al., 1990; Schwartzkopf and Shine, 1992).

Behavior also acts as an intermediate filter in the performance → fitness link in the context of habitat selection. For example, performance is clearly context-dependent—maximal capabilities are only valuable in habitats in which they can be used. Considering the thermal environment, Waldschmidt and Tracy (1983) demonstrated that *Uta stansburiana* utilize microhabitats that allow them to maintain body temperatures at which they can sprint fastest (cf. Adolph, 1990a;

Christian and Tracy, 1981; Grant, 1990; Grant and Dunham, 1988; Hertz et al., 1988). As the day proceeds, the thermal regimen of microenvironments changes, and the lizards tend to alter their position to maintain their temperature. The role of the structural environment is apparent in *Anolis* communities, where species tend not (but see Huey, 1983) to use habitats in which their locomotor abilities are compromised (see above, "Limb Length and Habitat Use").

Evolutionary adaptation is also mediated via habitat use (Dunson and Travis, 1991; Huey, 1991). In *Anolis* lizards, for example, the optimal temperature for locomotor performance has apparently evolved to match the thermal regimen of their environment (van Berkum, 1986), although some agamid, scincid, and *Sceloporus* lizards are more conservative (Crowley, 1985a; Garland et al., 1991; Hertz et al., 1983; van Berkum, 1988). Similarly, the habitat matrix model (see above) predicts that species utilizing different microhabitats will alter their behavior accordingly and that appropriate morphologies will evolve subsequently. Caribbean anoles exhibit all three components of the model: they shift habitat use in response to a number of factors, including climate and the presence of competitors (e.g., Moermond, 1986; Schoener and Schoener, 1971a, b; and see above); alter their locomotor behavior in different environments (Losos, 1990a; Moermond, 1979a, b; Pounds, 1988); and have evolved appropriate morphologies for their locomotor behavior (see above). Behavioral changes do not always precede changes in morphology and performance abilities, however (cf. Burggren and Bemis, 1990, p. 221). For example, Russell (1979) has shown that the enlarged lateral body folds used by some geckos in parachuting probably originally evolved to enhance crypsis (i.e., they are an exaptation, *sensu* Gould and Vrba, 1982). Similarly, the ability to flatten dorsoventrally in *Leiolepis belliani* may have evolved either as a means of thermoregulation or for social displays; the gliding ability that it entails seems of little significance to this beach-dwelling lizard (Losos et al., 1989). Similarly, the evolution of long legs for thermoregulatory purposes (cf. Muth, 1977; see above, "Limb Length and Habitat Use") might preadapt a lizard for the evolution of high sprint speed (cf. Gans, 1979, on "momentarily excessive construction"). Finally, the agamid *Calotes versicolor* has the ability to swim effectively, but when dropped in water, it swims effectively for a few moments and then becomes disoriented, cycles its limbs ineffectively, attempts to breathe while underwater, and eventually sinks and walks around aimlessly on the bottom (Gans, 1977)! Apparently, this lizard has the functional ability to swim, but not the appropriate behavior to use this ability.

Trade-offs and Correlations of Locomotor Abilities

Potential trade-offs and constraints are of major concern in evolutionary biology, behavioral ecology, and comparative morphology and physiology (e.g., Arnold et al., 1989; Barbault, 1988; Carrier, 1987, 1991; Congdon and Gibbons, 1987; Derrickson and Ricklefs, 1988; Feder and Londos, 1984; Garland and Huey, 1987; Grant and Dunham, 1988; Halliday, 1987; Maynard Smith et al., 1985; Miles and Dunham, 1992; Moermond, 1979b; Rose et al., 1987; Shine, 1988; Sinervo and Licht, 1991; Townsend and Calow, 1981). With respect to locomotion, performance of several functions may be correlated if the functions share a common mechanistic basis (cf. Emerson and Koehl, 1990). Such correlations may limit potential evolutionary pathways and prevent taxa from optimizing several performance abilities simultaneously (cf. Arnold, 1987, 1988; Brodie and Garland, 1993; Emerson and Arnold, 1989). However, predicting the existence of trade-offs based on first principles is not always easy. For example, contrary to simple models suggesting a necessary trade-off based on a dichotomy of fast versus slow muscle fiber types (cf. Gans et al., 1978; Pennycuick, 1991), speed and stamina are not negatively correlated at the level of individual variation (Garland, 1984, 1988; Garland and Else, 1987; Jayne and Bennett, 1990a; Secor et al., 1992), nor are they negatively genetically correlated in either *Thamnophis sirtalis* (Garland, 1988; see also Brodie, 1989b, 1992, on speed and distance crawled in *T. ordinoides*) or *Sceloporus occidentalis* (Tsuji et al., 1989).

Among species of *Anolis*, sprinting, jumping, and clinging ability are positively correlated, in part because all increase with body size (Losos, 1990b, c). With the effect of size removed by computing residuals, however, relative sprinting and jumping ability are still positively correlated, presumably due to the positive relationship of both to relative hindlimb length (Losos, 1990c). These correlations suggest that the evolution of sprinting and jumping ability is likely to be tightly linked. Consequently, the ability of *Anolis* to adapt in some ways to particular microhabitats may be constrained. Similarly, anole species adapt to utilize narrow surfaces such as twigs by evolving short limb length (Losos, 1990a; Williams, 1983), thus trading the ability to move quickly on broad surfaces for enhanced ability to move without difficulty on narrow surfaces (Losos and Sinervo, 1989). As a result, anoles cannot adapt to utilize narrow surfaces and still be able to make long jumps between perches.

Trade-offs may also occur between different types of performance. Chameleons, for example, have specialized for one locomotor task, grasping and moving upon extremely narrow surfaces, at the expense of another, sprint speed. Presumably, changes in the muscle architecture have played an important role in mediating this trade-off (see also Abu-Ghalyun et al., 1988).

Ecological circumstances may also dictate when correlations arise and when they are broken. For example, change in body size may be the path of least resistance when selection is for increased speed, generating among-population correlations of all traits that scale with size. Indeed, body size is often one of the most heritable traits within populations of many species (Falconer, 1989). In the case of arboreal or cliff-dwelling (cf. Kramer, 1951) species, however, an increase in body size might be selected against, resulting instead in an evolutionary increase in *relative* limb length.

CONCLUSIONS AND SUGGESTIONS FOR THE FUTURE

Our understanding of the ecological morphology of reptilian locomotor performance has increased greatly in the last decade. Many data have been gathered and a unified and general theoretical framework is coming into existence. This framework envisions measures of whole-animal performance as central for attempting to link morphological, physiological, or biochemical variation with behavior, fitness, or ecology. Nevertheless, many relatively simple ideas and hypotheses remain untested or understudied, for example, whether a trade-off between speed and stamina is ineluctable; whether and why limb length correlates with habitat characteristics; the selective importance of ability to recover from exhaustive exercise (cf. Gatten and Clark, 1989; Gatten et al., 1992; Gleeson, 1991; Pough et al., 1992); and possible relationships between locomotor abilities, vagility, and genetic variation (Gorman et al., 1977). Entire groups of reptiles are virtually unstudied (e.g., turtles; but see Dial, 1987; Janzen, 1993; Miller et al., 1987; Wyneken and Salmon, 1992; Zani and Clausen, in press). Studies of performance gradients, relating individual variation in morphology to performance, have been quite successful, but can become much more sophisticated (cf. work in mammalian exercise physiology). Fitness gradients, relating performance to fitness within populations, are starting to be quantified. However, quantitative data on field behavior, as derived from focal animal observations, are uncommon, making it difficult to identify possible selective agents responsible for correlations between laboratory locomotor performance and fitness/behavior/ecology. The lack of quantitative data on field behavior limits both comparisons among species and studies of natural selection (Bennett and Huey, 1990; Garland, 1993; Pough, 1989). For example, it has never even been documented that lizards with greater capabilities for sprinting (as measured in the lab) actually run faster in nature (cf. Garland, 1993; Hertz et al., 1988).

We predict that the next decade will continue to see exciting empirical and conceptual advances, aimed primarily at understanding the evolution of organismal performance (cf. Garland and Carter, 1994). Techniques for sophisticated measurements of locomotor performance and for quantitative genetic (Boake, 1994; Brodie and Garland, 1993) and phylogenetic analyses are now in place and awaiting integrated application. At present, these techniques are still novel enough to merit study in their own right. Soon, however, state-of-the-art studies will involve experimental, quantitative genetic, and phylogenetic analyses of measurements of morphology, performance, behavior, and fitness (or "adaptedness," when comparing populations or species).

In several cases, certain data sets already exist in partial form; these matrices can be filled in relatively easily. For example, relatively standardized techniques for measuring sprint speed have now been developed, and data on maximal speeds are now available for more than one hundred species of lizards; it is time to relate such data to morphology and to behavior and ecology (cf. Garland et al., 1988; Garland and Janis, 1993; Janis, in press; Losos, 1990b, c). In many cases, the most outstanding need is for quantitative data on field behavior. Comparisons would be greatly facilitated by standardization of methods for taking quantitative measurements of ecology and behavior (e.g., home ranges, movement rates, distances, speeds in nature; cf. Case, 1979; Cooper and Guillette, 1991; Garland, 1993; Huey and Pianka, 1981; Magnusson et al., 1985; Moermond, 1979a, b; Pietruszka, 1986). Another unresolved issue is the extent of behavioral compensation for changes in performance due to a full stomach, gravidity, or lowered body temperature. Behavioral adjustments related to changes in performance due to size and/or age also warrant study (cf. Grand, 1991; Pough and Kamel, 1984; Pounds et al., 1983; Taigen and Pough, 1981). How often does behavior shield performance from the direct effects of selection (fig. 10.2; Garland, 1994a; Garland et al., 1990b)?

Future comparative studies must deal directly with phylogenetic effects (Brooks and McLennan, 1991; Garland et al., 1991, 1992, 1993; Harvey and Pagel, 1991; Lauder, 1991; Losos and Miles, chap. 4, this volume; Lynch, 1991; Miles and Dunham, 1992; Pagel, 1993). As well, many existing studies warrant reanalysis with phylogenetically based methods. Computer programs are now available to do so (e.g., Garland et al., 1993; Lynch, 1991; Martins and Garland, 1991; A. Purvis, pers. comm.), and there is no excuse for not attempting to utilize whatever (even incomplete) phylogenetic information is available (Harvey and Pagel, 1991; Purvis and Garland, 1993). This historical context may be particularly important for investigations concerning the ecomorphology of reptilian locomotion, because many aspects of morphology, performance, and foraging mode seem strongly associated with phylogeny in both lizards and snakes (cf. Miles and Dunham, 1992). As well, many morphological bases—such as limblessness and toe fringes, and behavioral/ecological correlates such as active foraging—of locomotor performance have evolved many times independently in reptiles.

More studies of natural selection acting on individual differences within populations should be done. The available data base for reptilian locomotor performance is exceedingly small (see Bennett and Huey, 1990; Brodie, 1992; Sorci et al., in press). We need to know whether selection on performance is pervasive or rare, strong or weak, directional or stabilizing. (Unfortunately, quite large samples may be required to detect weak selection, although power can be enhanced by experimental manipulations [Sinervo et al., 1992].) Of particular interest would be studies in which traits at different levels of biological organization are measured simultaneously, such as antipredator display, speed, and scale counts in garter snakes or body mass, relative limb lengths, and endurance in lizards (cf. Lauder, 1990). Only these kinds of studies can tell us whether selection really is stronger for behavior and performance than for lower-level traits (cf. Garland et al., 1990b). Interestingly, Jayne and Bennett (1990b) report selection intensities on locomotor performance in snakes similar to those reported for morphometric traits in birds. Such studies can also tell us whether selection is correlational, favoring, for example, particular combinations of behavior and morphology/physiology (cf. Brodie, 1989b; Garland, 1988, 1994a, b; Huey and Bennett, 1987), and informing us about the form of the fitness function (cf. fig. 3.4 in Feder, 1987; Schluter, 1988). Finally, we might test whether selection ever acts *directly* on morphology (cf. fig. 10.4), although unmeasured characters could confound such analyses (Lande and Arnold, 1983; Mitchell-Olds and Shaw, 1987; Wade and Kalisz, 1990).

Although not an easy task, especially with the current difficulties in obtaining long-term funding, some studies of natural selection should be carried out for multiple years (cf. Grant, 1986; Grant and Grant, 1990; Huey et al., 1990). Detection of rare events may be difficult (Buffetaut, 1989; Weatherhead, 1986), but their potential selective and hence evolutionary importance (Grant, 1986; Wiens, 1977; Wiens and Rotenberry, 1980) dictates that substantial effort be expended to study them.

Another important direction for studies of natural selection is toward experimental manipulation to extend the range of natural variation and/or to diminish correlations between independent variables (e.g., performance and body size). Such manipulations can improve statistical power to detect selection acting on individual traits as well as allowing hypotheses suggested by comparative analyses to be experimentally probed (Baum and Larson, 1991; Mitchell-Olds and Shaw, 1987; Sinervo, 1990; Sinervo and Huey, 1990, pers. comm.). Hormonal or pharmacological manipulation, for example, could be used to change performance capacities of individuals, after which one could test for correlations between performance and survivorship in the field (Joos and John-Alder, 1990; John-Alder, 1990, pers. comm.; John-Alder et al., 1986b). Though probably impractical with reptiles, genetic perturbation experiments to detect selection have been successfully performed with invertebrates and with mice (Anderson et al., 1964; Hedrick, 1986).

Studies of natural selection should be integrated with estimates of heritabilities in nature (cf. Riska et al., 1989), to allow both prediction and reconstruction of microevolutionary phenomena (cf. S. J. Arnold, 1981, 1988). Estimation of field heritabilities (from free-living animals) is difficult but important, as field heritabilities may be considerably lower than those obtained in the laboratory (but see Hedrick, 1986, pp. 552–3).

Correlations between morphology and ecology, and trade-offs and constraints in performance, can be studied at multiple levels, yet integrative studies are uncommon (Emerson and Arnold, 1989; Sinervo, 1990; Sinervo and Licht, 1991). Given the growing information on interspecific variation, quantitative genetics, and physiological bases of locomotor performance, we anticipate studies in which physiological and biomechanical models are used to predict trade-offs in locomotor capacities, which are then tested, at complementary levels, by quantification of (1) genetic correlations within populations (e.g., Garland, 1988) and (2) "evolutionary correlations" through interspecific comparative studies (Garland et al., 1991, 1992; Garland and Janis, 1993; Harvey and Pagel, 1991; Lynch, 1991; Martins and Garland, 1991; Pagel, 1993). Such studies may also allow inferences concerning past patterns of (correlational) selection within populations (S. J. Arnold, 1981, 1988) and the direction of past evolutionary changes (e.g., Larson, 1984; Huey and Bennett, 1987). Of course, macroevolutionary processes, such as mass extinction (Buffetaut, 1989; Jablonski, 1986) and species selection (Arnold et al., 1989; Nee et al., 1992; Vrba, 1989; but see Williams, 1992), may also have influenced the covariation between ecology and morphology that exists for any given set of extant and/or extinct species (cf. Feder, 1987). If so, then patterns of among-population or among-species divergence predicted to have occurred based on known or hypothesized microevolutionary phenomena (e.g., individual selection, genetic correlation) may be obscured (cf. Emerson and Arnold, 1989; Garland and Carter, 1994).

The tools now exist to permit comprehensive studies of the ecological morphology of locomotor performance (cf. Bennett and Huey, 1990). Such studies will be greatly facilitated by interaction of morphologists, physiologists, ecologists, ethologists, geneticists, systematists, and evolutionary biologists. Particularly promising would seem to be collaborations between ecologists doing long-term field studies and morphologists or physiologists with laboratories equipped to measure ecologically relevant aspects of organismal performance (e.g., Huey et al., 1990).

ACKNOWLEDGMENTS

We thank Craig James for providing an unpublished portion of his thesis, A. F. Bennett and R. B. Huey for providing copies of unpublished manuscripts, and Eric Pianka for providing information. S. C. Adolph, R. A. Anderson, A. F.

Bennett, S. J. Bulova, M. R. Dohm, D. L. Gregor, K. L. Steudel, D. B. Miles, and the editors of this volume kindly reviewed the manuscript or provided helpful discussions. Supported in part by N.S.F. grants BSR-9006083 and BSR-9157268 to TG and BSR-8901205 to JBL and H. W. Greene.

REFERENCES

Abu-Ghalyun, Y., L. Greenwald, T. E. Hetherington, and A. S. Gaunt. 1988. The physiological basis of slow locomotion in chamaeleons. *J. Exp. Zool.* 245:225–231.

Adolph, S. C. 1990a. Influence of behavioral thermoregulation on microhabitat use by two *Sceloporus* lizards. *Ecology* 71:315–327.

Adolph, S. C. 1990b. Perch height selection by juvenile *Sceloporus* lizards: interspecific differences and relationship to habitat use. Journal of Herpetology 24:69–65.

Alexander, R. McN. 1968. *Animal Mechanics*. Seattle: University of Washington Press.

Ananjeva, N. B. 1977. Morphometrical analysis of limb proportions of five sympatric species of desert lizards (Sauria, *Eremias*) in the southern Balkhash Lake region. *Proc. Zool. Inst. Acad. Sci. USSR* 74:3–13.

Ananjeva, N. B., and S. M. Shammakov. 1985. Ecologic strategies and relative clutch mass in some species of lizard fauna in the USSR. *Sov. J. Ecol.* 16:241–247. (English translation of *Ekologiya* 4:58–65)

Anderson, P. K., L. C. Dunn, and A. B. Beasley. 1964. Introduction of a lethal allele into a feral house mouse population. *Amer. Nat.* 48:57–64.

Anderson, R. A., and W. H. Karasov. 1981. Contrasts in energy intake and expenditure in sit-and-wait and widely foraging lizards. *Oecologia* 49:67–72.

Anderson, R. A., and W. H. Karasov. 1988. Energetics of reproduction in a wide foraging lizard, *Cnemidophorus tigris,* and life history consequences of food acquisition mode. *Ecol. Mon.* 58:79–110.

Andren, C., and G. Nilson. 1981. Reproductive success and risk of predation in normal and melanistic colour morphs of the adder, *Vipera berus. Biol. J. Linnean Soc.* 15:235–246.

Andrews, R. M., F. H. Pough, A. Collazo, and A. de Queiroz. 1987. The ecological cost of morphological specialization: Feeding by a fossorial lizard. *Oecologia* 73:139–145.

Arnold, E. N. 1984a. Evolutionary aspects of tail shedding in lizards and their relatives. *J. Nat. Hist.* 18:127–169.

Arnold, E. N. 1984b. Ecology of lowland lizards in the eastern United Arab Emirates. *J. Zool.* (London) 204:329–354.

Arnold, E. N. 1988. Caudal autotomy as a defense. In *Biology of the Reptilia,* vol. 16, *Ecology B: Defense and Life Histories,* ed. C. Gans and R. B. Huey, 235–274. New York: Alan R. Liss.

Arnold, S. J. 1981. The microevolution of feeding behavior. In *Foraging Behavior: Ecological, Ethological and Psychological Approaches,* ed. A. Kamil and T. Sargent, 409–453. New York: Garland Press.

Arnold, S. J. 1983. Morphology, performance and fitness. *Amer. Zool.* 23:347–361.

Arnold, S. J. 1987. Genetic correlation and the evolution of physiology. In *New Directions in Ecological Physiology,* ed. M. E. Feder, A. F. Bennett, W. W. Burggren, and R. B. Huey, 189–215. Cambridge: Cambridge University Press.

Arnold, S. J. 1988. Quantitative genetics and selection in natural populations: Microevolution of vertebral numbers in the garter snake *Thamnophis elegans.* In *Proceedings of the Second International Conference on Quantitative Genetics,* ed. B. S. Weir, E. J. Eisen, M. J. Goodman, and G. Namkoong, 619–636. Sunderland, MA: Sinauer Associates.

Arnold, S. J., and A. F. Bennett. 1984. Behavioural variation in natural populations. III. Antipredator displays in the garter snake *Thamnophis radix. Anim. Behav.* 32:1108–1118.

Arnold, S. J., and A. F. Bennett. 1988. Behavioral variation in natural populations. V. Morphological correlates of locomotion in the garter snake *Thamnophis radix. Biol. J. Linnaen Soc.* 34:175–190.

Arnold, S. J., P. Alberch, V. Csanyi, R. C. Dawkins, S. B. Emerson, B. Fritzsch, T. J. Horder, J. Maynard Smith, M. J. Starck, E. S. Vrba, G. P. Wagner, and D. B. Wake. 1989. Group report. How do complex organisms evolve? In *Complex Organismal Functions: Integration and Evolution in Vertebrates,* ed. D. B. Wake and G. Roth, 403–433. Chichester: John Wiley and Sons.

Astrand, P. O., and K. Rodahl. 1986. *Textbook of Work Physiology,* 3d ed. New York: McGraw-Hill.

Atchley, W. R., T. E. Logsdon, D. E. Cowley, and E. J. Eisen. 1993. The uterine impact on adult skeletal morphology in the mouse: An embryo transplant study. *Evolution.* In press.

Autumn, K., R. B. Weinstein, and R. J. Full. 1994. Low cost of locomotion increases performance at low temperature in a nocturnal lizard. *Physiol. Zool.* 67. In press.

Avise, J. C., J. Arnold, R. M. Ball, E. Bermingham, T. Lamb, J. E. Neigel, C. A. Reeb, and N. C. Saunders. 1987. Intraspecific phylogeography: The mitochondrial DNA bridge between population genetics and systematics. *Ann. Rev. Ecol. Syst.* 18:489–522.

Baker, P. T. 1978. The adaptive fitness of high-altitude peoples. In *The Biology of High-altitude Peoples,* ed. P. T. Baker, 317–350. Cambridge: Cambridge University Press.

Ballinger, R. E. 1973. Experimental evidence of the tail as a balancing organ in the lizard, *Anolis carolinensis. Herpetologica* 29:65–66.

Ballinger, R. E., J. W. Nietfeldt, and J. J. Krupa. 1979. An experimental analysis of the role of the tail in attaining high running speed in *Cnemidophorus sexlineatus* (Reptilia: Squamata: Lacertilia). *Herpetologica* 35:114–116.

Barbault, R. 1988. Body size, ecological constraint, and the evolution of life-history strategies. *Evol. Biol.* 22:261–286.

Bartholomew, G. A., A. F. Bennett, and W. R. Dawson. 1976. Swimming, diving, and lactate production of the marine iguana, *Amblyrhynchus cristatus. Copeia* 1976:709–720.

Bauer, A. M. 1986. Saltation in the pygopodid lizard, *Delma tincta. J. Herpetol.* 20:462–463.

Baum, D. A., and A. Larson. 1991. Adaptation reviewed: A phylogenetic methodology for studying character macroevolution. *Syst. Zool.* 40:1–18.

Bauwens, D., and C. Thoen. 1981. Escape tactics and vulnerability to predation associated with reproduction in the lizard *Lacerta vivipara. J. Anim. Ecol.* 50:733–743.

Bauwens, D., T. Garland, Jr., A. M. Castilla, and R. Van Damme. In press. Evolution of sprint speed in lacertid lizards: Morphological, physiological, and behavioral coadaptation. *Evolution.*

Bauwens, D., R. Van Damme, D. Vanderstighelen, C. Thoen, D. Sanders, H. van Wezel, and R. F. Verheyen. 1987. Individuality in common lizards (*Lacerta vivipara*): A provisional review. In *Proceedings of the 4th Ordinary General Meeting of the Societas Europea Herpetologica,* ed. J. J. van Gelder, H. Strijbosch, and P. J. M. Bergers, 55–58. Nijmegen, The Netherlands: Faculty of Sciences.

Belkin, D. A. 1961. The running speeds of the lizards *Dipsosaurus dorsalis* and *Callisaurus draconoides. Copeia* 1961:223–224.

Bels, V. L., and J.-P. Theys. 1989. Mechanical analysis of the hind limb of *Anolis carolinensis* (Reptilia: Iguanidae) in jumping. In *Trends in Vertebrate Morphology: Proceedings of the 2d International Symposium on Vertebrate Morphology,* Vienna, 1986, ed. H. Splechtna and H. Hilgers, 608–611. Stuttgart: Gustav Fischer Verlag.

Bels, V. L., J.-P. Theys, M. B. Bennett, and L. Legrand. 1992. Biomechanical analysis of jumping in *Anolis carolinensis* (Reptilia: Iguanidae). *Copeia* 1992:492–504.

Benkman, C. W., and A. K. Lindholm. 1991. The advantages and evolution of a morphological novelty. *Nature* 349:519–520.

Bennett, A. F. 1980. The thermal dependence of lizard behaviour. *Anim. Behav.* 28:752–762.

Bennett, A. F. 1983. Ecological consequences of activity metabolism. In *Lizard Ecology: Studies of a Model Organism,* ed. R. B. Huey, E. R. Pianka, and T. W. Schoener, 11–23, 427–428, 487. Cambridge: Harvard University Press.

Bennett, A. F. 1987. Inter-individual variability: An underutilized resource. In *New Directions in*

Ecological Physiology, ed. M. E. Feder, A. F. Bennett, W. W. Burggren, and R. B. Huey, 147–169. Cambridge: Cambridge University Press.

Bennett, A. F. 1989. Integrated studies of locomotor performance. In *Complex Organismal Functions: Integration and Evolution in Vertebrates,* ed. D. B. Wake and G. Roth, 191–202. Chichester: John Wiley and Sons.

Bennett, A. F. 1990. The thermal dependence of locomotor capacity. *Amer. J. Physiol.* 259 (Regulatory Integrative and Comparative Physiology 28):R253–R258.

Bennett, A. F., and R. B. Huey. 1990. Studying the evolution of physiological performance. *Oxford Surveys Evol. Biol.* 7:251–284.

Bennett, A. F., and H. B. John-Alder. 1984. The effect of body temperature on the locomotory energetics of lizards. *J. Comp. Physiol.* B155:21–27.

Bennett, A. F., and J. A. Ruben. 1975. High altitude adaptation and anaerobiosis in sceloperine lizards. *Comp. Biochem. Physiol.* 50A:105–108.

Bennett, A. F., R. B. Huey, and H. B. John-Alder. 1984. Physiological correlates of natural activity and locomotor capacity in two species of lacertid lizards. *J. Comp. Physiol.* B154:113–118.

Bennett, A. F., R. S. Seymour, D. F. Bradford, and G. J. W. Webb. 1985. Mass-dependence of anaerobic metabolism and acid-base disturbance during activity in the salt-water crocodile, *Crocodylus porosus. J. Exp. Biol.* 118:161–171.

Bennett, A. F., K. M. Dao, and R. E. Lenski. 1990. Rapid evolution in response to high-temperature selection. *Nature* 346:79–81.

Bernardo, J. 1991. Manipulating egg size to study maternal effects on offspring traits. *Trends Ecol. Evol.* 6:1–2.

Bickler, P. E., and R. A. Anderson. 1986. Ventilation, gas exchange, and aerobic scope in a small monitor lizard, *Varanus gilleni. Physiol. Zool.* 59:76–83.

Birch, L. C., Th. Dobzhansky, P. O. Elliott, and R. C. Lewontin. 1963. Relative fitness of geographic races of *Drosophila serrata. Evolution* 17:72–83.

Boake, C. R. B., ed. 1994. *Quantitative Genetic Studies of Behavioral Evolution.* Chicago: University of Chicago Press. In press.

Bock, W. J., and G. von Wahlert. 1965. Adaptation and the form-function complex. *Evolution* 19:269–299.

Bolles, R. C. 1975. *Theory of Motivation,* 2d. ed. New York: Harper and Row.

Brodie, E. D., III. 1989a. Behavioral modification as a means of reducing the cost of reproduction. *Amer. Nat.* 134:225–238.

Brodie, E. D., III. 1989b. Genetic correlations between morphology and antipredator behaviour in natural populations of the garter snake *Thamnophis ordionoides. Nature* 342:542–543.

Brodie, E. D., III. 1992. Correlational selection for color pattern and antipredator behaviour in the garter snake *Thamnophis ordionoides. Evolution* 46:1284–1298.

Brodie, E. D. III. 1993. Homogeneity of the genetic variance-covariance matrix for antipredator traits in two natural populations of the garter snake. *Thamnophis ordionoides. Evolution* 44:844–854.

Brodie, E. D., III, and E. D. Brodie, Jr. 1990. Tetrodotoxin resistance in garter snakes: An evolutionary response of predators to dangerous prey. *Evolution* 44:651–659.

Brodie, E. D., III, and T. Garland, Jr. 1993. Quantitative genetics of snake populations. In *Snakes: Ecology and Behavior,* ed. R. A. Seigel and J. T. Collins, 315–362. New York: McGraw Hill.

Brooks, G. A., and T. D. Fahey. 1984. *Exercise Physiology: Human Bioenergetics and Its Applications.* New York: John Wiley and Sons.

Brooks, D. R., and D. A. McLennan. 1991. *Phylogeny, Ecology, and Behavior: A Research Program in Comparative Biology.* Chicago: University of Chicago Press.

Buffetaut, E., ed. 1989. Rare events, mass extinction and evolution. *Hist. Biol.* (special issue) 2:1–104.

Bulova, S. J. In press. Correlates of population and individual variation in antipredator behavior in two species of desert lizards. *Copeia.*

Burggren, W. W., and W. E. Bemis. 1990. Studying physiological evolution: Paradigms and pitfalls.

In *Evolutionary Innovations*, ed. M. H. Nitecki, 191–238. Chicago: University of Chicago Press.

Burghardt, G. M. 1984. On the origins of play. In *Play in Humans and Animals*, ed. P. K. Smith, 5–41. Oxford: Basil Blackwell.

Burnell, K. L., and S. B. Hedges. 1990. Relationships of West Indian *Anolis* (Sauria: Iguanidae): An approach using slow-evolving protein loci. *Carib. J. Sci.* 26:7–30.

Cannatella, D. C., and K. de Queiroz. 1989. Phylogenetic systematics of the anoles: Is a new taxonomy warranted? *Syst. Zool.* 38:57–69.

Carlquist, S. 1974. *Island Biology.* New York: Columbia University Press.

Carothers, J. H. 1986. An experimental confirmation of morphological adaptation: Toe fringes in the sand-dwelling lizard *Uma scoparia. Evolution* 40:871–874.

Carrier, D. R. 1987. The evolution of locomotor stamina in tetrapods: Circumventing a mechanical constraint. *Paleobiology* 13:326–341.

Carrier, D. R. 1991. Conflict in the hypaxial musculo-skeletal system: Documenting an evolutionary constraint. *Amer. Zool.* 31:644–654.

Case, T. J. 1979. Character displacement and coevolution in some *Cnemidophorus* lizards. In *Population Ecology.* Fortschr. Zoology 25, ed. U. Halbach and J. Jacobs, 235–282. Stuttgart: Gustav Fischer.

Castilla, A. M., and D. Bauwens. 1991. Observations on the natural history, present status, and conservation of the insular lizard *Podarcis hispanica atrata* on the Columbretes Archipelago, Spain. *Biol. Conserv.* 58:69–84.

Chappell, R. 1989. Fitting bent lines to data, with applications to allometry *J. Theor. Biol.* 138:235–256.

Christian, K. A., and C. R. Tracy. 1981. The effect of the thermal environment on the ability of hatchling Galapagos land iguanas to avoid predation during dispersal. *Oecologia* 49:218–223.

Christian, K. A., C. R. Tracy, and W. P. Porter. 1985. Inter- and intraindividual variation in body temperatures of the Galapagos land iguana (*Conolophus pallidus*). *J. Thermal Biol.* 10:47–50.

Colbert, E. H. 1967. Adaptations for gliding in the lizard *Draco. Amer. Mus. Nov.* 2283:1–20.

Congdon, J. D., and J. W. Gibbons. 1987. Morphological constraint on egg size: A challenge to optimal egg size theory? *Proc. Natl. Acad. Sci. USA* 84:4145–4147.

Cooper, W. E., Jr., and L. J. Guillette, Jr. 1991. Observations on activity, display behavior, coloration and androgen levels in the keeled earless lizard, *Holbrookia propinqua. Amphibia-Reptilia* 12:57–66.

Cooper, W. E., Jr., and L. J. Vitt. 1991. Influence of detectability and ability to escape on natural selection of conspicuous autotomic defenses. *Can. J. Zool.* 69:757–764.

Cooper, W. E., Jr., L. J. Vitt, R. Hedges, and R. B. Huey. 1990. Locomotor impairment and defense in gravid lizards (*Eumeces laticeps*): Behavioral shift in activity may offset costs of reproduction in an active forager. *Behav. Ecol. Sociobiol.* 27:153–157.

Cowley, D. E., D. Pomp, W. R. Atchley, E. J. Eisen, and D. Hawkins-Brown. 1989. The impact of maternal uterine genotype on postnatal growth and adult body size in mice. *Genetics* 122:193–203.

Crowley, S. R. 1985a. Thermal sensitivity of sprint-running in the lizard *Sceloporus undulatus:* Support for a conservative view of thermal physiology. *Oecologia* 66:219–225.

Crowley, S. R. 1985b. Insensitivity to desiccation of sprint running performance in the lizard, *Sceloporus undulatus. J. Herpetol.* 19:171–174.

Crowley, S. R. 1987. The effect of desiccation upon the preferred body temperature and activity level of the lizard *Sceloporus undulatus. Copeia* 1987:25–31.

Crowley, S. R., and R. D. Pietruszka. 1983. Aggressiveness and vocalization in the leopard lizard (*Gambelia wislizenii*): The influence of temperature. *Anim. Behav.* 31:1055–1060.

Daniels, C. B. 1983. Running: An escape strategy enhanced by autotomy. *Herpetologica* 39:162–165.

Daniels, C. B. 1985a. The effect of tail autotomy on the exercise capacity of the water skink, *Sphenomorphus quoyii. Copeia* 1985:1074–1077.

Daniels, C. B. 1985b. The effect of infection by a parasitic worm on swimming and diving in the water skink, *Sphenomorphus quoyii. J. Herpetol.* 19:160–162.

Daniels, C. B., S. P. Flaherty, and M. P. Simbotwe. 1986. Tail size and effectiveness of autotomy in a lizard. *J. Herpetol.* 20:93–96.

Daniels, C. B., and H. Heatwole. 1990. Factors affecting the escape behaviour of a riparian lizard. *Mem. Queensland Mus.* 29:375–387.

Darevskii, I. S. 1967. Rock lizards of the Caucasus: Systematics, ecology and phylogeny of the polymorphic lizards of the Caucasus of the subgenus *Archaeolacerta.* Izdatel'stvo "Nauka," Leningrad. (English translation by Indian National Scientific Documentation Centre, New Delhi, 1978)

Derrickson, E. M., and R. E. Ricklefs. 1988. Taxon-dependent diversification of life-history traits and the perception of phylogenetic constraints. *Func. Ecol.* 2:417–423.

Dial, B. E. 1987. Energetics and performance during nest emergence and the hatchling frenzy in loggerhead sea turtles (*Caretta caretta*). *Herpetologica* 43:307–315.

Dial, B. E., and L. C. Fitzpatrick. 1984. Predator escape success in tailed versus tailless *Scincella lateralis* (Sauria: Scincidae). *Anim. Behav.* 32:301–302.

Dial, B. E., R. E. Gatten, Jr., and S. Kamel. 1987. Energetics of concertina locomotion in *Bipes biporus* (Reptilia: Amphisbaenia). *Copeia* 1987:470–477.

Djawdan, M. 1993. Locomotor performance of bipedal and quadrupedal heteromyid rodents. *Func. Ecol.* 7:195–202.

Djawdan, M., and T. Garland, Jr. 1988. Maximal running speeds of bipedal and quadrupedal rodents. *J. Mammal.* 69:765–772.

Dodson, P. 1975. Relative growth in two sympatric species of *Sceloporus. Amer. Midland Nat.* 94:421–450.

Dunham, A. E., D. B. Miles, and D. N. Reznick. 1988. Life history patterns in squamate reptiles. *Biology of the Reptilia,* vol. 16, *Ecology B: Defense and Life Histories,* ed. C. Gans and R. B. Huey, 441–522. New York: Alan R. Liss.

Dunson, W. A., and J. Travis. 1991. The role of abiotic factors in community organization. *Amer. Nat.* 138:1067–1091.

Edwards, J. L. 1985. Terrestrial locomotion without appendages. In *Functional Vertebrate Morphology,* ed. M. Hildebrand, D. M. Bramble, K. F. Liem, and D. B. Wake, 159–172. Cambridge: Harvard University Press.

Eliott, J. P., I. McT. Cowan, and C. S. Holling. 1977. Prey capture by the African lion. *Can. J. Zool.* 55:1811–1828.

Emerson, S. B. 1985. Jumping and leaping. In *Functional Vertebrate Morphology,* ed. M. Hildebrand, D. M. Bramble, K. F. Liem, and D. B. Wake, 58–72. Cambridge: Harvard University Press.

Emerson, S. B., and S. J. Arnold. 1989. Intra- and interspecific relationships between morphology, performance, and fitness. In *Complex Organismal Functions: Integration and Evolution in Vertebrates,* ed. D. B. Wake and G. Roth, 295–314. Chichester: John Wiley and Sons.

Emerson, S. B., and M. A. R. Koehl. 1990. The interaction of behavioral and morphological change in the evolution of a novel locomotor type: "Flying" frogs. *Evolution* 44:1931–1946.

Endler, J. A. 1986. *Natural Selection in the Wild.* Princeton: Princeton University Press.

Faber, D. B., and J. R. Baylis. 1993. Body size effects on agonistic encounters between male jumping spiders (Araneae: Salticidae). *Anim. Behav.* 45:289–299.

Falconer, D. S. 1989. *Introduction to Quantitative Genetics,* 3d ed. London: Longman.

Feder, M. E. 1987. The analysis of physiological diversity: The prospects for pattern documentation and general questions in ecological physiology. In *New Directions in Ecological Physiology,* ed. M. E. Feder, A. F. Bennett, W. W. Burggren, and R. B. Huey, 38–70. New York: Cambridge University Press.

Feder, M. E., and P. L. Londos. 1984. Hydric constraints upon foraging in a terrestrial salamander, *Desmognathus ochrophaeus* (Amphibia: Plethodontidae). *Oceologia* 64:413–418.

Feder, M. E., A. F. Bennett, W. W. Burggren, and R. B. Huey, eds. 1987. *New Directions in Ecological Physiology*. New York: Cambridge University Press.

Felsenstein, J. 1985. Phylogenies and the comparative method. *Amer. Nat.* 125:1–15.

Ford, N. B., and G. A. Shuttlesworth. 1986. Effects of variation in food intake on locomotory performance of juvenile garter snakes. *Copeia* 1986:999–1001.

Formanowicz, D. R., Jr., E. D. Brodie, Jr., and P. J. Bradley. 1990. Behavioural compensation for tail loss in the ground skink, *Scincella lateralis. Anim. Behav.* 40:782–784.

Fox, S. F., E. Rose, and R. Myers. 1981. Dominance and the acquisition of superior home ranges in the lizard *Uta stansburiana. Ecology* 62:888–893.

Fox, W. 1948. Effect of temperature on development of scutellation in the garter snake, *Thamnophis elegans atratus. Copeia* 1948:252–262.

Fox, W., C. Gordon, and M. H. Fox. 1961. Morphological effects of low temperatures during the embryonic development of the garter snake, *Thamnophis elegans. Zoologica* 46:57–71.

Friedman, W. P., T. Garland, Jr., and M. R. Dohm. 1992. Individual variation in locomotor behavior and maximal oxygen consumption in mice. *Physiol. Behav.* 52:97–104.

Fuentes, E. R., and F. M. Jaksic'. 1980. Ecological species replacement of *Liolaemus* lizards along a habitat gradient. *Oecologia* 46:45–48.

Full, R. J. 1991. The concepts of efficiency and economy in land locomotion. In *Efficiency and Economy in Animal Physiology,* ed. R. W. Blake, 97–131. Cambridge: Cambridge University Press.

Full, R. J., D. A. Zuccarello, and A. Tullis. 1990. Effect of variation in form on the cost of terrestrial locomotion. *J. Exp. Biol.* 150:233–246.

Futuyma, D. J. 1986. *Evolutionary Biology.* 2d ed. Sunderland, MA: Sinauer Associates.

Gans, C. 1975. Tetrapod limblessness: Evolution and functional corollaries. *Amer. Zool.* 15:455–467.

Gans, C. 1977. Locomotor responses of *Calotes* to water (Agamidae: Sauria). *J. Bombay Nat. Hist. Soc.* 74:361–363.

Gans, C. 1979. Momentarily excessive construction as the basis for protoadaptation. *Evolution* 33:227–233.

Gans, C. 1988. Adaptation and the form-function relation. *Amer. Zool.* 28:681–697.

Gans, C. 1991. Efficiency, effectiveness, perfection, optimization: Their use in understanding vertebrate evolution. In *Efficiency and Economy in Animal Physiology,* ed. R. W. Blake, 1–11. Cambridge: Cambridge University Press.

Gans, C., H. C. Dessauer, and D. Baic. 1978. Axial differences in the musculature of uropeltid snakes: The freight-train approach to burrowing. *Science* 199:189–192.

Garland, T., Jr. 1982. Scaling maximal running speed and maximal aerobic speed to body mass in mammals and lizards. *Physiologist* 25:338.

Garland, T. Jr. 1983. Scaling the ecological cost of transport to body mass in terrestrial mammals. *Amer. Nat.* 121:571–587.

Garland, T., Jr. 1984. Physiological correlates of locomotory performance in a lizard: An allometric approach. *Amer. J. Physiol.* 247 (Regulatory Integrative and Comparative Physiology 16):R806–R815.

Garland, T., Jr. 1985. Ontogenetic and individual variation in size, shape and speed in the Australian agamid lizard *Amphibolurus nuchalis. J. Zool.* (London) A207:425–439.

Garland, T., Jr. 1988. Genetic basis of activity metabolism. I. Inheritance of speed, stamina, and antipredator displays in the garter snake *Thamnophis sirtalis. Evolution* 42:335–350.

Garland, T., Jr. 1993. Locomotor performance and activity metabolism of *Cnemidophorus tigris* in relation to natural behaviors. In *Biology of Whiptail Lizards (Genus* Cnemidophorus*),* ed. J. W. Wright and L. J. Vitt, 163–210. Norman: Oklahoma Museum of Natural History.

Garland, T., Jr. 1994a. Quantitative genetics of locomotor behavior and physiology in a garter snake. In *Quantitative Genetic Studies of Behavioral Evolution,* ed. C. R. B. Boake. Chicago: University of Chicago Press. In press.

Garland, T., Jr. 1994b. Phylogenetic analyses of lizard endurance capacity in relation to body size and body temperature. In *Lizard Ecology: Historical and Experimental Perspectives,* ed. L. J. Vitt and E. R. Pianka. Princeton: Princeton University Press. In press.

Garland, T., Jr., and S. C. Adolph. 1991. Physiological differentiation of vertebrate populations. *Ann. Rev. Ecol. Syst.* 22:193–228.

Garland, T., Jr., and S. J. Arnold. 1983. Effects of a full stomach on locomotory performance of juvenile garter snakes (*Thamnophis elegans*). *Copeia* 1983:1092–1096.

Garland, T., Jr., and A. F. Bennett. 1990. Quantitative genetics of maximal oxygen consumption in a garter snake. *Amer. J. Physiol.* 259 (Regulatory Integrative and Comparative Physiology 28):R986–R992.

Garland, T., Jr., and P. A. Carter. 1994. Evolutionary physiology. *Ann. Rev. Physiol.* 56:579–621.

Garland, T., Jr., and P. L. Else. 1987. Seasonal, sexual, and individual variation in endurance and activity metabolism in lizards. *Amer. J. Physiol.* 252 (Regulatory Integrative and Comparative Physiology 21):R439–R449.

Garland, T., Jr., and R. B. Huey. 1987. Testing symmorphosis: Does structure match functional requirements? *Evolution* 41:1404–1409.

Garland, T., Jr., and C. M. Janis. 1993. Does metatarsal/femur ratio predict maximal running speed in cursorial mammals? *J. Zool.* (London) 229:133–151.

Garland, T., Jr., P. L. Else, A. J. Hulbert, and P. Tap. 1987. Effects of endurance training and captivity on activity metabolism of lizards. *Amer. J. Physiol.* 252 (Regulatory Integrative and Comparative Physiology 21):R450–R456.

Garland, T., Jr., F. Geiser, and R. V. Baudinette. 1988. Comparative locomotor performance of marsupial and placental mammals. *J. Zool.* (London) 215:505–522.

Garland, T., Jr., E. Hankins, and R. B. Huey. 1990a. Locomotor capacity and social dominance in male lizards. *Func. Ecol.* 4:243–250.

Garland, T., Jr., A. F. Bennett, and C. B. Daniels. 1990b. Heritability of locomotor performance and its correlates in a natural population. *Experientia* 46:530–533.

Garland, T., Jr., R. B. Huey, and A. F. Bennett. 1991. Phylogeny and thermal physiology in lizards: A reanalysis. *Evolution.* 45:1969–1975.

Garland, T., Jr., P. H. Harvey, and A. R. Ives. 1992. Procedures for the analysis of comparative data using phylogenetically independent contrasts. *Syst. Biol.* 41:18–32.

Garland, T., Jr., A. W. Dickerman, C. M. Janis, and J. A. Jones. 1993. Phylogenetic analysis of covariance by computer simulation. *Syst. Biol.* 42:265–292.

Gatten, R. E., Jr. 1985. The uses of anaerobiosis by amphibians and reptiles. *Amer. Zool.* 25:945–954.

Gatten, R. E., Jr., and R. M. Clark. 1989. Locomotor performance of hydrated and dehydrated frogs: Recovery following exhaustive exercise. *Copeia* 1989:451–455.

Gatten, R. E., Jr., A. C. Echternacht, and M. A. Wilson. 1988. Acclimatization versus acclimation of activity metabolism in a lizard. *Physiol. Zool.* 61:322–329.

Gatten, R. E., Jr., J. D. Congdon, F. J. Mazzotti, and R. U. Fischer. 1991. Glycolysis and swimming performance in juvenile American alligators. *J. Herpetol.* 25:406–411.

Gatten, R. E., Jr., K. Miller, and R. Full. 1992. Energetics of amphibians at rest and during locomotion. In *Environmental Physiology of the Amphibia,* ed. M. E. Feder and W. W. Burggren, 314–377. Chicago: University of Chicago Press.

Gleeson, T. T. 1979a. Foraging and transport costs in the Galapagos marine iguana, *Amblyrhynchus cristatus. Physiol. Zool.* 52:549–557.

Gleeson, T. T. 1979b. The effects of training and captivity on the metabolic capacity of the lizard *Sceloporus occidentalis. J. Comp. Physiol.* 129:123–128.

Gleeson, T. T. 1991. Patterns of metabolic recovery from exercise in amphibians and reptiles. *J. Exp. Biol.* 160:187–207.

Gleeson, T. T., and A. F. Bennett. 1985. Respiratory and cardiovascular adjustments to exercise in reptiles. In *Circulation, Respiration, and Metabolism: Current Comparative Approaches,* ed. R. Gilles, 23–38. Berlin: Springer-Verlag.

Gleeson, T. T., and P. M. Dalessio. 1989. Lactate and glycogen metabolism in the lizard *Diposaurus dorsalis* following exhaustive exercise. *J. Exp. Biol.* 144:377–393.

Gleeson, T. T., and J. M. Harrison. 1988. Muscle composition and its relation to sprint running in the lizard *Diposaurus dorsalis*. *Amer. J. Physiol.* 255 (Regulatory Integrative and Comparative Physiology 24):R470–R477.

Gorman, G. C., Y. J. Kim, and Ch. E. Taylor. 1977. Genetic variation in irradiated and control populations of *Cnemidophorus tigris* (Sauria, Teiidae) from Mercury, Nevada with a discussion of genetic variability in lizards. *Theor. Appl. Genet.* 49:9–14.

Gould, S. J., and E. S. Vrba. 1982. Exaptation: A missing term in the science of form. *Paleobiology* 8:4–15.

Grand, T. I. 1991. Patterns of muscular growth in the African Bovidae. *Appl. Anim. Behav. Sci.* 29:471–482.

Grant, B. R., and P. R. Grant. 1990. *Evolutionary Dynamics of a Natural Population: The Large Cactus Finch of the Galapagos*. Chicago: University of Chicago Press.

Grant, B. W. 1990. Trade-offs in activity time and physiological performance for thermoregulating desert lizards, *Sceloporus merriami*. *Ecology* 71:2323–2333.

Grant, B. W., and A. E. Dunham. 1988. Thermally imposed time constraints on the activity of the desert lizard *Sceloporus merriami*. *Ecology* 69:167–176.

Grant, P. R. 1986. *Ecology and Evolution of Darwin's Finches*. Princeton: Princeton University Press.

Greene, H. W. 1986. Natural history and evolutionary biology. In *Predator-prey Relationships: Perspectives and Approaches from the Study of Lower Vertebrates*, ed. M. E. Feder and G. V. Lauder, 99–108. Chicago: University of Chicago Press.

Greer, A. 1989. *The Biology and Evolution of Australian Lizards*. Chipping Norton, N.S.W., Australia: Surrey Beatty & Sons.

Guyer, C., and J. M. Savage. 1986. Cladistic relationships among anoles (Sauria: Iguanidae). *Syst. Zool.* 35:509–531.

Guyer, C., and J. M. Savage. 1992. Anole systematics revisited. *Syst. Biol.* 41:89–110.

Hahn, W. E., and D. W. Tinkle. 1965. Fatbody cycling and experimental evidence for its adaptive significance to ovarian follicle development in the lizard *Uta stansburiana*. *J. Exp. Zool.* 158:79–86.

Hailey, A., and P. M. C. Davies. 1986. Effects of size, sex, temperature and condition on activity metabolism and defence behaviour of the viperine snake, *Natrix maura*. *J. Zool.* (London) A208:541–558.

Halliday, T. R. 1987. Physiological constraints on sexual selection. In *Sexual Selection: Testing the Alternatives*, ed. J. W. Bradbury and M. B. Andersson, 247–264. New York: John Wiley and Sons.

Hanken, J., and M. H. Wake. 1991. Introduction to the symposium: Experimental approaches to the analysis of form and function. *Amer. Zool.* 31:603–604.

Harvey, P. H., and M. D. Pagel. 1991. *The Comparative Method in Evolutionary Biology*. Oxford: Oxford University Press.

Hedrick, P. W. 1986. Genetic polymorphism in heterogenous environments: A decade later. *Ann. Rev. Ecol. Syst.* 17:535–566.

Heinrich, B., and G. A. Bartholomew. 1979. Roles of endothermy and size in inter- and intraspecific competition for elephant dung in an African dung beetle, *Scarabaeus laevistriatus*. *Physiol. Zool.* 52:484–496.

Henle, K. 1990. Population ecology and life history of three terrestrial geckos in arid Australia. *Copeia* 1990:759–781.

Hertz, P. E., R. B. Huey, and E. Nevo. 1982. Fight versus flight: Body temperature influences defensive responses of lizards. *Anim. Behav.* 30:676–679.

Hertz, P. E., R. B. Huey, and E. Nevo. 1983. Homage to Santa Anita: Thermal sensitivity of sprint speed in agamid lizards. *Evolution* 37:1075–1084.

Hertz, P. E., R. B. Huey, and T. Garland, Jr. 1988. Time budgets, thermoregulation, and max-imal locomotor performance: Are ectotherms Olympians or Boy Scouts? *Amer. Zool.* 28:927–938.

Herzog, H. H., Jr., and B. D. Bailey. 1987. Development of antipredator responses in snakes: II. Effects of recent feeding on defensive behaviors of juvenile garter snakes (*Thamnophis sirtalis*). *J. Comp. Psychol.* 101:387–389.

Hews, D. K. 1990. Examining hypotheses generated by field measures of sexual selection on male lizards, *Uta palmeri. Evolution* 44:1956–1966.

Heyer, W. R., and S. Pongsapipatana. 1970. Gliding speeds of *Ptychozoon lionatum* (Reptilia: Gek-konidae) and *Chrysopelea ornata* (Reptilia: Colubridae). *Herpetologica* 26:317–319.

Hildebrand, M., D. M. Bramble, K. F. Liem, and D. B. Wake. 1985. *Functional Vertebrate Mor-phology.* Cambridge: Belknap Press.

Hill, W. G., and A. Caballero. 1992. Artificial selection experiments. *Ann. Rev. Ecol. Syst.* 23:287–310.

Hill, W. G., and T. F. C. Mackay, eds. 1989. *Evolution and Animal Breeding: Reviews on Molecular and Quantitative Approaches in Honour of Alan Robertson.* Oxon, U.K.: Wallingford.

Hillman, S. S., and P. C. Withers. 1979. An analysis of respiratory surface area as a limit to activity metabolism in anurans. *Can. J. Zool.* 57:2100–2105.

Hochachka, P. W., and G. N. Somero. 1984. *Biochemical Adaptation.* Princeton: Princeton Univer-sity Press.

Huey, R. B. 1982. Phylogenetic and ontogenetic determinants of sprint speed performance in some diurnal Kalahari lizards. *Koedoe* 25:43–48.

Huey, R. B. 1983. Natural variation in body temperature and physiological performance in a lizard (*Anolis cristatellus*). In *Advances in Herpetology and Evolutionary Biology: Essays in Honor of Ernest E. Williams,* ed. A. G. J. Rhodin and K. Miyata, 484–490. Cambridge: Museum of Com-parative Zoology, Harvard University.

Huey, R. B. 1991. Physiological consequences of habit selection. *Amer. Nat.* (Supplement) 137:S91–S115.

Huey, R. B., and A. F. Bennett. 1987. Phylogenetic studies of coadaptation: Preferred temperatures versus optimal performance temperatures of lizards. *Evolution* 41:1098–1115.

Huey, R. B., and A. E. Dunham. 1987. Repeatability of locomotor performance in natural popula-tions of the lizard *Sceloporus merriami. Evolution* 41:1116–1120.

Huey, R. B., and P. E. Hertz. 1982. Effects of body size and slope on sprint speed of a lizard (*Stellio* (*Agama*) *stellio*). *J. Exp. Biol.* 97:401–409.

Huey, R. B., and P. E. Hertz. 1984a. Effects of body size and slope on acceleration of a lizard (*Stellio stellio*). *J. Exp. Biol.* 110:113–123.

Huey, R. B., and P. E. Hertz. 1984b. Is a jack-of-all-temperatures a master of none? *Evolution* 38:441–444.

Huey, R. B., and E. R. Pianka. 1981. Ecological consequences of foraging mode. *Ecology* 62:991–999.

Huey, R. B., and R. D. Stevenson. 1979. Integrating thermal physiology and ecology of ectotherms: A discussion of approaches. *Amer. Zool.* 19:357–366.

Huey, R. B., A. F. Bennett, H. B. John-Alder, and K. A. Nagy. 1984. Locomotor capacity and forag-ing behaviour of Kalahari lacertid lizards. *Anim. Behav.* 32:41–50.

Huey, R. B., A. E. Dunham, K. L. Overall, and R. A. Newman. 1990. Variation in locomotor perfor-mance in demographically known populations of the lizard *Sceloporus merriami. Physiol. Zool.* 63:845–872.

Huey, R. B., L. Partridge, and K. Fowler. 1991. Thermal sensitivity of *Drosophila melanogaster* responds rapidly to laboratory natural selection. *Evolution* 45:751–756.

Jablonski, D. 1986. Background and mass extinctions: The alternation of macroevolutionary re-gimes. *Science* 231:129–133.

Jackson, J. F. 1973a. The phenetics and ecology of a narrow hybrid zone. *Evolution* 27:58–68.

Jackson, J. F. 1973b. Distribution and population phenetics of the Florida scrub lizard, *Sceloporus woodi*. *Copeia* 1973:746–761.

Jackson, J. F., W. F. Ingram III, and H. W. Campbell. 1976. The dorsal pigmentation pattern of snakes as an antipredator strategy: A multivariate approach. *Amer. Nat.* 110:1029–1053.

Jackson, R. R., and S. E. A. Hallas. 1986. Capture efficiencies of web-building jumping spiders (Araneae, Salticidae): Is the jack-of-all-trades the master of none? *J. Zool.* (London) A209:1–7.

Jaksic', F. M., and H. Nuñez. 1979. Escaping behavior and morphological correlates in two *Liolaemus* species of Central Chile (Lacertilia: Iguanidae). *Oecologia* 42:119–122.

Jaksic', F. M., H. Nuñez, and F. P. Ojeda. 1980. Body proportions, microhabitat selection, and adaptive radiation of *Liolaemus* lizards in Central Chile. *Oecologia* 45:178–181.

James, C. D. 1989. Comparative ecology of sympatric scincid lizards (*Ctenotus*) in spinifex grasslands of Central Australia. Ph.D. diss., University of Sydney, Australia.

James, F. C. 1991. Complementary description and experimental studies of clinal variation in birds. *Amer. Zool.* 31:694–705.

James, F. C., R. F. Johnston, N. O. Wamer, G. J. Niemi, and W. J. Boecklen. 1984. The Grinnellian niche of the wood thrush. *Amer. Nat.* 124:17–30.

Janis, C. M. In press. Do legs support the arms race hypothesis in mammalian predator/prey relationships? In *Vertebrate Behaviour as Derived from the Fossil Record*, ed. J. R. Horner and K. Carpenter. New York: Columbia University Press.

Janzen, F. J. 1993. An experimental analysis of natural selection on body size of hatchling turtles. *Ecology* 74:332–341.

Jayne, B. C. 1985. Swimming in constricting (*Elaphe g. guttata*) and nonconstricting (*Nerodia fasciata pictiventris*) colubrid snakes. *Copeia* 1985:195–208.

Jayne, B. C. 1986. Kinematics of terrestrial snake locomotion. *Copeia* 1986:915–927.

Jayne, B. C. 1988a. Muscular mechanisms of snake locomotion: An electromyographic study of the sidewinding and concertina modes of *Crotalus cerastes*, *Nerodia fasciata* and *Elapha obsoleta*. *J. Exp. Biol.* 140:1–33.

Jayne, B. C. 1988b. Muscular mechanisms of snake locomotion: An electromyographic study of lateral undulation of the Florida banded water snake (*Nerodia fasciata*) and the yellow rat snake (*Elaphe obsoleta*). *J. Morphol.* 197:159–181.

Jayne, B. C., and A. F. Bennett. 1989. The effect of tail morphology on locomotor performance of snakes: A comparison of experimental and correlative methods. *J. Exp. Zool.* 252:126–133.

Jayne, B. C., and A. F. Bennett. 1990a. Scaling of speed and endurance in garter snakes: A comparison of cross-sectional and longitudinal allometries. *J. Zool.* (London) 220:257–277.

Jayne, B. C., and A. F. Bennett. 1990b. Selection on locomotor performance capacity in a natural population of garter snakes. *Evolution* 44:1204–1229.

Jayne, B. C., and J. D. Davis. 1991. Kinematics and performance capacity for the concertina locomotion of a snake (*Coluber constrictor*). *J. Exp. Biol.* 156:539–556.

John-Alder, H. B. 1984a. Reduced aerobic capacity and locomotory endurance in thyroid-deficient lizards. *J. Exp. Biol.* 109:175–189.

John-Alder, H. B. 1984b. Seasonal variations in activity, aerobic energetic capacities, and plasma thyroid hormones (T3 and T4) in an iguanid lizard. *J. Comp. Physiol.* B154:409–419.

John-Alder, H. B. 1990. Effects of thyroxine on standard metabolic rate and selected intermediary metabolic enzymes in field-active lizards *Sceloporus undulatus*. *Physiol. Zool.* 63:600–614.

John-Alder, H. B., and A. F. Bennett. 1981. Thermal dependence of endurance and locomotory energetics in a lizard. *Amer. J. Physiol.* 241:(Regulatory Integrative and Comparative Physiology 10):R342–R349.

John-Alder, H. B., C. H. Lowe, and A. F. Bennett. 1983. Thermal dependence of locomotory energetics and aerobic capacity of the Gila monster (*Heloderma suspectum*). *J. Comp. Physiol.* 151:119–126.

John-Alder, H. B., T. Garland, Jr., and A. F. Bennett. 1986a. Locomotory capacities, oxygen consumption, and the cost of locomotion of the shingle-back lizard (*Trachydosaurus rugosus*). *Physiol. Zool.* 59:523–531.

John-Alder, H. B., R. M. McAllister, and R. L. Terjung. 1986b. Reduced running endurance in gluconeogenesis-inhibited rats. *Amer. J. Physiol.* 251 (Regulatory Integrative and Comparative Physiology 20):R137–R142.

Joos, B., and H. B. John-Alder. 1990. Effects of thyroxine on standard and total metabolic rates in the lizard *Sceloporus undulatus. Physiol. Zool.* 63:873–885.

Joreskog, K. G., and D. Sorbom. 1978. LISREL IV: Analysis of linear structural relationships by the method of maximum likelihood. Chicago: International Educational Services.

Kamel, S., and R. E. Gatten, Jr. 1983. Aerobic and anaerobic activity metabolism of limbless and fossorial reptiles. *Physiol. Zool.* 56:419–429.

Karasov, W. H., and R. A. Anderson. 1984. Interhabitat differences in energy acquisition and expenditure in a lizard. *Ecology* 65:235–247.

Kaufmann, J. S., and A. F. Bennett. 1989. The effect of temperature and thermal acclimation on locomotor performance in *Xantusia vigilis*, the desert night lizard. *Physiol. Zool.* 62:1047–1058.

King, R. B. 1992. Lake Erie water snakes revisited: Morph- and age-specific variation in relative crypsis. *Evol. Ecol.* 6:115–124.

Kirkpatrick, B. W., and J. J. Rutledge. 1988. Influence of prenatal and postnatal fraternity size on reproduction in swine. *J. Anim. Sci.* 66:2530–2537.

Kramer, G. 1951. Body proportions of mainland and island lizards. *Evolution* 5:193–206.

LaBarbera, M. 1989. Analyzing body size as a factor in ecology and evolution. *Ann. Rev. Ecol. Syst.* 20:97–117.

Laerm, J. 1973. Aquatic bipedalism in the basilisk lizard: The analysis of an adaptive strategy. *Amer. Midland Nat.* 89:314–333.

Laerm, J. 1974. A functional analysis of morphological variation and differential niche utilization in basilisk lizards. *Ecology* 55:404–411.

Lamb, T., J. C. Avise, and J. W. Gibbons. 1989. Phylogeographic patterns in mitochondrial DNA of the desert tortoise (*Xerobates agassizi*), and evolutionary relationships among the North American gopher tortoises. *Evolution* 43:76–87.

Lande, R., and S. J. Arnold. 1983. The measurement of selection on correlated characters. *Evolution* 37:1210–1226.

Larson, A. 1984. Neontological inferences of evolutionary pattern and process in the salamander family Plethodontidae. *Evol. Biol.* 17:119–217.

Lauder, G. V. 1990. Functional morphology and systematics: Studying functional patterns in an historical context. *Ann. Rev. Ecol. Syst.* 21:317–340.

Lauder, G. V. 1991. An evolutionary perspective on the concept of efficiency: How does function evolve? In *Efficiency and Economy in Animal Physiology*, ed. R. W. Blake, 169–184. Cambridge: Cambridge University Press.

Lauder, G. V., and S. M. Reilly. 1988. Functional design of the feeding mechanism in salamanders: Causal bases of ontogenetic changes in function. *J. Exp. Biol.* 134:219–233.

Losos, J. B. 1990a. Concordant evolution of locomotor behaviour, display rate, and morphology in *Anolis* lizards. *Anim. Behav.* 39:879–890.

Losos, J. B. 1990b. Ecomorphology, performance capability, and scaling of West Indian *Anolis* lizards: An evolutionary analysis. *Ecol. Mon.* 60:369–388.

Losos, J. B. 1990c. The evolution of form and function: Morphology and locomotor performance in West Indian *Anolis* lizards. *Evolution* 44:1189–1203.

Losos, J. B. 1992. The evolution of convergent structure in Caribbean *Anolis* communities. *Syst. Biol.* 41:403–420.

Losos, J. B., and B. Sinervo. 1989. The effects of morphology and perch diameter on sprint performance of *Anolis* lizards. *J. Exp. Biol.* 145:23–30.

Losos, J. B., T. J. Papenfuss, and J. R. Macey. 1989. Correlates of sprinting, jumping, and parachuting performance in the butterfly lizard, *Leiolepis belliani. J. Zool.* (London) 217:559–568.

Losos, J. B., R. M. Andrews, O. J. Sexton, and A. L. Schuler. 1991. Behavior, ecology, and locomotor performance of the giant anole, *Anolis frenatus, Carib. J. Sci.* 27:173–179.

Losos, J. B., B. M. Walton, and A. F. Bennett. 1993. Trade-offs between sprinting and clinging ability in Kenyan chameleons. *Funct. Ecol.* 7:281–286.

Loumbourdis. N. S., and A. Hailey. 1985. Activity metabolism of the lizard *Agama stellio stellio. Comp. Biochem. Physiol.* 82A:687–691.

Luke, C. 1986. Convergent evolution of lizard toe fringes. *Biol. J. Linnaen Soc.* 27:1–16.

Lynch, M. 1991. Methods for the analysis of comparative data in evolutionary ecology. *Evolution* 45:1065–1080.

MacArthur, R. A. 1992. Foraging range and aerobic endurance of muskrats diving under ice. *J. Mammal.* 73:565–569.

Magnusson, W. E., L. J. de Paiva, R. M. da Rocha, C. R. Franke, L. A. Kasper, and A. P. Lima. 1985. The correlates of foraging mode in a community of Brazilian lizards. *Herpetologica* 41:324–332.

Marcellini, D. L., and T. E. Keefer. 1976. Analysis of the gliding behavior of *Ptychozoon lionatum* (Reptilia: Gekkonidae). *Herpetologica* 32:362–366.

Marsh, R. L. 1988. Ontogenesis of contractile properties of skeletal muscle and sprint performance in the lizard *Dipsosaurus dorsalis. J. Exp. Biol.* 137:119–139.

Marsh, R. L., and A. F. Bennett. 1985. Thermal dependence of isotonic contractile properties of skeletal muscle and sprint performance of the lizard *Dipsosaurus dorsalis. J. Comp. Physiol.* B155:541–551.

Marsh, R. L., and A. F. Bennett. 1986. Thermal dependence of sprint performance of the lizard *Sceloporus occidentalis. J. Exp. Biol.* 126:79–87.

Martins, E. P., and T. Garland, Jr. 1991. Phylogenetic analyses of the correlated evolution of continuous traits: A simulation study. *Evolution* 45:534–557.

Mautz, W. J., C. B. Daniels, and A. F. Bennett. 1992. Thermal dependence of locomotion and aggression in a xantusiid lizard. *Herpetologica* 48:471–479.

Mayer, G. C. 1989. Deterministic patterns of community structure in West Indian reptiles and amphibians. Ph.D. diss., Harvard University.

Maynard Smith, J., R. Burian, S. Kauffman, P. Alberch, J. Campbell, B. Goodwin, R. Lande, D. Raup, and L. Wolpert. 1985. Developmental constraints and evolution. *Q. Rev. Biol.* 60:265–287.

McLaughlin, R. L. 1989. Search modes of birds and lizards: Evidence for alternative movement patterns. *Amer. Nat.* 133:654–670.

Mertens, R. 1960. Gliding and parachuting flight among the amphibians and reptiles (translated by P. Gritis.) *Bull. Chicago Herpetol. Soc.* 21:42–46, 1985.

Michod, R. E. 1986. On fitness and adaptedness and their role in evolutionary explanation. *J. Hist. Biol.* 19:289–302.

Miles, D. B. 1987. Habitat related differences in locomotion and morphology in two populations of *Urosaurus ornatus. Amer. Zool.* 27:44A.

Miles, D. B. 1989. Selective significance of locomotory performance in an iguanid lizard. *Amer. Zool.* 29:146A.

Miles, D. B., and D. Althoff. 1990. Effects of differing substrates on sprint performance in sceloporine lizards. *Amer. Zool.* 30:18A.

Miles, D. B., and A. E. Dunham. 1992. Comparative analyses of phylogenetic effects in the life-history patterns of iguanid reptiles. *Amer. Nat.* 139:848–869.

Miller, K., G. C. Packard, and M. J. Packard. 1987. Hydric conditions during incubation influence locomotor performance of hatchling snapping turtles. *J. Exp. Biol.* 127:401–412.

Mitchell-Olds, T., and R. G. Shaw. 1987. Regression analysis of natural selection: Statistical and biological interpretation. *Evolution* 41:1149–1161.

Moermond, T. C. 1979a. The influence of habitat structure on *Anolis* foraging behavior. *Behaviour* 70:147–167.

Moermond, T. C. 1979b. Habitat constraints on the behavior, morphology, and community structure of *Anolis* lizards. *Ecology* 60:152–164.

Moermond, T. C. 1986. A mechanistic approach to the structure of animal communities: *Anolis* lizards and birds. *Amer. Zool.* 26:23–37.

Moore, F. R., and R. E. Gatten, Jr. 1989. Locomotor performance of hydrated, dehydrated, and osmotically stressed anuran amphibians. *Herpetologica* 45:101–110.

Moore, M. C., and C. A. Marler. 1987. Effects of testosterone manipulations on nonbreeding season territorial aggression in free-living male lizards, *Sceloporus jarrovi. Gen. Comp. Endocrinol.* 65:225–232.

Moore, M. C., and C. W. Thompson. 1990. Field endocrinology of reptiles: Hormonal control of alternative male reproductive tactics. In *Progress in Comparative Endocrinology,* ed. A. Epple, C. G. Scanes, and M. H. Stetson, 685–690. New York: Wiley-Liss.

Morey, S. R. 1990. Microhabitat selection and predation in the Pacific treefrog, *Pseudacris regilla. J. Herpetol.* 24:292–296.

Morgan, K. R. 1988. Body temperature, energy metabolism, and stamina in two neotropical forest lizards (*Ameiva,* Teiidae). *J. Herpetol.* 22:236–241.

Muth, A. 1977. Thermoregulatory postures and orientation to the sun: A mechanistic evaluation for the Zebra-tailed lizard, *Callisaurus draconoides. Copeia* 1977:710–720.

Nagy, K. A., R. B. Huey, and A. F. Bennett. 1984. Field energetics and foraging mode of Kalahari lacertid lizards. *Ecology* 65:588–596.

Nee, S., A. O. Mooers, and P. H. Harvey. 1992. Tempo and mode of evolution revealed from molecular phylogenies. *Proc. Natl. Acad. Sci. USA* 89:8322–8326.

Nie, N. H., C. H. Hull, J. G. Jenkins, K. Steinbrenner, and D. H. Bent. 1975. *Statistical Package for the Social Sciences,* 2d ed. New York: McGraw-Hill.

Norris, K. S. 1951. The lizard that swims in the sand. *Nat. Hist.* 60:404–407.

O'Brien, W. J., H. I. Browman, and B. I. Evans. 1990. Search strategies of foraging animals. *Amer. Sci.* 78:152–160.

Oliver, J. A. 1951. "Gliding" in amphibians and reptiles, with a remark on an arboreal adaptation on the lizard, *Anolis carolinensis carolinensis* Voigt. American Naturalist 85:171–176.

Oliver, J. A. 1955. *The natural history of North American amphibians and reptiles.* Princeton, D. Van Nostrand.

Osgood, D. W. 1978. Effects of temperature on the development of meristic characters in *Natrix fasciata. Copeia* 1978:33–47.

Packard, G. C., and T. J. Boardman. 1988. The misuse of ratios, indices, and percentages in ecophysiological research. *Physiol. Zool.* 61:1–9.

Pagel, M. 1993. Seeking the evolutionary regression coefficient: An analysis of what comparative methods measure. *J. Theor. Biol.* 164:191–205.

Pagel, M. D., and P. H. Harvey. 1988. The taxon-level problem in the evolution of mammalian brain size: Facts and artifacts. *Amer. Nat.* 132:344–359.

Payne, J. C., and R. E. Gatten, Jr. 1988. Thermal acclimation of activity metabolism in desert lizards (*Urosaurus graciosus* and *U. ornatus*). *J. Thermal Biol.* 13:37–42.

Pennycuick, C. J. 1991. Adapting skeletal muscle to be efficient. In *Efficiency and Economy in Animal Physiology,* ed. R. W. Blake, 33–42. Cambridge: Cambridge University Press.

Perry, G., I. Lampl, A. Lerner, D. Rothenstein, E. Shani, N. Sivan, and Y. L. Werner. 1990. Foraging mode in lacertid lizards: Variation and correlates. *Amphibia-Reptilia* 11:373–384.

Peterson, J. A. 1984. The locomotion of *Chamaeleo* (Reptilia: Sauria) with particular reference to the forelimb. *J. Zool.* (London) 202:1–42.

Pianka, E. R. 1966. Convexity, desert lizards, and spatial heterogeneity. *Ecology* 47:1055–1059.

Pianka, E. R. 1969. Sympatry of desert lizards (*Ctenotus*) in Western Australia. *Ecology* 50:1012–1030.

Pianka, E. R. 1986. *Ecology and Natural History of Desert Lizards.* Princeton: Princeton University Press.

Pianka, E. R., and H. D. Pianka. 1976. Comparative ecology of twelve species of nocturnal lizards (Gekkonidae) in the Western Australian desert. *Copeia* 1976:125–142.

Pietruszka, R. D. 1986. Search tactics of desert lizards: How polarized are they? *Anim. Behav.* 34:1742–1758.

Pond, C. M. 1981. Storage. In *Physiological Ecology: An Evolutionary Approach to Resource Use,* ed. C. R. Townsend and P. Calow, 190–219. Oxford: Blackwell Scientific.

Pough, F. H. 1976. Multiple cryptic effects of crossbanded and ringed patterns of snakes. *Copeia* 1976:834–836.

Pough, F. H. 1977. Ontogenetic change in blood oxygen capacity and maximum activity in garter snakes (*Thamnophis sirtalis*). *J. Comp. Physiol.* 116:337–345.

Pough, F. H. 1978. Ontogenetic changes in endurance in water snakes (*Natrix sipedon*): Physiological correlates and ecological consequences. *Copeia* 1978:69–75.

Pough, F. H. 1989. Organismal performance and Darwinian fitness: Approaches and interpretations. *Physiol. Zool.* 62:199–236.

Pough, F. H., and R. M. Andrews. 1985a. Use of anaerobic metabolism by free-ranging lizards. *Physiol. Zool.* 58:205–213.

Pough, F. H., and R. M. Andrews. 1985b. Energy cost of subduing and swallowing prey for a lizard. *Ecology* 66:1525–1533.

Pough, F. H., and S. Kamel. 1984. Post-metamorphic change in activity metabolism of anurans. *Oecologia* 65:138–144.

Pough, F. H., T. L. Taigen, M. M. Stewart, and P. F. Brussard. 1983. Behavioral modification of evaporative water loss by a Puerto Rican frog. *Ecology* 64:244–252.

Pough, F. H., with W. E. Magnusson, M. J. Ryan, K. D. Wells, and T. L. Taigen. 1992. Behavioral energetics. In *Environmental Physiology of the Amphibia*, ed. M. E. Feder and W. W. Burggren, 395–436. Chicago: University of Chicago Press.

Pounds, J. A. 1988. Ecomorphology, locomotion, and microhabitat structure: Patterns in a tropical mainland *Anolis* community. *Ecol. Mon.* 58:299–320.

Pounds, J. A., J. F. Jackson, and S. H. Shively. 1983. Allometric growth of the hind limbs of some terrestrial iguanid lizards. *Amer. Midland Nat.* 110:201–207.

Preest, M. R., and F. H. Pough. 1987. Interactive effects of body temperature and hydration state on locomotor performance of a toad, *Bufo americanus*. *Amer. Zool.* 27:5A.

Punzo, F. 1982. Tail autotomy and running speed in the lizards *Cophosaurus texanus* and *Uma notata*. *J. Herpetol.* 16:329–331.

Purvis, A., and T. Garland, Jr. 1993. Polytomies in comparative analyses of continuous characters. *Sys. Biol.* 42:569–575.

Putnam, R. W., and S. S. Hillman. 1977. Activity responses of anurans to dehydration. *Copeia* 1977:746–749.

Rand, A. S. 1964. Ecological distribution in anoline lizards of Puerto Rico. *Ecology* 45:745–752.

Rand, A. S. 1967. The ecological distribution of the anoline lizards around Kingston, Jamaica. *Breviora* 272:1–18.

Rand, A. S., and E. E. Williams. 1969. The anoles of La Palma: Aspects of their ecological relationships. *Breviora* 327:1–19.

Rayner, J. M. V. 1981. Flight adaptations in vertebrates. *Symp. Zool. Soc. Lond.* 48:137–172.

Rayner, J. M. V. 1987. Form and function in avian flight. *Curr. Ornithol.* 5:1–66.

Regal, P. J. 1983. The adaptive zone and behavior of lizards. In *Lizard Ecology: Studies of a Model Organism,* ed. R. B. Huey, E. R. Pianka, and T. W. Schoener, 105–118. Cambridge: Harvard University Press.

Ricqles, A., de. 1980. The gliding reptiles of the Paleozoic. *La Recherche* 107:75–77.

Rieser, G. D. 1977. A functional analysis of bipedalism in lizards. Ph.D. diss., University of California, Davis.

Riska, B. 1991. Regression models in evolutionary allometry. *Amer. Nat.* 138:283–299.

Riska, B., T. Prout, and M. Turelli. 1989. Laboratory estimates of heritabilities and genetic correlations in nature. *Genetics* 123:865–871.

Robinson, M. D., and A. B. Cunningham. 1978. Comparative diet of two Namib Desert sand lizards (Lacertidae). *Madoqua* 11:41–53.

Rose, M. R., P. M. Service, and E. W. Hutchinson. 1987. Three approaches to trade-offs in life-history evolution. In *Genetic Constraints on Adaptive Evolution,* ed. V. Loeschcke, 91–105. Berlin: Springer-Verlag.

Ruben, J. A., A. F. Bennett, and F. L. Hisaw. 1987. Selective factors in the origin of the mammalian diaphragm. *Paleobiology* 13:54–59.

Russell, A. P. 1979. The origin of parachuting locomotion in gekkonid lizards (Reptilia: Gekkonidae). *Zool. J. Linnaen Soc.* 65:233–249.

Russell, A. P., and A. M. Bauer. 1992. The *M. caudifemoralis longus* and its relationship to caudal autotomy and locomotion in lizards (Reptilia: Sauria). *J. Zool.* (London) 227:127–143.

Schall, J. J. 1986. Prevalence and virulence of a haemogregarine parasite of the Aruban whiptail lizard, *Cnemidophorus arubensis. J. Herpetol.* 20:318–324.

Schall, J. J. 1990. Virulence of lizard malaria: The evolutionary ecology of an ancient parasite-host association. *Parasitology* 100:535–552.

Schall, J. J., and M. D. Dearing. 1987. Malarial parasitism and male competition for mates in the western fence lizard, *Sceloporus occidentalis. Oecologia* 73:389–392.

Schall, J. J., A. F. Bennett, and R. W. Putnam. 1982. Lizards infected with malaria: Physiological and behavioral consequences. *Science* 217:1057–1059.

Scheibe, J. S. 1987. Climate, competition, and the structure of temperate zone lizard communities. *Ecology* 68:1424–1436.

Schiefflen, C. D., and A. de Queiroz. 1991. Temperature and defense in the common garter snake: Warm snakes are more aggressive than cold snakes. *Herpetologica* 47:230–237.

Schiotze, A., and H. Volsoe. 1959. The gliding flight of *Holapsis guentheri* Gray, a West African lacertid. *Copeia* 1959:259–260.

Schlager, G., and R. S. Weibust. 1976. Selection for hematocrit percent in the house mouse. *J. Hered.* 67:295–299.

Schluter, D. 1988. Estimating the form of natural selection on a quantitative trait. *Evolution* 42:849–861.

Schluter, D. 1989. Bridging population and phylogenetic approaches to the evolution of complex traits. In *Complex Organismal Functions: Integration and Evolution in Vertebrates,* ed. D. B. Wake and G. Roth, 79–95. Chichester: John Wiley and Sons.

Schoener, A., and T. W. Schoener. 1984. Experiments on dispersal: Short-term floatation of insular anoles, with a review of similar abilities in other terrestrial animals. *Oecologia* 63:289–294.

Schoener, T. W. 1971. Theory of feeding strategies. *Ann. Rev. Ecol. Syst.* 2:369–404.

Schoener, T. W., and A. Schoener. 1971a. Structural habitats of West Indian *Anolis* lizards. I. Lowland Jamaica. *Breviora* 368:1–53.

Schoener, T. W., and A. Schoener. 1971b. Structural habitats of West Indian *Anolis* lizards. II. Puerto Rican uplands. *Breviora* 375:1–39.

Schwartz, J. M., G. F. McCracken, and G. M. Burghardt. 1989. Multiple paternity in wild populations of the garter snake, *Thamnophis sirtalis. Behav. Ecol. Sociobiol.* 25:269–273.

Schwartzkopf, L., and R. Shine. 1992. Costs of reproduction in lizards: Escape tactics and susceptibility to predation. *Behav. Ecol. Sociobiol.* 31:17–25.

Secor, S. M., B. C. Jayne, and A. F. Bennett. 1992. Locomotor performance and energetic cost of sidewinding by the snake *Crotalus cerastes. J. Exp. Biol.* 163:1–14.

Seigel, R. A., M. M. Huggins, and N. B. Ford. 1987. Reduction in locomotor ability as a cost of reproduction in gravid snakes. *Oecologia* 73:481–485.

Seymour, R. S. 1982. Physiological adaptations to aquatic life. In *Biology of the Reptilia,* vol. 13, *Physiology D: Physiological Ecology,* ed. C. Gans and F. H. Pough, 1–51. London: Academic Press.

Seymour, R. S. 1989. Diving physiology: Reptiles. In *Comparative Pulmonary Physiology: Current Concepts,* ed. S. C. Wood, 697–720. New York: Marcel Dekker.

Shallenberger, E. W. 1970. Tameness in insular animals: A comparison of approach distances of insular and mainland iguanid lizards. Ph.D. diss., University of California, Los Angeles.

Shelford, R. 1906. A note on "flying" snakes. *Proc. Zool. Soc.* 1:227–230.

Shine, R. 1980. "Costs" of reproduction in reptiles. *Oecologia* 46:92–100.

Shine, R. 1986. Evolutionary advantages of limblessness: Evidence from the pygopodid lizards. *Copeia* 1986:525–529.

Shine, R. 1988. Constraints on reproductive investment: A comparison between aquatic and terrestrial snakes. *Evolution* 42:17–27.

Sinervo, B. 1990. The evolution of maternal investment in lizards: An experimental and comparative analysis of egg size and its effect on offspring performance. *Evolution* 44:279–294.

Sinervo, B., and S. C. Adolph. 1989. Thermal sensitivity of growth rate in hatchling *Sceloporus* lizards: Environmental, behavioral and genetic aspects. *Oecologia* 78:411–419.

Sinervo, B., and R. B. Huey. 1990. Allometric engineering: An experimental test of the causes of interpopulational differences in performance. *Science* 248:1106–1109.

Sinervo, B., and P. Licht. 1991. Proximate constraints on the evolution of egg size, number, and total clutch mass in lizards. *Science* 252:1300–1302.

Sinervo, B., and J. B. Losos. 1991. Walking the tight rope: A comparison of arboreal sprint performance among populations of *Sceloporus occidentalis* lizards. *Ecology* 72:1225–1233.

Sinervo, B., R. Hedges, and S. C. Adolph. 1991. Decreased sprint speed as a cost of reproduction in the lizard *Sceloporus occidentalis:* Variation among populations. *J. Exp. Biol.* 155:323–336.

Sinervo, B., P. Doughty, R. B. Huey, and K. Zamudio. 1992. Allometric engineering: A causal analysis of natural selection on offspring size. *Science* 258:1927–1930.

Slinker, B. K., and S. A. Glantz. 1985. Multiple regression for physiological data analysis: The problem of multicollinearity. *Amer. J. Physiol.* 249 (Regulatory Integrative and Comparative Physiology 18):R1–R12.

Snell, H. L., H. M. Snell, and C. R. Tracy. 1984. Variation among populations of Galapagos land iguanas (*Conolophus*): Contrasts of phylogeny and ecology. *Biol. J. Linnaen Soc.* 21:185–207.

Snell, H. L., R. D. Jennings, H. M. Snell, and S. Harcourt. 1988. Intrapopulation variation in predator-avoidance performance of Galapagos lava lizards: The interaction of sexual and natural selection. *Evol. Ecol.* 2:353–369.

Snyder, G. K., and W. W. Weather. 1977. Activity and oxygen consumption during hypoxic exposure in high altitude and lowland sceloporine lizards. *J. Comp. Physiol.* 117:291–301.

Sokal, R. R., and F. J. Rohlf. 1981. *Biometry,* 2d ed. San Francisco: W. H. Freeman and Co.

Sorci, G., M. Massot, and J. Clobert. In press. Maternal parasite load predicts offspring sprint speed in the philopatric sex. *Amer. Nat.*

Stamps, J. A. 1984. Rank-dependent compromises between growth and predator protection in lizard dominance hierarchies. *Anim. Behav.* 32:1101–1107.

Stefanski, M., R. E. Gatten, and F. H. Pough. 1989. Activity metabolism of salamanders: tolerance to dehydration. *J. Herpetol.* 23:45–50.

Stini, W. A. 1979. Adaptive strategies of human populations under nutritional stress. In *Physiological and Morphological Adaptation and Evolution,* ed. W. A. Stini, 387–407. The Hague: Mouton Publishers.

Strang, K. T., and K. Steudel. 1990. Explaining the scaling of transport costs: The role of stride frequency and stride length. *J. Zool.* (London) 221:343–358.

Taigen, T. L., and F. H. Pough. 1981. Activity metabolism of the toad (*Bufo americanus*): Ecological consequences of ontogenetic change. *J. Comp. Physiol.* 144:247–252.

Tenney, S. M. 1967. Some aspects of the comparative physiology of muscular exercise in mammals. *Circ. Res.* (Supplement I) 20–21:I7–I14.

Thompson, D. B. 1990. Different spatial scales of adaptation in the climbing behavior of *Peromyscus maniculatus:* Geographic variation, natural selection, and gene flow. *Evolution* 44:952–965.

Tokarz, R. R. 1985. Body size as a factor determining dominance in staged agonistic encounters between male brown anoles (*Anolis sagrei*). *Anim. Behav.* 33:746–753.

Townsend, C. R., and P. Calow. 1981. *Physiological Ecology: An Evolutionary Approach to Resource Use*. Sunderland, MA: Sinauer Associates.

Tracy, C. R., and K. A. Christian. 1985. Are marine iguana tails flattened? *Brit. J. Herpetol.* 6:434–435.

Tracy, C. R., and J. Sugar. 1989. Potential misuse of ANCOVA: Comment on Packard and Boardman. *Physiol. Zool.* 62:993–997.

Tsuji, J. S., R. B. Huey, F. H. van Berkum, T. Garland, Jr., and R. G. Shaw. 1989. Locomotor performance of hatchling fence lizards (*Sceloporus occidentalis*): Quantitative genetics and morphometric correlates. *Func. Ecol.* 3:240–252.

Turner, J. S., C. R. Tracy, B. Weigler, and T. Baynes. 1985. Burst swimming of alligators and the effect of temperature. *J. Herpetol.* 19:450–458.

Ultsch, G. R., R. W. Hanley, and T. R. Bauman. 1985. Responses to anoxia during simulated hibernation in northern and southern painted turtles. *Ecology* 66:388–395.

van Berkum, F. H. 1986. Evolutionary patterns of the thermal sensitivity of sprint speed in *Anolis* lizards. *Evolution* 40:594–604.

van Berkum, F. H. 1988. Latitudinal patterns of the thermal sensitivity of sprint speed in lizards. *Amer. Nat.* 132:327–343.

van Berkum, F. H., and J. S. Tsuji. 1987. Inter-familial differences in sprint speed of hatchling *Sceloporus occidentalis*. *J. Zool.* (London) 212:511–519.

van Berkum, F. H., R. B. Huey, and B. A. Adams. 1986. Physiological consequences of thermoregulation in a tropical lizard (*Ameiva festiva*). *Physiol. Zool.* 59:464–472.

van Berkum, F. H., R. B. Huey, J. S. Tsuji, and T. Garland, Jr. 1989. Repeatability of individual differences in locomotor performance and body size during early ontogeny of the lizard *Sceloporus occidentalis* (Baird & Girard). *Func. Ecol.* 3:97–105.

Van Damme, R., D. Bauwens, and R. F. Verheyen. 1989. Effect of relative clutch mass on sprint speed in the lizard *Lacerta vivipara*. *J. Herpetol.* 23:459–461.

Van Damme, R., D. Bauwens, D. Vanderstighelen, and R. F. Verheyen. 1990. Responses of the lizard *Lacerta* vivipara to predator chemical cues: The effects of temperature. *Anim. Behav.* 40:298–305.

Van Damme, R., D. Bauwens, F. Brana, and R. F. Verheyen. 1992. Incubation temperature differentially affects hatching time, egg survival, and hatchling performance in the lizard *Podarcis muralis*. *Herpetologica* 48:220–228.

Vitt, L. J. 1983. Tail loss in lizards: The significance of foraging and predator escape modes. *Herpetologica* 39:151–162.

Vitt, L. J. 1991. An introduction to the ecology of cerrado lizards. *J. Herpetol.* 25:79–89.

Vitt, L. J., and J. D. Congdon. 1978. Body shape, reproductive effort, and relative clutch mass in lizards: Resolution of a paradox. *Amer. Nat.* 112:595–608.

Vitt, L. J., and H. J. Price. 1982. Ecological and evolutionary determinants of relative clutch mass in lizards. *Herpetologica* 38:237–255.

Vleck, D., T. T. Gleeson, and G. A. Bartholomew. 1981. Oxygen consumption during swimming in Galapagos marine iguanas and its ecological correlates. *J. Comp. Physiol.* 141:531–536.

Vrba, E. S. 1989. What are the biotic hierarchies of integration and linkage? In *Complex Organismal Functions: Integration and Evolution in Vertebrates,* ed. D. B. Wake and G. Roth, 379–401. Chichester: John Wiley and Sons.

Vrba, E. S., and S. J. Gould. 1986. The hierarchical expansion of sorting and selection: Sorting and selection cannot be equated. *Paleobiology* 12:217–228.

Wade, M. J., and S. Kalisz. 1990. The causes of natural selection. *Evolution* 44:1947–1955.

Wainwright, P. C. 1991. Ecomorphology: Experimental functional anatomy for ecological problems. *Amer. Zool.* 31:680–693.

Waldschmidt, S., and C. R. Tracy. 1983. Interactions between a lizard and its thermal environment: Implications for sprint performance and space utilization in the lizard *Uta stansburiana*. *Ecology* 64:476–484.

Walls, G. L. 1942. *The Vertebrate Eye and Its Adaptive Radiation.* Cranbrook Institute of Science, Bloomfield Hills. Bulletin no. 19.

Walton, M., B. C. Jayne, and A. F. Bennett. 1990. The energetic cost of limbless locomotion. *Science* 249:524–527.

Ware, D. M. 1982. Power and evolutionary fitness of teleosts. *Can. J. Fish. Aq. Sci.* 39:3–13.

Weatherhead, P. J. 1986. How unusual are unusual events? *Amer. Nat.* 128:150–154.

Webb, P. W. 1976. The effect of size on the fast-start performance of rainbow trout *Salmo gairdneri*, and a consideration of piscivorous predator-prey interactions. *J. Exp. Biol.* 65:157–177.

Webster, D. B., and M. Webster. 1988. Hypotheses derived from morphological data: When and how they are useful. *Amer. Zool.* 28:231–236.

Wiens, J. A. 1977. On competition and variable environments. *Amer. Sci.* 65:590–597.

Wiens, J. A., and J. T. Rotenberry. 1980. Patterns of morphology and ecology in grassland and shrubsteppe bird populations. *Ecol. Mon.* 50:287–308.

Williams, E. E. 1972. The origin of faunas: Evolution of lizard congeners in a complex island fauna—a trial analysis. *Evol. Biol.* 6:47–89.

Williams, E. E. 1983. Ecomorphs, faunas, island size, and diverse end points. In *Lizard Ecology: Studies of a Model Organism,* ed. R. B. Huey, E. R. Pianka, and T. W. Schoener, 326–370. Cambridge: Harvard University Press.

Williams, E. E. 1989. A critique of Guyer and Savage (1986): Cladistic relationships among anoles (Sauria: Iguanidae). Are the data available to reclassify the anoles? In *Biogeography of the West Indies: Past, Present, and Future,* ed. C. A. Woods, 433–477. Gainesville, FL: Sandhill Crane Press.

Williams, G. C. 1992. *Natural Selection: Domains, Levels, and Challenges.* New York: Oxford University Press.

Wilson, B. S., and P. J. Havel. 1989. Dehydration reduces the endurance running capacity of the lizard *Uta stansburiana. Copeia* 1989:1052–1056.

Wilson, M. A., and R. E. Gatten, Jr. 1989. Aerobic and anaerobic metabolism of paired male lizards (*Anolis carolinensis*). *Physiol. Behav.* 46:977–982.

Wilson, M. A., R. E. Gatten, Jr., and N. Greenberg. 1989. Glycolysis in *Anolis carolinensis* during agonistic encounters. *Physiol. Behav.* 48:139–142.

Withers, P. C., and S. S. Hillman. 1988. A steady-state model of maximal oxygen and carbon dioxide transport in anuran amphibians. *J. Appl. Physiol.* 64:860–868.

Wyneken, J., and M. Salmon. 1992. Frenzy and postfrenzy swimming activity in loggerhead, green, and leatherback hatchling sea turtles. *Copeia* 1992:478–484.

Zani, P., and D. L. Clausen. In press. Voluntary and forced terrestrial locomotion in juvenile and adult painted turtles, *Chrysemys picta. Copeia.*

Zink, R. M. 1986. Patterns and evolutionary significance of geographic variation in the Schistacea group of the fox sparrow (*Passerella iliaca*). *Ornithol. Mon.* 40:1–119.

11

The Role of Physiological Capacity, Morphology, and Phylogeny in Determining Habitat Use in Mosquitoes

Timothy J. Bradley

This volume is intended to deal with ecomorphology, the capacity of morphological features to contribute to and indeed determine the ecological capacities of organisms. Morphological features can influence the ecology of organisms by changing performance, for example, through increased limb length or changes in bill shape (see Garland and Losos, chap. 10, this volume). The majority of morphological changes examined in this light influence behavioral aspects of organismal function such as locomotion, feeding, or sexual attraction.

An equally large number of morphological modifications influence functions which might most commonly be termed physiological or biochemical. The acquisition of novel cell types in the gut capable of the secretion of a strong salt solution certainly qualifies as a morphological change if these cells are morphologically unique. The consequence to the animal of this morphological change would not fall in the realm of what we would typically call a behavioral change, but would instead take the form of an increased capacity to produce concentrated excreta. The ecological consequences of this morphological change might be the capacity to invade highly saline environments previously unavailable to the species.

Before accepting the contention that such physiological changes are adaptive and of ecological significance, we would wish to observe such changes in organisms that had come to occupy a new ecological niche, but not in closely related organisms which retained the ancestral condition and which were excluded from the new niche. We should expect performance testing to demonstrate that the modified organisms possess physiological characteristics necessary for survival in the new niche and that the related forms possessing the ancestral condition could not perform adequately to survive in the niche (see Garland and Losos, chap. 10, this volume). Phylogenetic analysis should indicate that the physiological capability arose simultaneously with the occupation of the new niche (see Wainwright, chap. 3, this volume). Finally, the distribution of the organisms in the field should be in keeping with the contention that the physiological adaptation was necessary for survival in that niche. An analysis of this sort, therefore, requires physiological studies, phylogenetic analysis, performance analysis, and examination of the distribution and capabilities of the organisms in the field (Huey, 1987).

In this chapter I will describe studies we are carrying out which attempt to use these procedures to elucidate the physiological mechanisms and evolutionary pathways by which increased saline tolerance arose in a monophyletic group of insects. In particular, the chapter deals with the physiological mechanisms of ion transport and urine formation in mosquito larvae, and the role of these mechanisms in determining the ecological distribution of mosquito species.

The larvae of mosquitoes (family Culicidae) are aquatic. Important features of the aquatic habitat include the osmotic concentration, and the absolute concentrations and ratios of ionic species (e.g., salinity, hardness, pH, etc.). Mosquitoes can serve as a particularly valuable group of animals on which to conduct studies relating physiological capacity to ecological distribution for the following reasons.

First, mosquitoes are found in a wide array of aquatic habitats ranging from tiny containers of water such as are found in the leaf axils of bromeliads, to vast coastal marshes (O'Meara, 1976; Frank and Lounibos, 1983). The salinity of these habitats in which mosquito larvae can exist also spans the range from essentially distilled water derived from collected rainwater to desert saline pools several times the concentration of the open ocean (Bradley, 1987a). Mosquitoes as a group show a much larger range of salinity tolerance than any known vertebrate group, including fish.

Second, the physiology of salt transport and osmoregulation has been intensively studied in mosquitoes (reviewed by Bradley, 1987a). The ionic makeup and osmotic concentration of the hemolymph, gut contents, and urine can be fully characterized. The rates and directions of ion and fluid transport can be quantified using direct measurements, radioisotope studies, or electrophysiological techniques. The capacity of these transport systems to contribute to survival in differing habitats can be easily assessed using performance analysis. Performance analysis can include the analysis of survival times in saline stress, the response of hemolymph concentrations to external salt or osmotic stress, or even measurement of salt transport rates under conditions which maximize transport. The morphology of the osmoregulatory organs has been characterized using light and electron microscopy, including techniques for morphometrically differentiating cell types (Bradley, 1985). In short, the characters associated with osmoregulation can be fully described and easily compared and contrasted between species.

Finally, due to their medical importance, mosquitoes have been intensively studied from a taxonomic viewpoint. The insects commonly referred to as mosquitoes are all restricted to the family Culicidae and the relationships between the various genera are generally well accepted (Knight and Stone, 1977). The mosquitoes, therefore, represent a monophyletic group of insects which are taxonomically well studied and which occupy a broad range of ecological niches. They are an ideal group in which to seek information on behavioral and physiological adaptations while controlling for phylogeny (Bradshaw and Holzapfel, 1991, 1992; Sheplay and Bradley, 1982). In addition, having established the physiological mechanisms responsible for survival in media of varying salinity, we can take advantage of the extensive phylogenetic information available to attempt to reconstruct the evolutionary pathways by which these physiological processes arose.

I will restrict myself to those morphological and physiological mechanisms which are significant in determining the ecological distribution of mosquitoes with regard to osmotic niches and in elucidating the pathways by which increased salinity tolerance arose. Those readers interested in more detailed physiological information and a description of the experiments by which these were elucidated are referred to a review of these topics (Bradley, 1987a).

OBLIGATE FRESHWATER LARVAE

The osmoregulatory mechanisms used by the larvae of mosquitoes for surviving in fresh water have been examined in two species, *Aedes aegypti* and *Culex pipiens* (Wigglesworth, 1938; Bradley, 1987a). The morphological features associated with existence in fresh water, and the ion-transport mechanisms employed, are identical in these two species (Bradley, 1987a). In the absence of any information to the contrary, it is presumed that all larvae which are obligate freshwater forms employ the mechanisms described below.

All larvae of mosquitoes found in freshwater are osmoregulators, that is, they maintain relatively constant osmotic and ionic concentrations in the hemolymph, and these are relatively high compared to those in the external medium. As a result, the larvae experience water uptake across the external cuticle by osmosis. In addition, the ionic gradient between the hemolymph and the external medium favors the loss of ions.

Larvae are able to survive in freshwater by reducing drinking to a minimum, producing a copious, dilute urine and by actively transporting ions into the hemolymph both from the urine and the external medium. The organs employed for ion and water transport are the midgut, Malpighian tubules, rectum, and anal papillae (fig. 11.1).

Fluid or food entering the mouth passes down into the midgut. Here, osmotic equilibrium occurs, principally by entry of water into the hemolymph. Ions and nutrients are then actively transported into the hemolymph with water following.

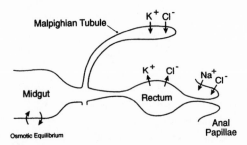

FIGURE 11.1 The relative positions of the midgut (MG), Malpighian tubules (MT), rectum (R), and anal papillae (AP) in freshwater mosquitoes. The Malpighian tubules produce the primary urine through the active transport of K^+ and Cl^-. This fluid, however, is always strictly isosmotic to the hemolymph and is therefore incapable of contributing to osmoregulation. The rectum resorbs K^+ and Cl^- to produce a very dilute urine and promote ionic and osmotic balance. The anal papillae also take up ions to promote ionic homeostasis. The rectum and anal papillae can only transport ions into the hemolymph, however, so these larvae cannot produce a concentrated urine and are restricted to media more dilute than the hemolymph.

To the extent that the food is mixed with the dilute external medium, eating and drinking both lead to the dilution of the body fluids.

The Malpighian tubules are used to produce the primary urine. Potassium and chloride ions are actively transported into the lumen of the Malpighian tubules and water follows. The primary urine produced in this fashion is strictly isosmotic to the hemolymph. If this fluid were excreted unmodified, it would cause a severe depletion of valuable ions from the hemolymph and make no contribution to osmotic homeostasis.

Modification of the urine to meet the ionic and osmotic regulatory needs of the larvae is carried out in the rectum. In this organ, potassium and chloride ions, as well as other ions such as sodium present in the feces, are actively returned to the hemolymph. The rectum is fairly osmotically tight and the resulting dilute urine is excreted. In this fashion, a very dilute urine is produced, allowing ions to be retained and excess water to be excreted.

When feeding rates are low, or in extremely dilute media, the larvae may require an additional source of ion uptake. The anal papillae, which are ion-transporting organs adjacent to the anal opening, can be used to actively transport sodium, chloride and to a lesser extent potassium from the external medium directly into the hemolymph.

It can be seen in figure 11.1 that with the exception of the Malpighian tubules, which are capable only of producing an isosmotic secretion, all the ion pumps in the gut and external epidermis are oriented to transport ions into the hemolymph and not out of it. As a result, the larvae can produce a dilute urine, but they cannot produce a concentrated secretion. These larvae must therefore be characterized as belonging to obligate freshwater species. They are restricted to media more dilute than their normal hemolymph concentration, about 350 milliosmoles/liter (mOsm). If the medium concentration rises above this critical level the larvae die (fig. 11.2) (Ramsay, 1950).

For the purposes of the topics we are discussing in this paper, the larvae of obligate freshwater mosquito species are characterized by the presence of a one-part rectum containing a single cell type, dilute urine, and failure to survive in media more concentrated than the normal hemolymph concentration (i.e., 350 mOsm). These morphological and physiological characteristics result in an osmotic strategy which firmly restricts larvae to hypoosmotic media.

THE EVOLUTION OF INCREASED SALINE TOLERANCE

The ancestral insect species from which the family Culicidae evolved is thought to have had obligate freshwater larvae. The evidence for this is the observations that 95% of all extant mosquito species are obligate freshwater forms and that the two families of insects most closely related to the mosquitoes, the Dixidae and the Chaoboridae, are both strictly freshwater.

A small number of species (roughly 125 out of the 2,500 known mosquito species) in nine separate genera have evolved a capacity to survive as larvae in waters more concentrated than 300 mOsm. Species from six genera have been

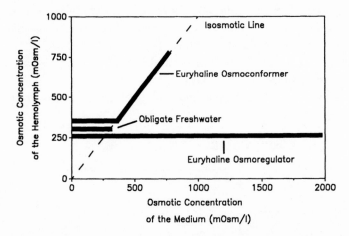

FIGURE 11.2 The response of the osmotic concentration of the larval hemolymph to the osmotic concentration of the medium. Larvae die of the effects of saline stress at the point where the lines end on the right. Obligate freshwater larvae osmoregulate at all times but cannot survive in water more concentrated than 300 mOsm. Euryhaline osmoconforming larvae osmoconform above 300 mOsm and show increased saline tolerance to salinities ranging up to 700 to 1000 mOsm, depending on the species. Euryhaline osmoregulators show osmoregulation in all media and can survive to 2000 mOsm as shown, or even higher (up to 4000 mOsm) in some species.

examined with regard to the mechanisms of ionic and osmotic regulation (Bradley, 1987a). The taxonomic relationships among these six genera are shown in figure 11.3. While all mosquitoes are in the family Culicidae, the *Anopheles* are in a separate subfamily and form an outgroup to the other genera shown. The genera *Aedes* and *Opifex* are sufficiently similar to be placed in one tribe, and the genera *Deinocerites* and *Culex* have similar affinities.

None of the saline-tolerant species is truly marine, since for reasons that are not clear, mosquitoes have not colonized open seawater. However, some species are found in tidal pools and coastal marshes which are ionically identical to sea-water and, following evaporation, can be even more concentrated than the open ocean. Saline-tolerant species are also found in inland saline waters including saline lakes and desert pools, which in some cases can have ionic ratios quite different from seawater.

Two distinct osmotic strategies have been elucidated in our studies of the physiology of osmoregulation in saline-tolerant mosquitoes (Bradley, 1987a). The following two sections review the physiological mechanisms employed by saline-tolerant mosquito larvae and the likely evolutionary relationships among larvae sharing a common osmoregulatory strategy.

Saline-Tolerant Osmoconforming Larvae

Culex tarsalis is an example of a mosquito species possessing larvae which are highly saline-tolerant. *Culex tarsalis* breed in inland waters, particularly in dry areas of the western portions of North America. The larvae are common pests around rice-growing areas and flooded pastures and, to a lesser extent, in coastal marshes. The larvae seem particularly well adapted to alkaline and saline seeps in the western deserts and the Great Basin.

Larvae of *Culex tarsalis* have no morphological specializations which would

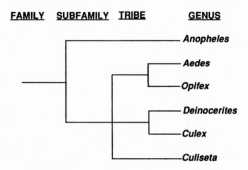

FIGURE 11.3 A dendrogram showing the taxonomic relationships among those genera of mosquitoes in which the mechanisms of saline tolerance have been examined.

allow them to survive in these saline waters. The Malpighian tubules and rectum are identical in gross morphology and cell type to those found in the obligate freshwater species *Culex pipiens*. When we examined the concentration of the hemolymph in larvae of *Culex tarsalis* acclimated to various concentrations of seawater, we found that the larvae could survive in waters ranging up to about 70% seawater (700 mOsm). Rather than regulating their hemolymph over this entire range, however, the larvae exhibit osmotic regulation of the hemolymph in waters more dilute than 300 mOsm but show osmoconformation in waters more concentrated than 300 mOsm. Measurements of sodium, potassium, chloride, calcium, and magnesium ions in the hemolymph demonstrate that the larvae regulate the concentrations of these ions in a narrow range over the full range of salinities in which they can survive. Since, the hemolymph osmotic concentration is increasing in higher salinities in parallel with the external medium, but ion levels in the hemolymph do not change, it is clear the hemolymph of these larvae contains substantial amounts of osmotically active compounds other than inorganic ions.

Garrett and Bradley (1987) examined the hemolymph of larvae acclimated to a dilute medium (5% seawater) and a concentrated medium (60% seawater) to search for compounds which might be acting as osmolytes in higher salinities. Using a variety of techniques to analyze organic compounds, they found that larvae accumulated proline, trehalose, and, to a lesser extent, serine when acclimated to higher salinities. Proline and trehalose have both been shown to be innocuous osmolytes in biological systems. It has been shown that, on the basis of their hydrogen bonding properties, these compounds do not denature most proteins and membranes even at high concentration (Yancey et al., 1982; Crowe et al., 1984).

It is apparent, therefore, that the strategy employed by osmoconforming mosquito larvae is one of accumulating, to extraordinarily high levels, two or three organic compounds which are present at lower levels in all insects. These compounds then serve as nontoxic osmolytes. No novel morphological features are known to be associated with this capacity. The changes relative to freshwater larval forms deal presumably with set points in biochemical pathways (Garrett and Bradley, 1987).

Larvae in the mosquito species *Culiseta inornata* have also been shown to osmoconform in saline waters more concentrated than 300 mOsm. These larvae also regulate hemolymph ion levels. Studies are presently underway to determine the organic compounds used as osmolytes in this species.

Deinocerites is a genus of mosquitoes containing only eight species, all located in the Caribbean region. Females in this genus oviposit in crab burrows. The larvae develop, therefore, in waters containing a substantial portion of seawater. Larvae in the genus *Deinocerites* demonstrate ionic regulation but osmoconformation in concentrated saline waters (J. K. Nayar, pers. comm.).

The strategy of accumulating organic compounds as a means of osmoconforming in waters more concentrated than 300 mOsm (fig. 11.2) is now known to occur in larvae from at least three genera (fig. 11.4). In view of the small number of species in the genera *Culex* and *Culiseta,* in which larvae are saline-tolerant, and the clear phylogenetic separation of these two genera, it seems reasonable to assume that the larvae in these two genera evolved these capacities independently. Since the genus *Deinocerites* is a small genus closely related to *Culex,* we cannot rule out, on the basis of present information, the possibility that these highly specialized mosquitoes may have evolved from a saline-tolerant, osmoconforming species of *Culex.* Even with this conservative assumption, we are left with the conclusion that the use of osmoconformation as a means of increasing saline-tolerance evolved at least twice independently in mosquitoes.

It should be pointed out that no "water pump" is known to exist in any biological system. The active movement of water across an epithelium is always the result of the generation of osmotic gradients by active ion transport. By adopting a strategy of osmoconforming to the external medium, larvae essentially eliminate water flux across the cuticle and therefore the need for a copious urine. Large fluxes of water through an animal, therefore, require large metabolic inputs for ion transport. In the case of freshwater larvae, the copious, dilute urine they produce is generated by transporting ions into the Malpighian tubules to produce the primary urine and then transporting them back out in the rectum to produce the dilute final urine. In the case of saline-water larvae producing a concentrated urine (see below) large ion gradients are produced by powerful ion pumps. Ion gradients still exist across the cuticle, but the epidermis of aquatic insects is usually quite impermeable to ions. A strategy of osmoconformation, therefore, substantially reduces the need for transporting ions as a means of moving water across epithelia.

Clearly there is a cost associated with accumulating proline and trehalose in the hemolymph since both are biologically useful, energy-rich compounds. Use of these compounds as osmolytes does not, in the long term, mean a loss of these compounds from the metabolic pool, however. The compounds serve as osmolytes in the larvae and, assuming that they are carried over into the adult, they are available as valuable, fully usable energy metabolytes. Therefore, there is no reason to believe that their synthesis and retention produces any net cost for the mosquito.

The increased saline tolerance demonstrated by certain larvae in the genera *Culex, Culiseta,* and *Deinocerites* enables the larvae to survive in waters approaching and in some cases including the salinity of seawater (see fig. 11.2

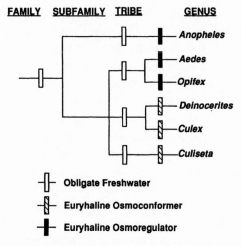

FIGURE 11.4 A dendrogram as in figure 11.3. The white box at the base of the dendrogram represents the ancestral, freshwater larval condition. The filled boxes indicate that all saline-tolerant larvae in these genera have proven to be osmoregulators. The striped boxes indicate that all saline-tolerant larvae in these genera have proven to be osmoconformers. It is important to note that in *Anopheles, Aedes, Culex,* and *Culiseta,* most of the larvae retain the ancestral, freshwater condition.

above). This permits the larvae to exploit inland saline ponds, alkaline waters around inland wetlands, estuaries, coastal marshes adjacent to sources of freshwater, and pools and burrows adjacent to marine waters. These sites are often highly biologically productive (O'Meara, 1976). In view of the presumed energetic efficiencies associated with a strategy of osmoconformation, it appears that osmoconforming species have evolved the capacity to exploit highly productive sites at relatively low physiological cost.

Saline-Tolerant Osmoregulating Larvae

Some species of mosquito have larvae which are among the most saline-tolerant aquatic animals known. These larvae are not found in the genera *Culex, Culiseta,* and *Deinocerites,* but rather in *Aedes, Opifex,* and *Anopheles.* Research conducted on *Aedes taeniorhynchus* illustrates the physiological processes involved in osmoregulation in saline-tolerant *Aedes.*

Larvae of *Aedes taeniorhynchus* can survive in salinities ranging essentially from distilled water to 350% seawater. Over this full range of salinities, the larvae strictly regulate the osmotic and ion concentrations of the hemolymph, with the hemolymph concentration being maintained at about 350 mOsm (see fig. 11.2 above). In media more dilute than this, the larvae hyperregulate the hemolymph, actively taking up ions and producing a dilute urine in a manner identical to obligate freshwater larvae (Bradley and Phillips, 1977). In media more concentrated than the hemolymph, the larvae actively transport ions out through the production of a concentrated urine (Bradley and Phillips, 1975, 1977). It clear on the basis of this physiological performance alone that the larvae of *Aedes taeniorhynchus* are very different from those of osmoconforming species. Ramsay (1950) pointed out a very distinct morphological character which differentiates osmoregulating, saline-tolerant mosquitoes from all other forms (fig. 11.5). The larvae of saline-tolerant *Aedes* have an additional rectal segment not found in the rectum of other species. The anterior rectal segment in these larvae is identical in ultrastructure and cell type to the rectal cells in freshwater species. The posterior portion contains cells with deep membrane folds closely associated with numerous mitochondria. It is this posterior rectal segment which allows the saline-tolerant osmoconforming larvae to produce a concentrated urine.

Bradley and Phillips (1977) demonstrated that the posterior rectal segment (fig. 11.5) is the site of active ion transport out of the larvae. This segment produces a concentrated fluid which enters the urine immediately prior to excretion. Depending on the medium in which the larvae are reared, the posterior rectal cells can actively transport sodium, chloride, magnesium, and potassium out of the hemolymph to produce a hyperosmotic urine. The capacity of these larvae to osmoregulate in a wide variety of media through the production of a dilute or highly concentrated urine is largely dependent on the powerful ion transporting capacities of the rectum. However, if larvae of *Aedes taeniorhynchus* are reared in saline waters rich in sulfate, the Malpighian tubules can actively transport sulfate ions into the urine as a way of ridding the hemolymph of these ions (fig. 11.5).

A two-part rectum and the capacity to osmoregulate by production of a concentrated urine are characteristics found in all the euryhaline (broadly saline-tolerant) species of *Aedes* examined to date. Most of these larvae are placed in the subgenus *Ochlerotatus*. On the basis of our current knowledge of the evolution of this group, it is possible although unlikely that these mosquitoes represent the decendents of a single euryhaline osmoregulating ancestor (see "Studies of the Evolution of Saline-Tolerance in Mosquitoes at the Species Level" below). A separate subgenus of *Aedes, Finlaya,* is separated from *Ochlerotatus* on the basis of a number of morphological features of the larvae and adults (Knight and Stone, 1977). *Aedes togoi,* an osmoregulating species in the subgenus *Finlaya,* has an anterior rectal segment in the normal rectal region and a posterior salt-secreting cell type in the anal canal. This morphological difference, coupled with the phylogenetic evidence, argues that its capacities to produce a concentrated urine evolved separately from those of the *Aedes* in the subgenus *Ochlerotatus*.

The monotypic genus *Opifex* contains the species *Opifex fuscus,* found only in New Zealand. Nicholson and Leader (1974) demonstrated that this species can

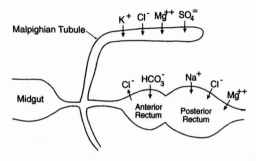

FIGURE 11.5 The organs contributing to ionic regulation in saline-water *Aedes*. The posterior rectal segment (PR) can produce a hyperosmotic fluid; the anterior rectal segment (AR) carries out the functions of the rectum in freshwater, larvae but also contributes bicarbonate to hyperosmotic urine in alkaline waters; the Malpighian tubules (MT) are the sole site of active sulphate transport; midgut (MG).

produce a concentrated urine using the rectum. It is not known if the larvae of this species have a two-part rectum. Since the genus *Opifex* is closely related to *Aedes*, we cannot rule out the possibility that *Opifex fuscus* evolved from a species of saline-water *Aedes* that became isolated in New Zealand and underwent sufficient specialization to warrant being placed in a separate genus. Our most conservative estimate would be, therefore, that the strategy of producing a concentrated urine in the rectum as a means of osmoregulating in highly saline environments evolved at least twice in the genera *Aedes* and *Opifex*.

There is another group of highly saline-tolerant mosquitoes which has not yet been discussed, those in the genus *Anopheles* (see fig. 11.3 above). My laboratory has been investigating the mechanisms contributing to saline-tolerance in the species *Anopheles* sp. nr. *salbai* (Bradley, 1987b), a species whose larvae are found in desert pools near the Red Sea. Our studies have shown that these larvae have a two-part rectum which is morphologically distinct from that found in *Aedes*. The rectal segment resembling that of freshwater *Anopheles* species is located on the ventral portion of the rectum of *A.* sp. nr. *salbai*, while the "new" rectal portion containing cells with deep membrane folds and numerous mitochondria is located dorsally. This pattern is clearly different from that in *Aedes*, where the two segments are located anteriorly and posteriorly. The larvae of *A.* sp. nr. *salbai* can survive in water ranging from 10% seawater to 400% seawater. They survive over this tremendous range of salinities by strict osmoregulation in both dilute and concentrated media. The rectum is the site of urine concentration.

The osmoregulatory strategy employed by *A.* sp. nr. *salbai* appears to be physiologically identical to but not homologous to that of *Aedes*. The larvae produce concentrated urine and osmoregulate the hemolymph through the use of a second rectal segment found only in saline-water forms. The morphology of this two-part rectum in *Anopheles* is very different from that of *Aedes*. On the basis of phylogenetic information (see fig. 11.4 above), it is clear that osmoregulating larvae in *Anopheles* and *Aedes* evolved similar physiological functions independently.

Of the six genera so far examined, representing at a minimum five separate evolutionary transitions to saline-tolerance, it is noteworthy that only two strategies have been observed. In one strategy, innocuous osmolytes are accumulated in the hemolymph to eliminate osmotic gradients. In the absence of large water fluxes, ionic regulation is much reduced. Presumably some adaptation of the cells to reduced water activity is required. In the other strategy, powerful ion pumps located in a unique cell type have evolved in the posterior rectal region. These pumps produce a urine more concentrated than the external medium. It is not clear why these two strategies have evolved repeatedly to the exclusion of others. It is tempting to attribute this to "phylogenetic constraints" in mosquitoes which channel evolutionary movement in one of these two directions. I find such an explanation merely descriptive of the syndrome and in no way explanatory of the mechanism. In the case of the osmoconformers using organic osmolytes, the evolutionary step to osmoconformation is simplified by the presence in all mosquitoes of the major osmolytes, proline and trehalose. Adjustment of the levels of these osmolytes to osmotically significant levels might only require changes in the set points of regulatory enzymes. The larvae also require systems which transport ions outward, systems which are entirely lacking in their freshwater ancestors. Until we know more about the mechanisms and location of these transport mechanisms we cannot comment intelligently on the reasons that some mosquitoes have apparently been channeled into this strategy.

The modifications required to achieve osmoregulation would seem to be more extreme than those required for osmoconformation. For osmoregulation, a new cell type must arise in the rectal region, possessing the properties of producing an osmotically tight epithelium which can pump a multitude of ion types out of the hemolymph at a high rate. In the absence of knowledge about genes affecting segmentation in this region of the gut and the control of ion pump differentiation, speculation about the ease and frequency of such genetic modifications is again teleological and post hoc.

STUDIES OF THE EVOLUTION OF SALINE-TOLERANCE IN MOSQUITOES AT THE SPECIES LEVEL

The above studies demonstrate that we can use highly defined and quantifiable characters of morphology, physiological performance, organic and ionic concentrations, and ion transport mechanisms to rapidly and accurately characterize the osmotic strategy and mechanisms used by mosquitoes for invading saline habitats. Using these characters and our knowledge of mosquito phylogeny, we should be able to chart the pathways by which these mechanisms and increased tolerance evolved. Described below is our initial progress along these lines.

My discussion to this point has dealt with an attempt to understand the evolution of saline-tolerance in mosquitoes from the perspective of an overview of the family. Evolution does not, of course, occur in this fashion. Instead, presumably, individual mosquitoes or perhaps small populations of mosquitoes evolved mechanisms which permitted them to exploit a new habitat, that is, saline waters. Eventually, these mosquitoes became reproductively isolated, perhaps due to their isolation in newly exploited habitats. This led to new saline-tolerant species or perhaps clusters of closely related species which evolved from the original pioneering species.

The pattern of saline-tolerant species in *Aedes* would seem to support such an evolutionary scenario. Rohlf (1963) used larval and adult morphological characters to produce a phylogenetic tree for species of *Aedes* in the subgenus *Ochlerotatus*. His methods produced one pair of closely related saline-tolerant species (*A. taeniorhynchus* and *A. sollicitans*) in one part of the tree and a second pair of sibling species (*A. campestris* and *A. dorsalis*) in another, with many freshwater species in between. More recently, Schultz et al. (1986) have analyzed the phylogenetic relationships in the saline-tolerant *Aedes* using isozyme analysis and arrived at similar conclusions, that is, that these two species clusters, each showing saline-tolerance, osmoregulation, and a two-part rectum, are not as closely related to each other as they are to some freshwater species. These species provide an ideal opportunity to study in detail the morphological and physiological mechanisms promoting saline tolerance in these species clusters and comparing them to the primitive condition observed in the closely related freshwater species.

We have undertaken such a study using the cluster of saline water species (*Aedes caspius, A. dorsalis, A. campestris*) shown in figure 11.6 and comparing these to the most closely related freshwater species, *A. melanimon. Aedes melanimon* occurs in California in montane flooded meadows and pastures. It is described in the literature as a freshwater species with a slight preference and increased tolerance for alkaline waters. We found that *A. melanimon* has a two-part rectum and is capable of surviving in full-strength seawater. Clearly, *A. melanimon* is highly saline-tolerant and has morphological features and physiological capabilities identical to the cluster of saline-water species.

Two different interpretations of these results are possible. In the first scenario, one could conclude that *A. melanimon* and the other closely related saline-tolerant species evolved from a common ancestor with a two-part rectum and eu-

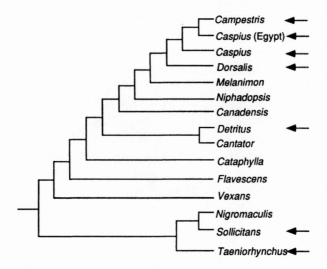

FIGURE 11.6 A phylogeny of the saline-water *Aedes* and related *Aedes* species based on Schultz et al. (1986). Those species which have previously been shown to produce a concentrated urine and to possess a two-part rectum are marked with arrowheads. Note that the positions of the saline-tolerant species and the locations of the intervening, purported freshwater forms argues for repeated independent evolution of the euryhaline osmoregulating condition in this group of *Aedes*.

ryhaline osmoregulatory capabilities. Therefore, *A. melanimon* would appear to be a saline-tolerant species that returned to a freshwater habitat. One is still left, however, with the fact that the freshwater condition is ancestral and that the genus *Aedes* originally arose as a freshwater species whose larvae contained a one-part rectum. The fact that *A. melanimon* has a two-part rectum and is found in fresh water could be interpreted to mean that a two-part rectum, powerful ion pumps, and the capacity to survive in highly saline waters evolved in freshwater species and confers some other advantage outside of saline waters. Indeed, a strict parsimony interpretation of the cladistic data would demand such an explanation.

Such a scenario parallels in some ways the observations of Bradshaw and Holzapfel (1991), who showed that a species of *Anopheles* avoided competition with other mosquito species, and therefore showed increased fitness in treehole environments, due to behavioral characteristics common to all anophelines and which evolved outside of the treehole environment.

At this early stage in our studies of the evolution of saline tolerance in mosquito species, perhaps the most important point to make is that we possess the tools to resolve these conflicting scenarios. The morphological, physiological, and performance characteristics of saline-tolerant forms in this group are well described and highly quantifiable. The cladistics of the group are also well in hand. Further studies should, therefore, enable us to determine the evolutionary pathway by which saline tolerance arose in these mosquitoes in the genus *Aedes,* subgenus *Ochlerotatus.* A particularly important and interesting question arising from this particular study is whether the morphological and physiological characters found in present-day saline-water species arose simultaneously with the invasion of saline-water habitats. Analysis at this level of detail is only possible through simultaneous application of careful morphological, physiological, and cladistic procedures.

SUMMARY

Casual analysis of the osmoregulatory capabilities of larvae in the mosquitoes (family Culicidae) reveals that while most species are restricted to freshwater, the larvae of about 5% of all species possess the capability of living in waters more concentrated than 35% seawater (350 mOsm). These larvae occur in nine separate genera.

Detailed studies of the morphology and physiology of the larvae from six of these genera revealed two general strategies for survival in saline waters: (1) osmoconformation through the accumulation of innocuous organic osmolytes in the hemolymph or (2) the acquisition of a powerful salt-secreting organ in the hindgut which permitted hemolymph osmoregulation in waters as concentrated as 300 to 400% seawater. Cladistic analysis revealed that the morphological and physiological traits associated with these strategies were not always homologous. Osmoconformation arose at least twice in the mosquitoes, and hypoosmoregulation at least three times. Further, ongoing studies using morphological, physiological, and performance characters in the context of phylogenetic analysis will permit us to characterize the pathways by which these evolutionary changes occurred. These studies will enable us to determine the causal and temporal relationships between morphological, physiological, and performance changes, and the ecological distribution of a monophyletic but ecologically diverse group of animals.

REFERENCES

Bradley, T. J. 1985. The excretory system: Structure and physiology. In *Comprehensive Insect Physiology, Biochemistry and Pharmacology,* ed. G. A. Kerkut and L. I. Gilbert, 421–465. Oxford: Oxford Pergamon Press.

Bradley, T. J. 1987a. Physiology of osmoregulation in mosquitoes. *Ann. Rev. Entomol.* 36:439–462.

Bradley, T. J. 1987b. Evidence for hypo- and hyperosmotic regulation in the larvae of an anopheline mosquito. *Amer. Zool.* 27(4):130A.

Bradley, T. J., and J. E. Phillips. 1975. The secretion of hyperosmotic fluid by the rectum of a saline-water mosquito larva, *Aedes taeniorhynchus. J. Exp. Biol.* 63:331–342.

Bradley, T. J., and J. E. Phillips. 1977. The location and mechanism of hyperosmotic fluid secretion in the rectum of the saline-water mosquito larvae, *Aedes taeniorhynchus. J. Exp. Biol.* 66:111–126.

Bradshaw, W. E., and C. M. Holzapfel. 1991. Fitness and habitat segregation of British treehole mosquitoes. *Ecol. Entomol.* 16:133–144.

Bradshaw, W. E., and C. M. Holzapfel. 1992. Resource limitation, habitat segregation, and species interactions of British tree-hole mosquitoes in nature. *Oecologia* 90:227–237.

Crowe, J. H., M. Crowe, and D. Chapman. 1984. Preservation of membranes in anhydrobiotic organisms: The role of trehalose. *Science* 223:701–703.

Frank, J. H., and L. P. Lounibos, eds. 1983. *Phytotelmata: Terrestrial Plants as Hosts of Aquatic Insect Communities*. Medford, NJ: Plexus Publishing.

Garrett, M. A., and T. J. Bradley. 1987. Accumulation of proline, serine and trehalose in the hemolymph of osmoconforming brackish-water mosquitoes. *J. Exp. Biol.* 129:231–238.

Huey, R. B. 1987. Phylogeny, history, and the comparative method. In *New Directions in Ecological Physiology,* ed. M. E. Feder, A. F. Bennett, W. W. Burggren, and R. B. Huey, Cambridge: Cambridge University Press.

Knight, K. L., and A. Stone. 1977. *A Catalog of the Mosquitoes of the World,* 2d ed. College Park, MD: Ent. Soc. Amer.

Nicholson, S. W., and J. P. Leader. 1974. The permeability to water of the cuticle of the larva of *Opifex fuscus* (Hutton) (Diptera: Culicidae). *J. Exp. Biol.* 60:593–603.

O'Meara, G. F. 1976. Saltmarsh mosquitoes (Diptera: Culicidae). In *Marine Insects,* ed. L. Chang. New York: North Holland.

Ramsay, J. A. 1950. Osmotic regulation in mosquito larvae. *J. Exp. Biol.* 27:145–157.

Rohlf, F. J. 1963. Classification of *Aedes* by numerical taxonomic methods (Diptera: Culicidae). *Ann. Entomol. Soc. Amer.* 56:798–804.

Schultz, J. H., P. G. Meier, and H. D. Newson. 1986. Evolutionary relationships among the salt marsh *Aedes* (Diptera: Culicidae). *Mosquito Syst.* 18(2):145–180.

Sheplay, A. W., and T. J. Bradley. 1982. A comparative study of magnesium sulphate tolerance in saline-water mosquito larvae. *J. Insect Physiol.* 28(7):641–646.

Wigglesworth, V. B. 1938. The regulation of osmotic pressure and chloride concentration in the hemolymph of mosquito larvae *J. Exp. Biol.* 15:235–247.

Yancey, P. H., M. E. Clark, S. C. Hank, R. D. Bowlus, and G. N. Somero. 1982. Living with water stress: Evolution of osmolyte systems. *Science* 217:1214–1222.

12

The Ecological Morphology of Metamorphosis: Heterochrony and the Evolution of Feeding Mechanisms in Salamanders

Stephen M. Reilly

INTRODUCTION

Morphological evolution proceeds through the alteration of developmental processes underlying morphogenesis (Wake and Roth, 1989a; Raff et al., 1990; Nitecki, 1990; Müller and Wagner, 1991). Heterochrony, or changes in the timing and/or rates of development, is our best source of data to study how developmental processes are shifted relative to each other in evolution. A major goal of heterochronic analyses is to understand how the forces of selection (ecological factors) affect developmental responses and ontogenetic variability within species. On this intraspecific level, heterochrony refers to the processes that affect the plasticity of ontogenetic development among individuals within populations or species. For example, it is well known that variation in the timing of metamorphosis in amphibians is an important adaptive response to environmental variability (Istock, 1967; Wilbur, 1980; Hanken, 1992; Travis, chap. 5, this volume).

Another major goal of heterochronic analyses is to understand how patterns of morphology among species reveal the ways that the forces of selection have produced evolutionary change. On this phyletic level, heterochrony refers to processes that have altered the ontogeny of species so that features of descendant species are displaced in time relative to the ancestral ontogeny (Gould, 1977). Phyletic differences in ontogenetic patterns for ancestor and descendant lineages can be compared, and the mechanisms by which species have come to differ can be inferred (Alberch et al., 1979). Thus, heterochrony can be thought of in terms of interspecific differences in ontogeny or intraspecific differences in ontogeny within species (whether heritable or environmental). One major point of this paper is that it is critical to distinguish these two levels of heterochronic variation in ecological and evolutionary studies.

Alberch et al. (1979) formalized a model that attempts to relate patterns of morphological change to the processes of interspecific heterochrony. The model compares the ontogenetic trajectories of ancestor and descendant species to detect differences in the timing or rate of development. Perturbations in the growth rate and the onset and offset of growth during ontogeny define six major heterochronic processes that can produce a phyletic change in ontogeny relative to the ancestral condition. Developmental perturbations can produce a terminal shape in the descendant species that falls short of the ancestral species shape (paedomorphosis), or transcends the ancestral species shape (peramorphosis). Paedomorphosis, which describes descendant species that resemble earlier ontogenetic stages of the ancestral species (Garstang, 1922; Gould, 1977; Alberch et al., 1979) is by far the most common type of heterochrony observed in evolution (McNamara, 1988).

The Alberch et al. (1979) model, though generally accepted (but see Alberch, 1985), has seen relatively little empirical use because little is known about the actual timing of developmental and life history events in most species (Emerson, 1986; Jones, 1988). Empirical applications that are available show that age information (rather than size) must be available to distinguish between heterochronic rate and timing processes (Shea, 1983, 1988; Emerson, 1986; Jones, 1988; but see Strauss and Fuiman, 1985). And, because ancestral trajectories are rarely known, the phylogenetic outgroup condition must be used as the best estimate of the ancestral ontogeny (Fink, 1982, 1988).

Thus, to recognize heterochronic change one must first identify character transformations of interest in the context of a phylogenetic hypothesis of relevant taxa. The second step is to quantify ontogeny of features relative to age. The third step is to attempt to categorize the change by comparing the ancestral (or outgroup) trajectory to that of the descendant(s) of interest. Given that a phylogeny is available and that the model accurately categorizes heterochronic processes, the first and last steps are relatively straightforward. It is the second step, the combination of ecological (age, life history) and detailed morphological data to quantify ontogenetic trajectories, that is not well integrated into analyses of heterochrony (Gould, 1988; Jones, 1988).

The purpose of this chapter is to illustrate the insights to be gained by integrating ecology and morphology in the study of heterochronic variation and evolution. To do this I will compare ontogenetic trajectories of head development through metamorphosis in salamanders. Head shape ontogeny is based on large-scale changes that occur in the feeding mechanism (skull and hyobranchial apparatus) over time. Case studies illustrating interspecific heterochrony will be contrasted to cases of intraspecific heterochrony. The ecomorphological approach in this paper involves the comparison of morphological aspects of metamorphosis of the feeding mechanism to changes in design, performance, function, and environment during metamorphosis. The integration of these data reveals that heterochronic changes in morphological development (whether phyletic, phenotypic, or environmental) do not necessarily coincide with functional changes predicted from the life history stage or external morphology alone.

Salamanders are especially suited to the study of heterochrony for several reasons. First, heterochrony is a powerful evolutionary process in this group. Among vertebrates the greatest known expression of heterochrony is in salamanders (Gould, 1988) and the most common event in both fossil and extant groups is paedomorphosis involving the retention of pre-metamorphic features of the gills and head (Duellman and Trueb, 1986; McNamara, 1988, Wake, 1966). Second, the hormonal control of sexual maturation is strongly dissociated from somatic development (Lofts, 1974; Gould, 1988). Thus, heterochronic changes in somatic development are not associated with changes in the time of sexual maturation. Third, metamorphosis generally involves drastic ontogenetic changes in ecology (Duellman and Trueb, 1986), morphology (Reilly, 1986, 1987, 1990; Lauder and Reilly, 1990; Reilly and Lauder, 1990a), and function (Lauder and Reilly, 1988, 1990; Reilly and Lauder, 1990b, c, 1991a, b); therefore, evolutionary changes can be interpreted in terms of environmental selection on a large suite of quantifiable characters.

THE HEAD SHAPE AXIS

Analysis of the ontogeny of head morphology through normal metamorphosis reveals a major morphological "shape" axis on which heterochronic processes can be compared. The shape axis is based on the overall morphological condition of numerous features of the head for several ontogenetic samples of individuals at a known age for each species (Reilly, 1986, 1987, 1990; Lauder and Reilly, 1990; Reilly and Lauder, 1990a). Because of the nature of head development through metamorphosis the position of individual specimens along the shape axis can be qualitatively located by examination of individual features and overall cranial shape.

If we consider the ontogenetic development of the head in salamanders that metamorphose, in terms of changes in morphology and shape along a head shape axis versus time (fig. 12.1), there is rapid head shape change through embryogenesis until the salamanders hatch out in the aquatic larval form (Harrison, 1969; references in Duellman and Trueb, 1986; table 5-6). The larval form is defined by a large suite of morphological characters describing the shape of the head, hyobranchial apparatus, and skeletal features of the cranium. These include characteristics of pure shape of these components (Reilly, 1990; Reilly and Lauder, 1990a) as well as the presence, degree of ossification, and relative size of various individual elements (Wake, 1966; Reilly, 1986, 1987; Lauder and Reilly, 1990).

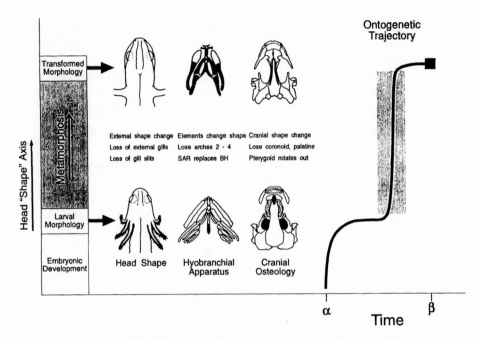

FIGURE 12.1 Head shape axis used to quantify ontogenetic trajectories of head development in salamanders. The Y-axis, representing qualitative ontogenetic changes in head shape from the embryo through metamorphosis to the terminal shape (■) when shape remains constant, is based on numerous morphological characters that can be traced through ontogeny. The shaded boxes represent metamorphosis. Three examples of metamorphic changes within each of the three major components of the head are given. The X-axis is based on known age of timing of various developmental and life history events. Together these two axes produce the ontogenetic trajectory of head shape development from the onset of development (α) through metamorphosis to the offset of shape change (β). Comparisons to a standard metamorphic trajectory can identify both phyletic patterns of heterochrony and developmental plasticity exhibited within species.

Salamanders retain the basic larval shape but grow in size for several months until the onset of metamorphosis. At this time they undergo rapid ontogenetic changes in most aspects of the head as they metamorphose to the terrestrial form. Metamorphosis, indicated in figures 12.1 to 12.4 by gray boxes, involves numerous major morphological changes in external head shape, the hyobranchial apparatus, and the cranium that have been studied in detail (Reilly, 1986, 1987, 1990; Lauder and Reilly, 1990; Reilly and Lauder, 1990, 1991b) and can be traced through the process (fig. 12.1).

Examples of these morphological transformations include (1) changes in external head shape and loss of the gills and gill slits; (2) changes in shape in the hyobranchial apparatus and its various elements, loss of branchial arches two through four, and the loss and movement of hyobranchial muscles; (3) changes in overall cranial shape; (4) changes in muscles involved in tongue projection; and (5) changes in the shape of individual bones, or their disappearance or relocation. Metamorphosed salamanders retain the terminal "transformed" morphology and shape but continue to grow in size throughout their lives.

By quantifying shape change through time one can construct an ontogenetic trajectory of head shape change for a particular taxon based on changes along this head shape axis plotted from the onset of growth, through key life history events, through the terminal shape when shape remains constant. Because the onset of maturity in salamander species usually occurs after the terminal shape has been attained and the time of maturity does not change among closely related salamanders, I will use the time (age) of terminal shape as the offset of the ontogenetic trajectory (*sensu* Fink, 1982). Using this head shape trajectory for complete metamorphosis (the primitive condition) as a standard, ontogenetic trajectories of head shape development versus age in different taxa can be compared to quantify the degree to which shape is truncated relative to complete metamorphosis and to examine the heterochronic processes that produce the changes in ontogeny. In the following section, I present interspecific heterochronic case studies that illustrate the two basic developmental pathways by which heterochrony has produced paedomorphosis in the heads of salamanders.

INTERSPECIFIC HETEROCHRONY IN CRANIAL DEVELOPMENT

The first case study is an example of paedomorphosis in the family Ambystomatidae, illustrated by the ontogeny of the axolotl, *Ambystoma mexicanum* (Rusconi, 1823; Altman and Dittmer, 1962; Giles and Stapleton, 1966; Brandon, 1989). The axolotl is known for its permanent complete retention of larval morphology relative to its sister taxon, *Ambystoma tigrinum,* and other outgroup taxa from the genus *Ambystoma* which exhibit complete metamorphosis (fig. 12.2*A*). If we plot the ontogenetic trajectory, based on head shape change over time, for the ancestral condition represented by *A.. tigrinum* (Shaffer, 1984), we see that it is a curve composed of embryonic, larval, metamorphic, and transformed periods (fig. 12.2*B:* solid line). The axolotl follows the ancestral embryonic and larval trajectories, then the head develops no further and sexual maturity occurs at the same time as in the ancestral condition (Brandon, 1977; Pfingsten and Downs, 1989). In the case of the axolotl we definitely have paedomorphosis, a permanent truncation of head shape development in this species relative to the ancestral condition (fig. 12.2*B:* arrow indicating truncation of shape). Using the time of the terminal shape as the end point of the ontogenetic trajectory we find that the axolotl has an earlier offset or cessation of shape development relative to the ancestral condition observed in *A. tigrinum* (fig. 12.2*B*). Because there is a negative perturbation of the shape offset, the heterochronic process observed in

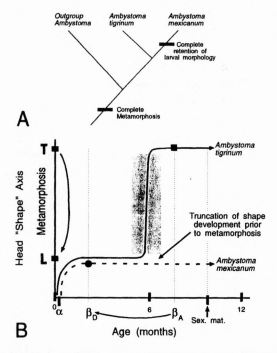

FIGURE 12.2 Interspecific heterochronic patterns of cranial paedomorphosis in ambystomatid salamanders (A). The axolotl, *Ambystoma mexicanum* (dashed trajectory), is known for its permanent complete retention of larval morphology relative to the primitive condition (solid trajectory) exemplified by its metamorphosing sister taxon *Ambystoma tigrinum* and outgroups (B). Paedomorphosis in the axolotl indicated by the shape truncation on the Y-axis (arrow) is a result of an early offset of head shape development relative to the ancestral trajectory represented by A. *tigrinum* (hypomorphosis, indicated by the arrow on the x-axis). Note that permanent truncation of development in the axolotl occurs prior to the onset of metamorphosis. L = larval morphology; T = transformed morphology; β_D = offset of descendent; β_A = offset of ancestor; (●) axolotl terminal shape, (■) ancestral terminal shape (A. *tigrinum*). Sex. mat. = onset of sexual maturity. (Tree adapted from Shaffer, 1984.)

the axolotl would be categorized as hypomorphosis (Shea, 1988; same as progenesis of Alberch et al., 1979). In addition, the cessation of axolotl shape development occurs prior to the onset of metamorphosis. Because there appears to be genetic (Tompkins, 1978; Brad Shaffer, pers. comm.) and hormonal (Dodd and Dodd, 1976; White and Nicoll, 1981) control of metamorphosis, selection seems to have acted to permanently block the onset of the entire developmental process of metamorphosis.

In the second case study we see a different pattern of hypomorphosis in the family Cryptobranchidae (Smith, 1907, 1912; Fukada, 1928; Aoyama, 1930; Noble, 1931; De Beer, 1937; Bishop, 1941; Nickerson and Mays, 1973;

Kuwabara et al., 1989). In the hellbender, *Cryptobranchus allegheniensis,* the retention of larval characters is limited to gill structures only, relative to the complete metamorphosis seen in its sister taxon, *Andrias japonicus,* the Asian giant salamander, and fossil outgroups which metamorphose completely (fig. 12.3*A;* Estes, 1981).

When we examine the ontogenetic trajectories for head shape versus age, for the ancestral condition represented by *A. japonicus,* we find that the hellbender follows the embryonic and larval trajectories and most of the metamorphic trajectory and then shape development terminates before gill structures have completely transformed (fig. 12.3*B*). The hellbender exhibits paedomorphosis (y-axis arrow) via hypomorphosis (x-axis arrow), indicated by the early and permanent offset of shape relative to the ancestral condition, not by blocking the

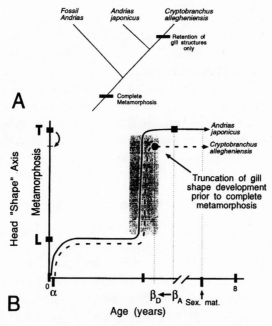

FIGURE 12.3 Interspecific heterochronic patterns of cranial paedomorphosis in cryptobranchid salamanders (*A*). The hellbender, *Cryptobranchus allegheniensis* (dashed trajectory), retains gill structures only relative to the primitive condition (solid trajectory) exemplified by its sister taxon, *Andrias japonicus,* and fossil outgroups (*B*). The heterochronic process producing paedomorphosis (y-axis arrow) in the hellbender is a result of early offset of gill shape development (hypomorphosis, indicated by the x-axis arrow). Abbreviations as in fig. 12.2. Note: following Shea (1988) the more appropriate terms *hypomorphosis* (for *progenesis*) and *deceleration* (for *neoteny*) are substituted to avoid confusion with other historical definitions of the terms used in the Alberch et al. (1979) model. (●) Hellbender terminal shape, (■) ancestral terminal shape (*Andrias japonicus*).

onset of metamorphosis but rather by affecting the ability of *some tissues* to finish metamorphosis. Other aspects of the head are metamorphosed and only branchial tissues fail to completely transform, and thus hypomorphosis is limited to gill structures. This may be due to fluctuations in the level of secretion of hormones producing metamorphosis or changes in the sensitivity of target issues. Whatever the cause, cranial morphology in this paedomorphic pattern differs from the pattern observed in the axolotl. Both the axolotl and the hellbender exhibit hypomorphosis, but the timing of shape offset occurs at different times and to different degrees relative to the onset and completion of metamorphosis.

A preliminary survey of the ontogenetic trajectories of many salamander taxa reveals that paedomorphic cranial shape trajectories (interspecific heterochronic patterns) conform to one of the two basic hypomorphic pathways shown in these two case studies (shape is offset either before or during metamorphosis as in figs. 12.2 and 12.3). Shape offset prior to metamorphosis, as in the axolotl, results from the loss of metamorphosis by the developmental truncation of a large suite of coupled transformations. Metamorphosis simply fails to occur and is evolutionarily lost (e.g., *Haideotriton, Necturus, Siren*). Shape offset during metamorphosis, as in the hellbender, results from incomplete metamorphosis or the decoupling and truncation of development of some structures prior to complete metamorphosis. Metamorphosis occurs in most cranial components, but some structures fail to metamorphose completely (e.g., *Amphiuma*). Thus, we see that paedomorphosis in the heads of salamanders seems to be channeled into one of two basic developmental pathways: loss of metamorphosis or failure to complete metamorphosis.

INTRASPECIFIC HETEROCHRONY IN CRANIAL DEVELOPMENT

In the previous section I referred to heterochrony, or changes in the timing and/or rates of development, in the evolutionary context in which it is usually defined and used (Gould, 1977, 1992), that is, interspecific heterochrony as in the two case studies just described. Interspecific heterochrony involves phyletic variation in the onset or timing of development in which the terminal shape of descendant species is offset relative to ancestral species. Interspecific heterochrony refers to heterochronic differences in development among species reflecting phyletic offsets in shape, defined as either paedomorphosis or peramorphosis (Alberch et al., 1979). It is the product of evolution and is identified by ancestor-descendant species comparisons (Alberch et al., 1979).

In a different context heterochronic development occurs on an individual level. It involves intraspecific variation in the onset or timing of development where individuals within species exhibit temporary perturbations of normal ontogenetic development en route to the species terminal shape. All individuals eventually will (can) arrive at the species terminal shape, and thus no phyletic change in shape occurs. Intraspecific heterochrony appears to be a strategy for coping with variable environments and may reflect plasticity in response to environmental heterogeneity, or it may be the result of some genetic predisposition for alternate or variant life history pathways (Smith-Gill and Berven, 1979; Wilbur, 1980). The following two case studies illustrate examples of intraspecific heterochrony in cranial development in salamanders.

The first case is intraspecific heterochrony in the mole salamander, *Ambystoma talpoideum* (fig. 12.4A). The normal species trajectory for head shape in *A. talpoideum* (fig. 12.4A: solid line, which is also the ancestral trajectory) is the standard metamorphosis trajectory (Reilly, 1987). Mole salamanders undergo embryogenesis, develop into the larval morphology, maintain this shape while growing for a few months, then metamorphose into the terrestrial form, and maturity is attained by the end of the first year of life. Within *A. talpoideum,* some individuals exhibit an alternate developmental pathway through metamorphosis (fig. 12.4A: dashed line). These individuals have the same embryonic and larval trajectories as their metamorphosing cohorts but delay the onset of metamorphosis for one or more years (with no change in the timing of sexual maturity), before undergoing normal metamorphosis to the terrestrial form (Semlitsch, 1985; Reilly, 1987). All individuals eventually metamorphose (or will when the pond dries), and thus, all attain the species' terminal shape. Comparison of the two ontogenetic trajectories reveals that the heterochronic process operating here is the temporary delay of the onset of metamorphosis.

The second case study is intraspecific heterochrony in the eastern newt, *Notophthalmus viridescens* (fig. 12.4B). The normal species trajectory for *N. viridescens* is the standard metamorphic trajectory through metamorphosis to the terrestrial "eft" stage with sexual maturity occurring after approximately three years (fig. 12.4B: solid line; Gill, 1978; Reilly, 1987).

Within this species, some individuals exhibit a variant "branchiate" trajectory (fig. 12.4B: dashed line). They go through normal larval development and metamorphosis with their cohorts, except that the metamorphic resorption of gill structures is temporarily and variably delayed, often well beyond the onset of sexual maturity (Brandon and Bremer, 1966; Reilly, 1986, 1987). Other aspects of transformation are complete and the tardy gill structures complete transformation with time or when the pond dries (Albert, 1967; Gates, 1978; Reilly, 1987), and thus all individuals eventually reach the species terminal shape. Comparison of the two ontogenetic trajectories of cranial development shows that the heterochronic process operating in *N. viridescens* is the delay of completion of branchial metamorphosis that is decoupled from otherwise normal transformation of the rest of the head (Reilly, 1986, 1987).

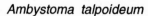

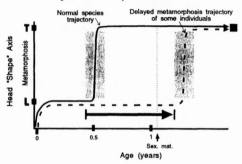

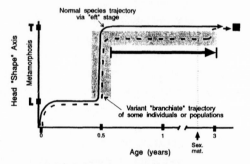

FIGURE 12.4 Case studies in intraspecific heterochrony illustrating intraspecific developmental plasticity within two metamorphosing species. (A) Some individuals of *Ambystoma talpoideum* (dashed trajectory) exhibit temporal plasticity (delay) in the onset of metamorphosis (large arrow) relative to the normal species trajectory of metamorphosis in the first year of life (solid trajectory). Head development is temporarily delayed (offset) prior to metamorphosis. All individuals can eventually proceed through normal transformation to the species terminal shape (■: T on y-axis). (B) Some individuals of *Notophthalmus viridescens* exhibit delayed branchial metamorphosis (large arrow) via temporal plasticity in the completion of metamorphosis (dashed trajectory) relative to the normal species trajectory of complete metamorphosis (solid trajectory). As branchial transformation proceeds (large arrow) all individuals eventually attain the species terminal shape (■: T on y-axis). Note that because there is no change in the species terminal shape for either case this cannot be termed paedomorphosis. Individuals within both species exhibit "temporary hypermorphosis" but the species differ in the degree of morphological transformation. (See text for discussion.)

Several points are obvious from these two intraspecific heterochronic patterns. First, there is no permanent shape offset of the terminal shape of individuals of either of these species (no phyletic shape change), thus by definition (Alberch et al., 1979), this is not paedomorphosis (there is no provision in the model for temporary perturbations in development that do not affect the species terminal shape). Many workers, when discussing individuals or species exhibiting these variant trajectories incorrectly refer to them as paedomorphs (e.g., Gould, 1977; Harris et al., 1990; Semlitsch et al., 1990).

Second, when fit to the six developmental perturbations categorized by the Alberch et al. (1979) model (ignoring the requirement for species comparisons and terminal shape shifts) the intraspecific patterns are categorized as the opposite developmental process that would produce paedomorphosis when it becomes permanent in a species. For example, using terminal shape regardless of sexual maturity as the endpoint, the model would categorize the pattern of delayed metamorphosis in *A. talpoideum* as "temporary hypermorphosis" (actually temporarily delayed offset), because shape offset of the variants is temporarily delayed relative to the species normal condition. Thus, in this case, perturbations introduced into the ontogeny of individuals that temporarily *extend* the offset point (temporary hypermorphosis, fig. 12.4), appear as *truncations* of development when the perturbation becomes permanently fixed within species (hypomorphosis, figs. 12.2 and 12.3).

Third, the actual heterochronic processes operating here to produce these intraspecific patterns are obvious from comparisons of variant trajectories to the species normal trajectories. Some *A. talpoideum* individuals exhibit temporary plasticity in timing of the onset of metamorphosis. During the delay the animals remain fully larval in morphology and feeding function. Individual *N. viridescens* exhibit temporary plasticity in the timing of completion of metamorphosis. At the time they depart from the normal metamorphic trajectory, most aspects of the body are metamorphosed except for the incomplete resorption of some posterior hyobranchial elements and external gill structures. Thus, some individuals within some populations of these species (and others such as *A. gracile, A. tigrinum*) exhibit plasticity in the timing of metamorphosis. This might be an example of "adaptive plasticity" as discussed by Travis (chap. 5, this volume) that differs in that the polymorphism (the presence of transformed and neotenic adults) is produced by the simple delayance of the onset of metamorphosis in some individuals and thus the polymorphism results from environmentally induced punctuated ontogeny. Such intraspecific variation in the timing of metamorphosis is widespread in amphibians.

Fourth, the two temporary intraspecific developmental perturbations (fig. 12.4) involve the same perturbations observed to be permanently fixed during the cranial development of paedomorphic taxa (figs. 12.2 and 12.3). Thus, we see that patterns of temporary developmental plasticity occurring presently in metamorphosing species are the same ontogenetic pathways (points of perturbation) that have become evolutionarily fixed in paedomorphic taxa.

Intra- versus Interspecific Heterochrony in Metamorphosis

The four case studies discussed above provide the basis for comparisons of the two forms of heterochrony (fig. 12.5). In intraspecific heterochrony, some individuals express variation in the timing or rate of development; these individuals

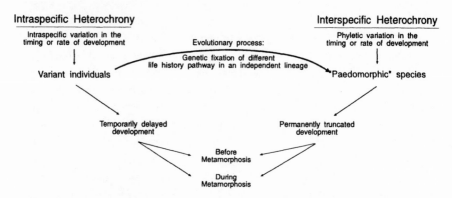

FIGURE 12.5 Intra- versus interspecific heterochrony. Heterochronic variation within species is contrasted with patterns among species. It is important to distinguish these two levels of heterochrony because interspecific patterns reflect phyletic differences in ontogeny while intraspecific patterns reveal heterochronic responses to local ecological conditions. Identical points of perturbation occur in both forms of heterochrony, indicating that cranial paedomorphosis may have evolved from the evolutionary fixation of variant developmental pathways observed within metamorphosing species today.

exhibit plasticity in the timing of metamorphosis, which is temporarily delayed either before or during metamorphosis. Intraspecific heterochronic patterns, whether environmental or heritable, reveal developmental responses to environmental conditions (Wilbur, 1980; Travis, chap. 5, this volume). In interspecific heterochrony, phyletic variation in the timing or rate of development in paedomorphic species, is exhibited as permanent truncation of development, either before or during metamorphosis. Interspecific heterochronic patterns reveal the differences that have become evolutionarily fixed. If the intraspecific heterochronic patterns became permanent and genetically isolated one would observe the two patterns of interspecific heterochrony seen in paedomorphic taxa today.

It is the genetic fixation of different life history pathways in an independently evolving lineage that separates intraspecific from interspecific patterns of heterochrony and is the evolutionary process leading to paedomorphosis. Thus, our interest in mechanisms of evolution is focused on the causal bases of these variant intraspecific patterns, occurring here and now in species that metamorphose, and how they become evolutionarily fixed. It is not known whether variation in the frequency of these variant intraspecific heterochronic pathways reflects genetic differences resulting from different selective regimes that individuals encounter in local environments, or if environmental factors lead to local or geographic variation in the phenotypic expression of these alternate pathways. A series of studies specifically designed to examine the genetic versus the environmental basis of these alternate developmental pathways has not demonstrated a clear genetic basis for these patterns and leads to the conclusion that some individuals within metamorphic species exhibit developmental plasticity in the timing of metamorphosis (Semlitsch et al., 1990).

METAMORPHOSIS OF AQUATIC FEEDING FUNCTION

From a functional morphological perspective the two developmental pathways (shape offset before or during metamorphosis) are very different in terms of aquatic feeding function and, therefore, success in exploiting aquatic food resources. Analyses of aquatic feeding behavior of salamanders (Lauder and Shaffer, 1986; Reilly and Lauder, 1988, 1992; Lauder and Reilly, 1993) have shown that prey capture performance decreases drastically at metamorphosis due to changes at several levels of functional design (Lauder and Reilly, 1990, 1993). Normal metamorphosis involves a transformation in the hydrodynamic design of the feeding mechanism from a unidirectional flow system in larvae to a bidirectional aquatic feeding system after metamorphosis (Lauder and Shaffer, 1986). In fully larval (nontransformed) salamanders, water containing the prey enters the mouth and exits posteriorly through the gill slits. After metamorphosis the gill openings have closed and water entering the mouth during aquatic prey capture must reverse direction and exit via the mouth. The change in hydrodynamic design alone is sufficient to reduce aquatic feeding performance due to the effects of hydrodynamics on the negative and positive pressure waveforms in the buccal cavity during feeding (Lauder and Reilly, 1988). In addition, metamorphic changes in the hyobranchial apparatus and the size of muscles that generate negative buccal pressure contribute significantly to reduced aquatic feeding performance in transformed individuals (Lauder and Reilly, 1990, 1993).

Thus, the two developmental pathways, whether intraspecifically plastic or phyletic, fall into two functional categories related to aquatic feeding ability. Those taxa that delay metamorphosis entirely (or offset development prior to metamorphosis), such as *A. talpoideum* and the axolotl, remain completely larval in morphology, hydrodynamic design, and aquatic feeding performance and thus are well adapted to exploit aquatic food resources. Those that retain only gill structures, such as *N. viridescens* and the hellbender, have effectively metamorphosed feeding mechanisms. The temporary retention of hyobranchial elements in "branchiate" *N. viridescens* affords no aquatic feeding performance advantage over completely metamorphosed *N. viridescens* (Reilly and Lauder, 1988) and thus, they are functionally transformed and less effective in capturing aquatic prey than larvae. In hellbenders, although a tiny single gill opening is retained permanently, bidirectional (metamorphosed) hydrodynamic flow is used during aquatic feeding (Reilly and Lauder, unpubl.). Prey capture performance in hellbenders may be secondarily enhanced, however, by an increase in buccal capacity and unique modulatory feeding behaviors and associated novelties in morphology (Cundall et al., 1987; Reilly and Lauder, 1992). Because *N. viridescens* and hellbenders are functionally metamorphosed, they are faced with a significant reduction in feeding ability when feeding in the water.

These conclusions are good examples of the power of an ecomorphological approach. By integrating of morphological development with ecological and functional data, insights are gained that could not be known from a purely ecological perspective. For example, based on external morphology alone, one might assume that hellbenders and branchiate *N. viridescens* are "larval" and therefore possess larval feeding mechanisms, and that selection favors the retention of larval gill morphology because it affords better aquatic feeding performance. Neither of these ideas are borne out by functional or ontogenetic data.

Consideration of the functional differences in the feeding mechanisms of these two pathways raises questions concerning the nature of the selective or evolutionary scenarios producing these patterns. Does the strategy of delayed metamorphosis in which individuals retain a high performance, unidirectional aquatic feeding mechanism allow animals to remain in the aquatic situation to exploit rich aquatic resources as appears to be the case in *A. talpoideum* (Semlitsch, 1987)? Or has strong selection against terrestrials produced the permanent loss of metamorphosis, as appears to be the case in many paedomorphic cave plethodontids (Sprules, 1974)?

In the other strategy of incomplete or partial metamorphosis, *N. viridescens* and hellbenders have functionally transformed bidirectional aquatic feeding systems. Because the retention of gill structures appears to be unrelated to feeding function (Reilly and Lauder, unpubl.), perhaps different selective scenarios or historical constraints are operating where aquatic resources are exploited by essentially metamorphosed forms. The inability of branchial tissues to complete metamorphosis may be a correlated response to selection on some other aspect of metamorphosis, such as the hormonal control of land-drive behavior that causes metamorphosed salamanders to leave the aquatic habitat (Reilly, 1986). Thyroxin, besides being responsible for inducing morphological metamorphosis also induces land-drive behavior (Grant and Cooper, 1965). Therefore, selection for reduced thyroxin secretion or decreased target tissue sensitivity could result in the retention of both larval morphology (gill structures) and larval behavior (aquatic habitat preference) while otherwise normal metamorphosis occurs (Marangio, 1978). Likewise, selection against terrestrials could eliminate the entire process of metamorphosis via selection against the high thyroxin levels that cause the onset of metamorphosis. Much more integration of hormonal and ecological aspects of metamorphosis is needed before we will understand the different processes acting to produce these intraspecific patterns of development and how they have become evolutionarily fixed.

ECOMORPHOLOGY AND THE ANALYSIS OF HETEROCHRONY

Given these inferences from empirical data from salamanders what can we conclude about the ecological morphological approach to the analysis of heterochrony? First, the integration of ecological (life history) and morphological data is necessary to identify and understand heterochronic processes. The age at which developmental events occur and the age of maturity are needed to construct ontogenetic trajectories. In addition, detailed ontogenetic morphological data are needed to determine the nature of differences present in organisms being compared. For example, without detailed morphological data one could commonly wrongly assume that branchiate *N. viridescens* are completely larval in head morphology. The integration of morphology, functional performance, life history data, and ecology are necessary to form a better understanding of ecological and evolutionary patterns and processes. Tests of the functional bases of differences in morphology provide valuable insights into patterns of morphological change (Lauder, 1990) but are more important in identifying ecologically relevant aspects of morphological and functional changes related to resource use. One could not know that branchiate *N. viridescens* retain only some gill structures and are functionally metamorphosed without tests of performance and function. Because branchiate *N. viridescens* have poor aquatic feeding performance one might predict that they feed on different aquatic prey than larval *N. viridescens*.

Second, intra- and interspecific patterns of heterochrony must be distinguished before heterochronic phenomena can be defined. For example, individuals exhibiting intraspecific heterochronic patterns as seen in *A. talpoideum* and *N. viridescens*, which have no terminal shape offset, are not paedomorphic. Failure to distinguish between intraspecific variability and phyletic change in ontogeny leads to conceptual confusion. For example, Harris et al. (1990) characterize the pattern of temporarily delayed metamorphosis exhibited by individuals of *A. talpoideum* as "an adaptation to variable aquatic environments" (phenotypic plasticity), on one hand, and "a macroevolutionary pattern" (transpecific change) on the other hand. Interspecific heterochrony reflects patterns of phyletic change that have become fixed in independent phylogenetic lineages. Intraspecific heterochrony involves tokogenetic relationships among semaphoronts and provides a view into the processes affecting environmental, phenotypic, and genetic variation (of individuals) in response to environmental change. Comparison of intra- and interspecific patterns of development leads to the conclusion that paedomorphic species appear to have evolved through the evolutionary fixation of variant patterns (offset points) observed within metamorphosing species. For example, it is probable that *A. mexicanum* evolved by the permanent loss of metamorphosis along a variant ontogenetic trajectory similar to that exhibited by *A. talpoideum*. Only after a speciation event has occurred can alternative intra-specific heterochronic patterns be considered in phyletic terms such as paedomorphosis. Confusion also arises with the use of the terms "facultative" and "obligate" paedomorphosis. Because paedomorphosis, by definition (Gould, 1977, 1992, pers. comm; Alberch et al., 1979; Kluge and Strauss, 1985; Wake and Roth, 1989b), can only be obligate (i.e., evolutionarily fixed, a phyletic difference among species), the term "facultative paedomorphosis" is meaningless, the term "obligate paedomorph" is a tautologism, and neither should be used to describe the intraspecific heterochronic patterns such as those exhibited by *N. viridescens* and *A. talpoideum*. Studies bridging intra- and interspecific heterochrony are necessary to make meaningful hypotheses relating ecological and genetic variation to transpecific evolutionary change. The Alberch et al. (1979) framework was formalized to categorize phylogenetic heterochronic processes and has no provision for categorizing developmental heterochrony that does not affect the species terminal shape (intraspecific heterochrony). Great care should be taken when applying the model to categorize ontogenetic perturbations of intraspecific heterchronic patterns. Its implementation to identify paedomorphic or peramorphic phenomena should be restricted only to comparisons of phyletic differences in development (interspecific heterochrony) where more or less complete ontogenetic trajectories are available.

Third, heterochronic analysis can reveal major ontogenetic pathways through which development is channeled or decoupled by the nature of genetic and developmental systems or in response to environmental selection. As illustrated in the case studies, developmental plasticity in metamorphosis appears to be channeled into two basic pathways. One involves the total truncation of metamorphic development while the other involves the decoupling of development of some tissues during metamorphosis. This important difference would not be known without an integrative approach. Much more detailed analyses are needed to quantify the developmental envelope and decouplings that encompass the ontogenetic trajectories for all of the many transformations that occur during metamorphosis.

ACKNOWLEDGMENTS

I thank Al Bennett, Ron Brandon, Jim Collins, Alistair Cullum, Bill Duellman, Zoe Eppley, Bill Fink, Alice Gibb, Steve Gould, Joe Holomuzki, Bruce Jayne, George Lauder, Dan Meinhardt, Don Miles, Scott Moody, Rich Strauss, Linda Trueb, Miriam Ashley-Ross, Peter Wainwright, Dave Wake, and Ed Wiley for many discussions of the concepts of heterochrony and how best to describe them. This research was supported by the Graduate School of Southern Illinois University at Carbondale, NSF grants BSR 8520305 and DCB 8710210 to G. Lauder, and an Ohio University Research Challenge grant. Generous support from the Cocos Foundation for the symposium "Ecological morphology: Integrative approaches in organismal biology" presented at the American Society of Zoologists meetings in San Antonio (1990) is gratefully acknowledged.

References

Alberch, P. 1985. Problems with the interpretation of developmental sequences. *Syst. Zool.* 34:46–58.

Alberch, P., S. J. Gould, G. F. Oster, and D. B. Wake. 1979. Size and shape in ontogeny and phylogeny. *Paleobiology* 5:296–317.

Albert, E. H. 1967. Life history of neotenic newts *Notophthalmus viridescens louisianensis* (Wolterstorff), in southern Illinois. Master's thesis, Southern Illinois University, Carbondale.

Altman, P. L., and D. S. Dittmer. 1962. *Biological Handbooks, II. Growth Including Reproduction and Morphological Development.* Washington, D.C.: Federation of American Societies for Experimental Biology.

Aoyama, F. 1930. Die Entwicklungsgeschichte des Kopfskelettes des *Cryptobranchus japonicus. Z. Anat.* 93:109–181.

Bishop, S. C. 1941. *The Salamanders Of New York.* New York State Mus. Bull. 324:1–365.

Brandon, R. A. 1977. Interspecific hybridization among Mexican and United States salamanders of the genus *Ambystoma* under laboratory conditions. *Herpetologica* 33:133–152.

Brandon, R. A. 1989. Natural history of the axolotl and its relationship to other ambystomatid salamanders. In *Developmental Biology of the Axolotl,* ed. J. B. Armstrong and G. M. Malacinski, 13–21. Oxford: Oxford University Press.

Brandon, R. A., and D. J. Bremer. 1966. Neotenic newts *Notophthalmus viridescens louisianensis* (Wolterstorff), in southern Illinois. *Herpetologica* 22:213–217.

Cundall, D., J. Lorenz-Elwood, and J. D. Groves. 1987. Asymmetric suction feeding in primitive salamanders. *Experientia* 43:1229–1231.

De Beer, G. R. 1937. *The Development of the Vertebrate Skull.* Oxford: Clarendon Press.

Dodd, M. H. I., and J. M. Dodd. 1976. The biology of metamorphosis. In *Physiology of the Amphibia,* vol. 3., ed. B. Lofts, 467–599. New York: Academic Press.

Duellman, W. E., and L. Trueb. 1986. *Biology of Amphibians.* New York: McGraw Hill.

Emerson, S. B. 1986. Heterochrony and frogs: The relationship of a life history trait to morphological form. *Amer. Nat.* 127:167–183.

Estes, R. 1981. Gymnophiona, Caudata. *Handb. Paläherpetol.* 2:1–115.

Fink, W. L. 1982. The conceptual relationship between ontogeny and phylogeny. *Paleobiology* 8:254–264.

Fink, W. L. 1988. Phylogenetic analysis and the detection of ontogenetic patterns. In *Heterochrony in Evolution,* ed. M. L. McKinney, 71–91. New York: Plenum Publishing Corp.

Fukuda, Y. 1928. Zur Morpholgie und Entwicklungsgeschichte des Hyobranchialskelettes von *Megalobatrachus japonicus. Folia Anat. Japonica* 6:327–374.

Garstang, W. 1922. The theory of recapitulation: A critical restatement of the biogenetic law. *J. Linnean Soc. Zool.* 35:81–101.

Gates, D. W. 1978. Relation of prolonged retention of larval characters to sex, size, age, and season in newts (*Notophtalmus viridescens*). Master's thesis, Southern Illinois University, Carbondale.

Giles, R. A., and M. L. Stapleton. 1966. The eventual variations in size (weight, length) and maturity of the Mexican axolotl (*Siredon mexicanum*) resulting from various feeding schedules. *Proc. Penn. Acad. Sci.* 39:78–81.

Gill, D. E. 1978. The metapopulation ecology of the red-spotted next, *Notophthalmus viridescens* (Rafinesque). *Ecol. Mon.* 48:145–166.

Gould, S. J. 1977. *Ontogeny and Phylogeny.* Cambridge: Harvard University Press.

Gould, S. J. 1988. The uses of heterochrony. In *Heterochrony in Evolution*, ed. M. L. McKinney, 1–13. New York: Plenum Publishing Corp.

Gould, S. J. 1992. Heterochrony. In *Keywords in Evolutionary Biology*, ed. E. F. Keller and E. A. Lloyd, 158–165. Cambridge: Harvard University Press.

Grant, W. C., and G. Cooper. 1965. Behavioral and integumentary changes associated with induced metamorphosis in *Diemictylus*. *Biol. Bull.* 129:510–522.

Hanken, J. 1992. Life history and morphological evolution. *J. Evol. Biol.* 5:549–557.

Harris, R. N., R. D. Semlitsch, H. M. Wilbur, and J. E. Fauth. 1990. Local variation in the genetic basis of paedomorphosis in the salamander *Ambystoma talpoideum*. *Evolution* 44:1588–1603.

Harrison, R. G. 1969. Harrison stages and description of the normal development of the spotted salamander, *Amblystoma punctatum* (Linn.). In *Organization and Development of the Embryo*, ed. R. G. Harrison, 44–66. New Haven: Yale University Press.

Istock, C. A. 1967. The evolution of complex life cycle phenomena: An ecological perspective. *Evolution* 21:592–605.

Jones, D. G. 1988. Sclerochronology and the size versus age problem. In *Heterochrony in Evolution*, ed. M. L. McKinney, 93–108. New York: Plenum Publishing Corp.

Kluge, A. G., and R. E. Strauss. 1985. Ontogeny and systematics. *Ann. Rev. Ecol. Syst.* 16:247–268.

Kuwabara, K., N. Suzuki, F. Wakabayashe, H. Ashikaga, T. Inoue, and J. Kobara. 1989. Breeding the Japanese giant salamander *Andrias japonicus*. *Int. Zoo Yb.* 28:22–31.

Lauder, G. V. 1990. Functional morphology and systematics: Studying functional patterns in an historical context. *Ann. Rev. Ecol. Syst.* 21:317–340.

Lauder, G. V., and S. M. Reilly. 1988. Functional design of the feeding mechanism in salamanders: Causal bases of ontogenetic changes in function. *J. Exp. Biol.* 143:219–233.

Lauder, G. V., and S. M. Reilly. 1990. Metamorphosis of the feeding mechanism in tiger salamanders (*Ambystoma tigrinum*) *J. Zool.* (London) 222:59–74.

Lauder, G. V., and S. M. Reilly. 1994. Amphibian feeding behavior: Comparative biomechanics and evolution. In *Biomechanics and Feeding in Vertebrates*, ed. R. Gilles. Berlin: Springer-Verlag. In press.

Lauder, G. V., and H. B. Shaffer. 1986. Functional design of the feeding mechanism in lower vertebrates: Unidirectional and bidirectional flow systems in the tiger salamander. *Zool. J. Linnean Soc.* 88:277–290.

Lofts, B., ed. 1974. *Physiology of the Amphibia*, vol. 2. New York: Academic Press.

McNamara, K. J. 1988. The abundance of heterochrony in the fossil record. In *Heterochrony in Evolution*, ed. M. L. McKinney, 287–325. New York: Plenum Publishing Corp.

Marangio, M. S. 1978. The occurrence of neotenic rough-skinned newts (*Taricha granulosa*) in montane lakes of southern Oregon. *Northwest Sci.* 52:343–350.

Müller, G. B., and G. P. Wagner. 1991. Novelty in evolution: Restructuring the concept. *Ann. Rev. Ecol. Syst.* 22:229–256.

Nickerson, M. A. and C. E. Mays. 1973. *The Hellbenders.* Milwaukee Pub. Mus. Biol. Geol. no. 1.

Nitecki, M. H., ed. 1990. *Evolutionary Innovations.* Chicago: University of Chicago Press.

Noble, G. K. 1931. *The Biology of the Amphibia.* New York: McGraw-Hill.

Pfingsten, R. A., and F. L. Downs, eds. 1989. *Salamanders of Ohio.* Ohio Biol. Surv. Bull. New Series vol. 7 no. 2.

Raff, R. A., A. L. Parks, and G. A. Wray. 1990. Heterochrony and other mechanisms of radical change in early development. In *Evolutionary Innovations*, ed. M. H. Nitecki, 71–98. Chicago: University of Chicago Press.

Reilly, S. M. 1986. Ontogeny of cranial ossification in the eastern newt, *Notophthalmus viridescens* (Caudata: Salamandridae), and its relationship to metamorphosis and neoteny. *J. Morphol.* 188:215–326.

Reilly, S. M. 1987. Ontogeny of the hyobranchial apparatus in the salamanders *Ambystoma talpoideum* (Ambystomatidae) and *Notophthalmus viridescens* (Salamandridae): The ecological morphology of two neotenic strategies. *J. Morphol.* 191:205–214.

Reilly, S. M. 1990. Comparative ontogeny of cranial shape in salamanders using Resistant Fit Theta Rho analysis. In *Proceedings of the Michigan Morphometrics Workshop*, ed. F. J. Rohlf and F. L. Bookstein, 311–321. Ann Arbor: University of Michigan Press.

Reilly, S. M., and G. V. Lauder. 1988. Ontogeny of aquatic feeding performance in the eastern newt *Notophthalmus viridescens* (Salamandridae). *Copeia* 1988:87–91.

Reilly, S. M., and G. V. Lauder. 1990a. Metamorphosis of cranial design in the tiger salamander (*Ambystoma tigrinum*): A morphometric analysis of ontogenetic change. *J. Morphol.* 204:121–127.

Reilly, S. M., and G. V. Lauder. 1990b. Evolution of tetrapod feeding behavior: Kinematic homologies in prey transport. *Evolution* 44:1542–1557.

Reilly, S. M. and G. V. Lauder. 1990c. The strike of the salamander: Quantitative electromyography and muscle function during prey capture. *J. Comp. Physiol.* A167:827–939.

Reilly, S. M., and G. V. Lauder. 1991a. Prey transport in the tiger salamander: Quantitative electromyography and muscle function in tetrapods. *J. Exp. Zool.* 260:1–17.

Reilly, S. M., and G. V. Lauder. 1991b. Experimental morphology of the feeding mechanism in salamanders. *J. Morphol.* 210:33–44.

Reilly, S. M., and G. V. Lauder. 1992. Morphology, behavior and evolution: Comparative kinematics of aquatic feeding in salamanders. *Brain Behav. Evol.* 40:182–196.

Rusconi, M. 1823. Observations on the natural history and structure of the aquatic salamander, and on the development of the larva of these animals from the egg up to the perfect animal. *Edinburgh Phil. J.* 9:107–118.

Semlitsch, R. D. 1985. Reproductive strategy of a facultatively paedomorphic salamander *Ambystoma talpoideum*. *Oecologia* 65:305–313.

Semlitsch, R. D. 1987. Paedomorphosis in *Ambystoma talpoideum:* Effects of density, food, and pond drying. *Ecology* 68:994–1002.

Semlitsch, R. D., R. N. Harris, and H. M. Wilbur. 1990. Paedomorphosis in *Ambystoma talpoideum:* Maintenance of population variation and alternative life-history pathways. *Evolution* 44:1604–1613.

Shea, B. T. 1983. Allometry and heterochrony in the African apes. *Amer. J. Phys. Anthrop.* 62:275–289.

Shea, B. T. 1988. Heterochrony in primates. In *Heterochrony in Evolution*, ed. M. L. McKinney, 237–266. New York: Plenum Publishing Corp.

Shaffer, H. B. 1984. Evolution of a paedomorphic lineage. I. An electrophoretic analysis of the Mexican ambystomatid salamanders. *Evolution* 38:1194–1206.

Smith, B. G. 1907. The life history and habits of *Cryptobranchus alleghaniensis*. *Biol. Bull.* 13:291–303.

Smith, B. G. 1912. The embryology of *Cryptobranchus alleghaniensis* including comparisons with some other vertebrates. *Biol. Bull.* 23:455–579.

Smith-Gill, S. J., and K. A. Berven. 1979. Predicting amphibian metamorphosis. *Amer. Nat.* 133:563–585.

Sprules, W. G. 1974. The adaptive significance of paedogenesis in North American species of *Ambystoma* (Amphibia: Caudata): An hypothesis. *Can. J. Zool.* 52:393–400.

Strauss, R. E., and L. A. Fuiman. 1985. Quantitative comparisons of body form and allometry in larval and adult Pacific sculpins (Teleostei: Cottidae). *Can. J. Zool.* 63:1582–1589.

Tompkins, R. 1978. Genic control of axolotl metamorphosis. *Amer. Zool.* 18:313–319.

Wake, D. B. 1966. Comparative osteology and evolution of the lungless salamanders, family Plethodontidae. *Mem. S. Calif. Acad. Sci.* 4:1–111.

Wake, D. B., and G. Roth, eds. 1989a. *Complex Organismal Functions: Integration and Evolution in Vertebrates.* Chichester: John Wiley and Sons.

Wake, D. B., and G. Roth, eds. 1989b. The linkage between ontogeny and phylogeny in the evolution of complex systems. In *Complex Organismal Functions: Integration and Evolution in Vertebrates,* ed. D. B. Wake and G. Roth, 361–377. Chichester: John Wiley and Sons.

White, B. A., and C. S. Nicoll. 1981. Hormonal control of metamorphosis in amphibians. In *Metamorphosis: A Problem in Developmental Biology,* ed. L. I. Gilbert and E. Frieden, 363–396. New York: Plenum Press.

Wilbur, H. 1980. Complex life cycles. *Ann. Rev. Ecol. Syst.* 11:67–93.

13

Conclusion: Ecological Morphology and the Power of Integration

Stephen M. Reilly and Peter C. Wainwright

Ecological morphology is broadly concerned with making connections between how organisms are constructed and the ecological and evolutionary consequences of that design. Three avenues of research have been particularly prominent in the development of ecological morphology, and in various combinations they form the conceptual framework for the chapters of this book. First, perhaps the greatest long-term success in morphological research has been in analyzing how organisms function (Alexander, 1975; Gans, 1985; Denny, 1988; Lauder, 1990, 1991; Niklas, 1992; Reilly and Lauder, 1992; Vogel, 1981; Wainwright et al., 1976; Wake and Liem, 1985). This line of investigation continues to be extremely productive and has led to a broad understanding of morphological diversity within a functional context. Second, a perspective introduced by some ecologists has been to assume that organisms are reasonably well adapted to their environment, to hypothesize that ecological traits of species can be inferred from morphological traits, and to test this hypothesis through community-level analyses of the correlation between measured morphological and ecological attributes (Karr and James, 1975; Miles and Ricklefs, 1984; Miles et al., 1987; Ricklefs and Travis, 1980; Winemiller, 1991). This approach has also provided several testable predictions about the morphological distribution of species in communities (Karr and James, 1975; Hutchinson, 1959; Ricklefs and Miles, chap. 2, this volume). Third, much work is aimed at asking how morphological and functional diversity evolves. In this area are both population studies of natural selection of focal traits (Endler, 1986; Lande and Arnold, 1983) and phylogenetic studies that seek to establish the role of evolutionary history in shaping the form of the organism and their relationship with the environment (Brooks and McLennan, 1991; Lauder and Liem, 1989).

The overriding theme of this volume has been the integration of these three research axes toward a deeper understanding of the ecological and evolutionary consequences of organismal design. Thus, through this volume the contributors advocate an approach to morphology in which one asks how the design functions, what the consequences are of its use as the organism interacts with its environment, and how the organism came to have that design. This is a multidimensional task. Accomplishing this level of thoroughness requires the integration of techniques and viewpoints that have traditionally been separate.

We began by posing several questions to each author: What approaches are used to understand ecomorphological relationships? How can the knowledge of morphology and function improve the understanding of ecology and vice versa? How can the different approaches be better integrated into a common general framework for studying ecological morphology? What other factors must be accounted for in making inferences about the links between morphology, function, and ecology? We hope that the authors' answers to these questions suggest some valuable directions that integrative morphological research can take and illustrate some of its successes. Here we offer a discussion of the levels of analysis that may be integrated in ecomorphology, a summary of the messages generated by the authors in response to the issues posed above, and some implications of those views.

LEVELS OF ANALYSIS

Biological diversity is a product of evolution and any attempt to understand diversity must recognize the relationships between morphology and ecology. But, how do we make the connection between the form of the organism and the ecological and evolutionary consequences of that form? To expand the discussion of the ecomorphological approach a generalized integrative scheme for analyzing ecomorphological relationships is presented in figure 13.1 as a series of interconnected levels of inquiry. This representation allows the discussion of several approaches to the study of ecomorphology and illustrates a hierarchical relationship between areas of traditional focus. We note that variations on this diagram are discussed throughout this book (chaps. 2, 3, and 10) and by Arnold (1983). The levels of analysis are arranged from the sub-organismal traits of interest to their functional roles, performance consequences, to their effects on resource use and fitness. To facilitate the discussion of ecomorphological approaches we first provide a general description of each level and how they are interconnected.

Morphology: Ecologically Relevant Structures

Morphology is the lowest level and it represents the description of any aspect of the structure of the phenotypic traits to be related to ecology. This level is the analysis of anatomy, form, or shape of specific traits from which ecological inferences are to be made. These can be of any level of organismal design from molecular and cellular structures to the various components of complex morphological systems of behavioral mechanisms.

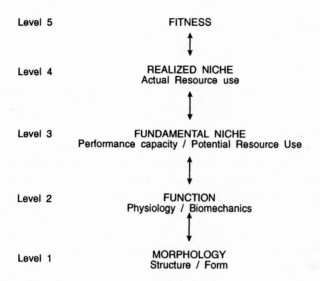

FIGURE 13.1 Hierarchical relationships of interconnected levels of analysis in ecomorphological studies. Function and performance provide the key linkages in relating morphological and ecological patterns of variation. The view that integrating ecology (realized niche, resource use, and fitness), performance capacity (and inferred fundamental niche), and functional morphology is necessary to accurately interpret the ecological role and evolution of organismal design has emerged as a central paradigm in ecomorphology.

Function: How Does the System Work?

The second level moves from the structures of interest to how they are used by the organism. The "function" of a structure may be defined as how it is used by the organism in natural behaviors and it may encompass physiological to biomechanical levels. On a structural level, studies of material properties and biomechanics have revealed physical principles that underlie the functional consequences of morphological design (Vogel, 1981, 1988; Wainwright, 1988; Wainwright et al., 1976). Building from these biomechanical principles, the rigorous analysis of function and its morphological basis is the next step for a sound understanding the higher ecological and evolutionary consequences of form. It is useful to distinguish between the function of a structure in a particular morphological system (biceps flexes forearm) and its "biological role" (Bock and van Wahlert, 1965), which refers to how the structure is used in the day-to-day life of the organism (forearm elevates fly-rod). Analysis of function, however, need not be made within the context of the biological role. Indeed, a rich tradition exists in inferring function by comparing structures against theoretical design optima (Alexander, 1982; Dullemeijer, 1974; Sibly and Calow, 1986). In such cases, the biological role is usually greatly simplified in order to facilitate the modeling.

Performance

The key concept conveyed in figure 13.1 is the role of organismal performance as a mechanistic link between organismal design and the ecological and evolutionary consequences of design. This idea has been introduced by several authors (e.g., Huey and Stevenson, 1979; Bock and von Wahlert, 1965; Bartholomew, 1982, 1986), although Arnold (1983) is generally credited with the popularization of the concept and with connecting the scheme to experimental paradigms in quantitative genetics and ecology. By performance, Arnold meant standardized measures of organismal ability that represent integrative measures of the sum of the functional and physiological processes and morphological properties that produce a particular behavior. His focus was to relate variation in morphology to variation in survival by measuring natural selection on some measure of whole organism performance, and in this way the morphological basis of performance could be related to fitness. Performance, or how well a particular behavior or function is performed, is the key to linking morphology to the ability of the individual to perform ecologically relevant tasks, and thus it has become a central component of research programs that seek a causal connection between organismal design and ecological patterns (Bennett, 1989; Emerson and Arnold, 1989; Garland, 1985; Lauder, 1991; Wainwright, 1987, 1991). It is useful to divide "performance" into two levels that can be related to "potential" and "actual" resource use. Borrowing from Hutchinson's (1957) niche terminology, the performance capacity (maximum measured range of performance) can be used to predict the potential resource limits of the functional system (the fundamental niche), which can in turn be compared to actual performance and resource use (the realized niche).

Fundamental Niche: Inferring Potential Resource Use from Performance Capacity

An individual's performance capacity places fundamental limits on the range of possible activities, and ultimately the resources that can be obtained. The performance capacity (fig. 13.1), whether actually measured or predicted from morphology, generally reflects the maximum range of abilities of functioning morphological systems. For example, the range of potential prey for a mollusk-eating fish will be limited to those with shells that the fish is able to crush. In such a case prey-crushing performance measured in the laboratory (performance capacity) can be clearly interpreted as indicative of the range of mollusk prey (along a prey-hardness axis) the fish can potentially eat (Wainwright, 1987). Thus, the performance capacity of a sample of fishes could be used to infer the fundamental niche or potential prey that could be exploited.

Ultimately it is the relationship between morphology and function that determines the performance capacity of the individual and thus, its potential resource use. So our ability to understand the causal connection between morphology and performance influences our confidence in the morphological basis of the performance capacity being measured and potential resource use being inferred. What the animal and morphological system do in the real world and actual resource use is our second performance level.

Realized Niche

The measurement of actual performance or resource use under natural conditions is known as the realized niche. The realized niche space determined by the particular morphology and function under study involves the quantification of ranges of behavior (performance) in the field or patterns of resource use (i.e., good, space, habitat use). In the example of the mollusk-crushing fish, the realized niche is the range of prey types used by free-living individuals. Comparing the fundamental niche to the realized niche identifies what portion of performance capacity is utilized by the organism in nature and how well performance capacity matches resource use.

Organisms may utilize only part of their performance capacity in nature or use only part of the resources that they could potentially exploit. However, when resource use predicted from performance capacity closely matches resource use in natural populations, a clear link is established in which performance capacity can be thought of as representing the maximum range of resources that the individual will use (Wainwright, 1987). For example, if wave forces limit habitat use on rocky intertidal substrata (Denny, chap. 8, this volume), and a given body morphology can only resist wave forces up to some clear limit, then the potential habitat of the individual, as a function of the animal's ability to cling to the substrate under the stresses of waves, can be inferred directly from laboratory performance tests of its clinging ability. In this case the realized niche (actual habitat use) matches the fundamental niche (potential habitat use) predicted from performance capacity (clinging ability). Comparing potential resource use estimated from performance studies to actual resource use measured in the field is an important link to be considered before performance capacity (and thus morphology) can be meaningfully related to the next level of inquiry: fitness (fig. 13.1).

Fitness: Selection on Realized Performance

The ultimate test of the adaptive significance of morphological features through their functional roles, potential, and realized performance is their effect on the individual's fitness. This involves the demonstration of whether the traits of interest are heritable and the nature of selection on their performance through the analysis of quantitative genetics, survivorship, and the number of offspring surviving. This is usually done by comparing the distribution of phenotypes in a population before and after periods of selection (Lande and Arnold, 1983; Emerson and Arnold, 1989). Studies of natural selection in natural populations attempt to elucidate the current ecological importance of the traits under study.

Viewing performance gradients in two levels (relating performance capacity to potential and actual resource use) sets up an obvious question that probes the ultimate connection between morphology and ecology. Do differences between species, or individuals within species, in performance capacity account for differences in resource use? Since many factors other than behavioral capability work to further refine patterns of resource use and survival (e.g., competition, environmental variability, energetic considerations, threat of predation, size), it cannot be assumed that differences between organisms in resource use patterns are due directly to the functional capabilities of those individuals. Indeed, few studies have partitioned the relative roles of behavioral capacity, actual resource use, and other factors in shaping observed patterns of resource distribution and fitness. In most cases this is because resource use is inferred from morphology, rather than being directly measured or estimated from performance capacity. Individual fitness depends on the ability of the individual to obtain resources, survive, and maximize reproductive output. However, fitness may be most dependent on performance capacity during a few short life-or-death interactions with predators that ultimately influence reproductive success. Integrative efforts between functional morphologists, natural historians, and ecologists are needed to identify and quantify the interconnections between performance capacity, ecological patterns, and individual fitness. Quantification of these various links will indeed provide new insights and spawn conceptual advances in the understanding of adaptation, fitness and evolution.

ECOMORPHOLOGICAL APPROACHES

Ecomorphology has many meanings to many people. Historically, the focus of several subdisciplines in comparative biology has been to describe and compare the morphology of organisms indicated by level 1 in figure 13.1. Data from this level of analysis is often used to directly *infer* the function of morphological traits, or its is combined with experimentally measured functional data to determine the function, physiology, or biomechanics of complex morphological systems (level 2). Information on morphology and function are then used to make inferences about the evolution of morphological systems and their anatomical components. Lauder and Liem (1989) and Reilly and Lauder (1992) outline multilevel approaches to the analysis of functional morphology employed in determining the mechanistic bases of behavioral performance. Levels might include central nervous system circuitry, central nervous system structure, motor pattern, muscle physiology, and peripheral morphology. One or more of these levels may evolve as behavioral performance adapts to changes in the environment. This approach can be inferentially powerful, especially when individual and taxonomic variation is accounted for and the role of phylogenetic history in constraining form and function is considered (Lauder, 1981, 1982, 1990, 1991; Losos and Miles, chap. 4, this volume).

Often, functional morphologists take an ecomorphological approach that attempts to relate design features and performance in functional systems to specific ecological traits such as diet (levels 1 through 4). A major strength of this approach is that detailed biomechanical or experimental analyses can pinpoint the specific morphological traits that can be expected to have the greatest impact on performance and thus, resource use (Wainwright, chap. 3). When variation in these traits is compared to relevant aspects of the organism's ecology, then convincing conclusions can be drawn regarding the adaptive roles of morphology, function, and performance in shaping ecology and fitness (Wainwright, chap. 3; Denny, chap. 8).

Another major approach in ecomorphology has often been employed by ecologists to match species to environments and to examine the morphological diversification in communities of species in relation to habitat use (Ricklefs and Miles, chap. 2). Thus, the distribution of morphological types in ecological space is determined by connecting levels 1 and 4 (fig. 13.1). This provides comparisons (usually multivariate) of the nature of the morphospace and ecospace occupied by species in a community. A major assumption (usually untested) of this approach is that variation in the measured morphological traits reflects environmentally induced natural selection and the historical affects of species interactions on those traits. In chapter 2, Ricklefs and Miles review the utility of this approach in producing generalizations about patterns of convergence, community organization, species-habitat associations, foraging ecology, and ecological relationships measured independently of the habitat context. Many studies using this approach have found a directionality to the relationship between morphology and ecology (morphology predicts ecology better than ecology predicts morphology), however much of the variance in both types of variables is often left unexplained. Norberg uses this approach in chapter 9 to relate the morphological basis of flight performance to feeding behavior in bats to show that aerodynamic flight capabilities predicted from morphology closely match the foraging behavior and associated diets observed in a large cross-section of morphotypes. Norberg estimates the flight-performance consequences of bat wing morphology, using aerodynamic theory developed in engineering research. She makes clear predictions about how interspecific differences in wing morphology will influence aspects of flight performance such as flight speed, hovering ability, and agility. A good correlation exists between morphological patterns and habitat use that follow from the expected performance consequences of wing design. Our confidence in the causal relationship between wing morphology and habitat use in bats is fundamentally dependent upon our confidence that actual flight performance reflects the biomechanical and aerodynamic principles assumed in the analysis of wing design. These predictions are being tested through the analysis of actual performance from filmed flight trials.

Arnold's (1983) thesis that performance is a key link between the traits of interest and the fitness of individuals has emerged as a central paradigm in ecomorphology. An important corollary of his thesis is that to fully understand the causal relationship between morphology and fitness one must identify the morphological basis of behavioral performance (demonstrate causal links between levels 1 through 3; Lauder, 1990, 1991) and then establish how fitness is effected by the ecological limits of performance (by adding level 4). This view forms a central message from the chapters of this book: integrating ecology (realized niche, resource use, and fitness), performance capacity (and inferred fundamental niche), and functional morphology is necessary to accurately interpret the evolution and ecological role of organismal design.

INTEGRATION IN ECOMORPHOLOGICAL STUDIES

A theme in most of the chapters of this book and in works by previous proponents of performance testing is that much can be gained by the integration of several levels of analysis in ecomorphology. It is easy to see the power of integrating all of the levels of analysis depicted in figure 13.1. One advantage of integrating levels of analysis is that it allows prediction with understanding. In the ideal situation one could determine the functions of specific morphological traits in performing some ecologically relevant task, measure maximal performance in the lab and realized performance in nature, and then demonstrate how ecological variation in realized performance relates to survival and fitness of individuals in a population over time. One could then track the functional significance of morphological traits on performance to the ecological importance of performance on fitness. To date no study has implemented such a comprehensive integrative approach employing all of the levels of analysis (fig. 13.1); however, the studies in this book that integrate two or more of these levels emphasize two major points that call for integration in ecomorphology.

First, inferences made about the effects of one level on another depend on the assumption (or demonstration) that the traits under study are causally relevant to the next level of analysis. Ricklefs and Miles (chap. 2) examine the assumption that morphological traits have some adaptive significance that determines patterns of resource allocation and discuss one problem that weakens inferences about the ecomorphological associations: the function of many morphological traits that are used are too poorly understood to infer their functional role in either the fundamental or realized niche. Studies of function are necessary to identify the morphological traits most important in producing the behavior of interest (Wainwright, chap. 3; Bradley, chap. 11). Once this link has been made, analysis of fitness correlates of morphological and performance traits can be used to reveal those aspects that are most important for survival and thus, are under selection in that particular environment or community. In addition, it is important that experimental performance measures accurately reflect the presumed ecological role as closely as possible.

Second, the adaptive significance of many ecologically important traits has not been demonstrated. In reviewing research on locomotion in reptiles, Garland and Losos (chap. 10) point out that although the physiological and functional bases of locomotor performance have been extensively studied and are reasonably well understood, we know comparatively little about the ecological consequences or adaptive significance of differences in locomotor capacities among individuals or species. Ecological traits used to estimate the realized niche should be carefully selected to establish or estimate the ecological role of the system in determining resource use. In addition, the adaptive value of traits will likely vary among individuals, among populations, and temporally within populations; thus, care must be taken to quantify ecological characteristics of the populations under study.

Obviously there is a need for better integration between ecologists, functional biologists, and morphologists and their data, but this is not a new idea. Bock (1977) argued for integration and collaboration and points out that many earlier workers have suggested that functional and ecological observations of animals were essential to understand the adaptiveness and evolution of morphological features. So how can these levels be better integrated? The best approach seems to be to integrate the laboratory analysis of descriptive and functional morphology and organismal performance capacity with field investigations of the realized performance niche, biological roles, and ecological factors. All of these types of information are needed to carry out an integrative study that convincingly interrelates all of the levels outlined in figure 13.1.

Another approach is to better integrate the literature. Morphologists often ignore fairly extensive bodies of literature that could be used to identify pertinent ecological traits that could be related to function or performance. Ecologists too, frequently fail to use the often extensive morphological and functional literature from which they could better judge specific morphological traits to relate to ecological data. Workers from both disciplines need to discuss how traits are chosen, how they are expected to function, how this function or behavior is going to be related to its ecology, and how performance is specifically related to its biological role. Performance must be tested in its biological role before it can be related to variation in morphology or ecology. The obvious solution to the necessity to integrate all of these levels of analysis is collaborative research among workers in these different subdisciplines.

OTHER FACTORS TO CONSIDER

Biologists often attribute ecomorphological associations to the process of adaptation. Many studies, however, have shown that ecomorphological associations may be affected by a variety of other factors which may have either strong or negligible effects on the matching of the organism to the environment. These factors that may constrain or explain ecomorphological associations and their affects on the relationship between organismal design and performance must be considered before we can have a full picture of the interaction between morphology and the environment.

Accounting for Variation

All links to be inferred in ecomorphology depend on our success measuring characteristics about the organism, its abilities, and its niche. When sampling these traits it is important that we quantify the variability of each measure. Interindividual variation in morphology, function, performance, and ecology is often very large, so great as to cloud or prevent our interpretation of the differences between individuals, populations, or species. Some measure of the variability in traits among individuals or populations is necessary to be able to empirically relate a value for a trait to its functional, ecological, and evolutionary consequences. Accounting for variability is necessary at all levels of analysis: morphological variation and its relation to functional variation, repeatability of performance measures, variation in resource use, and the effects of variation on the resolution of measures of fitness.

Interpopulational variation is also shown to be great, so great as to make comparisons among populations potentially hazardous. This precludes the substitution of data from populations other than those on which the other levels of analysis have been studied. Due to the nature of variation, the substitution of data at any level from other populations or individuals, however similar, can effectively negate any conclusions rendered from those ecomorphological comparisons. Workers should strive to use the same specimens, populations, species, and localities when gathering data from each level of inquiry. Viewed in another light, however, variation is central to all levels of ecomorphological analysis. It is the relationship between individual variation in morphology, performance, and fitness that is the focus of studies of natural selection. Similarly, community level analyses look for the concordance between morphological variation among species and variation in their use of the environment (Ricklefs and Miles, chap. 2).

Phenotypic and Environmental Plasticity

Phenotypic and environmental plasticity illustrates the bidirectional nature of forces in figure 13.1. A given genotype can produce different phenotypes in response to cues in the environment, thus phenotypic traits are rarely (never?) constant within species. Some traits have lows levels of variation while others exhibit interindividual, interpopulational, or generational variation. Variation may be due to genetic variation or environmental effects on development. As discussed by Travis in chapter 5, evaluating the nature and adaptive role of morphological plasticity is a major challenge for ecomorphology and illustrates the need to distinguish the effects of variation *in* the genotype from variation in environmental effects *on* a common genotype (a point also made by Reilly, chap. 12). Travis points out that phenotypic plasticity itself is typically adaptive and not simply a passive property of the organism with no influence on the individual's fitness. The discreteness of polymorphisms and the possibility of "adaptive plasticity" must be accounted for through analyses of variation. Ideally, the same consideration should be given to each level of analysis, including variation in function, performance, and resource use. Herein lies another rich area for integration, in this case between functional biologists and evolutionary biologists. It has been known for some time that physiological properties of organisms are remarkably environment specific, and that many tissues can undergo rapid remodeling in response to changes in the relevant environmental parameters. Recent advances have seen the incorporation of developmental plasticity into evolutionary theory, and progress is rapidly being made toward the integration of physiology, development, and the evolution of plasticity.

Size and Scaling

Effects of scale are ubiquitous in biology (Calder, 1984; Schmidt-Nielsen, 1984), yet ecomorphological studies often do not adequately account for the potentially confounding effects of body size. Emerson, Greene, and Charnov, in chapter 7, illustrate in their discussion of predator-prey interactions that allometric relationships potentially underlie all components of ecomorphological relationships. An understanding of these relationships can be gained in several ways. First, functionally relevant variables documenting the size of the organism and the functional morphological unit under study should be measured if samples vary in size or age. Second, allometric models can be used to account for size prior to testing ecomorphological hypotheses about the causal links among levels. In addition, size should be accounted for when interpreting the functional basis of performance and resource use.

Integrating Phylogeny

The preceding discussion focuses primarily on extrinsic influences on the evolution of organismal design (Lauder, 1981, 1982). Extrinsic factors may be thought of generally as all possible environmental forces that shape design through natural selection and phenotypic plasticity. Extrinsic factors are contrasted with intrinsic factors: features inherited from ancestors that limit, canalize, and promote particular transformations in design. Examples of intrinsic factors that have been proposed to influence evolution include genetic correlations (Cheverud, et al., 1983, 1989; Cheverud, 1984), developmental shifts (Alberch, 1980; Fink, 1982; Reilly, chapter 12), and functional couplings where the same structures being used in two or more functions (Lauder, 1982; Liem, 1988; Wake and Larson, 1987). Understanding the nature of extrinsically versus intrinsically determined ecomorphological patterns is the focus of chapters 4 (Losos and Miles), 5 (Travis), 11 (Bradley), and 12 (Reilly).

Efforts to examine the role of intrinsic factors rely upon the use of a phylogenetic hypothesis of the group in question, and the past decade has seen an explosion of interest in the use of the comparative method as a framework for asking questions about the evolution of almost any aspect of an organism's biology (Brooks and McClennan, 1991; Lauder, 1990, 1991; Harvey and Pagel, 1991). Recent developments in this area, especially as they relate to testing ecomorphological hypotheses, are reviewed in chapter 4 by Losos and Miles. More than simply provide a caveat for the adaptationist program, the phylogenetic perspective has produced a battery of approaches and techniques that can be used to ask questions about the correlation between morphology and ecology while accounting for phylogenetic constraints. Losos and Miles illustrate a number of methods that are useful for addressing specific questions in ecological morphology. While they note that a few studies have been able to incorporate detailed phylogenetic information into their analyses, one major roadblock to the comparative method is the lack of well-corroborated, independently derived phylogenetic hypotheses for the vast majority of taxa.

Another major influence of integrating phylogeny is in greatly enhancing our understanding of the evolution of morphology and ecology. Comparing intraspecific and interspecific patterns can reveal evolutionary mechanisms and intrinsic constraints (Emerson and Arnold, 1989). Analyzing multiple levels with a comparative phylogenetic approach makes it possible to describe historical patterns of ecomorphological relationships and allows one to test hypotheses about the conditions surrounding specific evolutionary transformations. Bradley's chapter (11) is a good example of the explanatory power of a phylogeny in ecomorphological analyses because it illustrates how confusing patterns of morphology, physiology, performance, and ecology take on a predictive nature when mapped onto a phylogenetic hypothesis for the organisms of interest. In addition, using the comparative phylogenetic approach patterns of conservatism and congruence of morphology, function, and ecology can be inferred and used to identify key constraints and interactions among different levels of biological design that play an important role in the origin and transformation of behavioral novelties and performance capacities underlying the diversification of ecological niches. A similar approach becomes a powerful inferential tool for paleontologists to understand ecomorphological patterns in fossil organisms and communities by comparisons to experimental functional analyses of similar taxa and communities living today (Van Valkenburgh, chap. 7).

CONCLUSION

The aim of this project was to bring together diverse approaches and perspectives to discuss the ecology and evolution of form and function. The contributors have tried to show how various ecomorphological links can be forged, how interrelationships among levels of analysis can be interpreted, and why various factors must be considered in partitioning variation in traits at each level of analysis. Together, the contributors and previous workers argue for a strong integrative effort among research disciplines traditionally distinct methodologically and conceptually. In some cases this effort can be sufficiently streamlined and the questions focused so as to permit a single individual to undertake all aspects of the project. In other instances the task will require the collaboration of individuals with differing expertise.

The potential for reciprocal illumination of morphology and ecology is strong. To fully understand biomechanical and physiological design are we not compelled to ask what the ecological consequences are of particular design features? Similarly, to understand the ecological ramifications of a phenotype in a particular environment should we not ask what the causal connection is between the phenotype and fitness? We ask not only *if* but also *how* variation in the phenotype affects fitness. The framework of such questions spans several levels of inquiry (fig. 13.1), and a dominant theme of the volume is that behavioral performance provides the key link between morphological and ecological levels of investigation. There is a natural tendency in functional research to become increasingly focused on smaller parts of the whole organism. Analyzing function at one level of organization typically generates questions about function at lower levels as the search for causal explanations leads inexorably to ever lower levels of design. While this proclivity for reductionism has beneficial consequences, it nevertheless is important to maintain a constant reference to the whole organism if ecological and evolutionary relevance are sought. Evolution acts at the level of the whole organism. Individual organisms die or survive, and individuals use environmental resources. The organism is the ultimate integrator of functional morphology and physiology on the one hand and ecology and evolution on the other.

We have emphasized integration, and yet we recognize there are several factors that have tended to restrict integrative research. Not least of these factors is that the structure of major funding agencies like the National Science Foundation tends to fund research that emphasizes one level of analysis over others in a given research program, even when trying to promote "integrative" studies. In addition, high-quality interdisciplinary research requires mastery of methodology and the literature in at least two areas. But in weighing the practical barriers against the fruits of integrative research we hope that the barriers can be seen simply as challenges and the fruits as desirable enough to justify the effort.

Understanding the diversity of complex organisms and their ecologies is a multifaceted problem. Evolution is not simply a problem for the geneticist, nor is morphology completely understood in a physiological context. We hope that the chapters of this book encourage morphologists and ecologists to consider the position of their work in the context of the hierarchy in figure 13.1 and ask what could be gained by moving up or down through the levels of inquiry. It is precisely such forays into new areas that will be needed if we are to realize a more ecological morphology.

REFERENCES

Alberch, P. 1980. Ontogenesis and morphological diversification. *Amer. Zool.* 20:653–667.

Alexander, R. M. 1975. *Biomechanics*. London: Chapman and Hall.

Alexander, R. M. 1982. *Optima for Animals*. London: Edward Arnold.

Arnold, S. J. 1983. Morphology, performance and fitness. *Amer. Zool.* 23:347–361.

Bartholomew, G. A. 1982. Scientific innovation and creativity: A zoologist's point of view. *Amer. Zool.* 22:227–235.

Bartholomew, G. A. 1986. The role of natural history in contemporary biology. *Bioscience* 36:324–329.

Bennett, A. F. 1989. Integrated studies of locomotor performance. In *Complex Organismal Functions: Integration and Evolution in Vertebrates*, ed. D. B. Wake and G. Roth. Chichester: John Wiley and Sons.

Bock, W. J., and G. von Wahlert. 1965. Adaptation and the form-function complex. *Evolution* 19:269–299.

Bock, W. J. 1977. Toward an ecological morphology. *Die Vogelwarte* 29:127–135.

Brooks, D. R., and D. A. McLennan. 1991. *Phylogeny, Ecology, and Behavior*. Chicago: University of Chicago Press.

Calder, W. A. III. 1984. *Size, Function, and Life History*. Cambridge: Harvard University Press.

Cheverud, J. M. 1984. Quantitative genetics and developmental constraints on evolution by selection. *J. Theor. Biol.* 110:155–171.

Cheverud, J. M., J. J. Rutledge, and W. R. Atchley, 1983. Quantitative genetics of development: Genetic correlations among age-specific trait values and the evolution of ontogeny. *Evolution* 37:895–905.

Denny, M. W. 1988. *Biology and the Mechanics of the Wave-swept Environment.* Princeton: Princeton University Press.

Dullemeijer, P. 1974. *Concepts and Approaches in Animal Morphology.* The Netherlands: Van Gorcum.

Emerson, S. B., and S. J. Arnold. 1989. Intra- and interspecific relationships between morphology, performance and fitness. In *Complex Organismal Functions: Integration and Evolution in Vertebrates,* ed. D. B. Wake and G. Roth, 295–314. London: John Wiley and Sons.

Endler, J. 1986. *Natural Selection in the Wild.* Princeton: Princeton University Press.

Fink, W. L. 1982. The conceptual relationship between ontogeny and phylogeny. *Paleobiology* 8:254–264.

Gans, C. 1985. Vertebrate morphology: Tale of a phoenix. *Amer. Zool.* 25:689–694.

Garland, T., Jr. 1985. Ontogenetic and individual variation in size, shape and speed in the Australian agamid lizard *Amphibolurus nuchalis.* J. Zool (London) A207:425–439.

Harvey, P. H., and M. D. Pagel. 1991. *The Comparative Method in Evolutionary Biology.* Oxford: Oxford University Press.

Huey, R. B., and R. D. Stevenson. 1979. Integrating thermal physiology and ecology of ectotherms: A discussion of approaches. *Amer. Zool.* 19:357–366.

Hutchinson, G. E. 1957. Concluding remarks. *Cold Spring Harbor Symp. Quant. Biol.* 22:415–427.

Hutchinson, G. E. 1959. Homage to Santa Rosalia, or why are there so many kinds of animals? *Amer. Nat.* 93:145–159.

Karr, J. R., and F. C. James. 1975. Ecomorphological configurations and convergent evolution in species and communities. In *Ecology and Evolution of Communities,* ed. M. L. Cody and J. M. Diamond, 258–291. Cambridge: Belknap Press.

Lande, R., and S. J. Arnold. 1983. The measurement of selection on correlated characters. *Evolution* 37:1210–1226.

Lauder, G. V. 1981. Form and function: Structural analysis in evolutionary morphology. *Paleobiology* 7:430–442.

Lauder, G. V. 1982. Historical biology and the problem of design. *J. Theor. Biol.* 97:57–67.

Lauder, G. V. 1990. Functional morphology and systematics: Studying functional patterns in an historical context. *Ann. Rev. Syst. Zool.* 21:317–340.

Lauder, G. V. 1991. Biomechanics and evolution: Integrating physical and historical biology in the study of complex systems. In *Biomechanics and Evolution,* ed. J. M. V. Rayner and R. J. Wootton. Cambridge: Cambridge University Press.

Lauder, G. V., and K. F. Leim. 1989. The role of historical factors in the evolution of complex organismal functions. In *Complex Organismal Functions: Integration and Evolution in Vertebrates,* ed. D. B. Wake and G. Roth, 63–78. London: John Wiley and Sons.

Liem, K. F. 1988. Form and function of lungs: The evolution of air breathing mechanisms. *Amer. Zool.* 28:739–759.

Miles, D. B., and R. Ricklefs. 1984. The correlation between ecology and morphology in deciduous forest passerine birds. *Ecology* 65:1629–1640.

Miles, D. B., R. E. Ricklefs, and J. Travis. 1987. Concordance of eco-morphological relationships in three assemblages of passerine birds. *Amer. Nat.* 129:347–364.

Niklas, K. J. 1992. *Plant Biomechanics: An Engineering Approach to Plant Form and Function.* Chicago: University of Chicago Press.

Reilly, S. M., and G. V. Lauder. 1992. Morphology, behavior and evolution: Comparative kinematics of aquatic feeding in salamanders. *Brain Behav. Evol.* 40:182–196.

Ricklefs, R. E., and J. Travis. 1980. A morphological approach to the study of avian community organization. *Auk* 97, 321–338.

Schmidt-Nielsen, K. 1984. *Scaling: Why Is Animal Size so Important?* Cambridge: Cambridge University Press.

Sibly, R. M., and P. Calow. 1986. *Physiological Ecology of Animals: An Evolutionary Approach.* Oxford: Blackwell Scientific.

Vogel, S. 1981. *Life in Moving Fluids.* Princeton: Princeton University Press.

Vogel, S. 1988. *Life's Devices.* Princeton: Princeton University Press.

Wainwright, P. C. 1987. Biomechanical limits to ecological performance: Mollusk-crushing by Caribbean hogfish, *Lachnolaimus maximus* (Labridae). *J. Zool.* (London) 213:283–297.

Wainwright, P. C. 1991. Ecomorphology: Experimental functional anatomy for ecological problems. *Amer. Zool.* 31:680–693.

Wainwright, S. A. 1988. *Axis and Circumference.* Cambridge: Harvard University Press.

Wainwright, S. A., W. D. Biggs, J. D. Curry, and J. M. Gosline. 1976. *Mechanical Design on Organisms.* London: Edward Arnold.

Wake, D. B., and A. Larson. 1987. Multidimensional analysis on an evolving lineage. *Science* 238:42–48.

Wake, D. B., and K. F. Liem. 1985. Morphology: Current approaches and concepts. In *Functional Vertebrate Morphology,* ed. M. Hildebrand, D. M. Bramble, K. F. Liem, D. B. Wake. Cambridge: Belknap Press.

Winemiller, K. O. 1991. Ecomorphological diversification in lowland freshwater fish assemblages from five biotic regions. *Ecol. Mon.* 61:343–365.

Contributors

Timothy J. Bradley
  Department of Ecology and
  Evolutionary Biology
  University of California
  Irvine, California 92717

Eric L. Charnov
  Department of Biology
  University of Utah
  Salt Lake City, Utah 84112

Mark W. Denny
  Department of Biological Sciences
  Stanford University
  Hopkins Marine Station
  Pacific Grove, California 93950

Sharon B. Emerson
  Department of Biology
  University of Utah
  Salt Lake City, Utah 84112

Theodore Garland, Jr.
  Department of Zoology
  University of Wisconsin
  Madison, Wisconsin 53706

Harry W. Greene
  Museum of Vertebrate Zoology
  University of California
  Berkeley, California 94720

Jonathan B. Losos
  Department of Biology
  Washington University
  St. Louis, Missouri 63130

Donald B. Miles
  Department of Biological Sciences
  Ohio University
  Athens, Ohio 45701

Ulla Norberg
  Department of Zoology
  University of Göteborg
  S-41390 Göteborg
  Sweden

Stephen M. Reilly
  Department of Biological Sciences
  Ohio University
  Athens, Ohio 45701

Robert Ricklefs
  Department of Biology
  University of Pennsylvania
  Philadelphia, Pennsylvania 19104

Joseph Travis
  Department of Biological Science
  Florida State University
  Tallahassee, Florida 32306

Blaire Van Valkenburgh
  Department of Biology
  University of California
  Los Angeles, California 90024

Peter C. Wainwright
  Department of Biological Science
  Florida State University
  Tallahassee, Florida 32306

Index